Integrated Waste Management:
A Sustainable Approach

Integrated Waste Management:
A Sustainable Approach

Editor: Dave Whittaker

CALLISTO REFERENCE

www.callistoreference.com

Callisto Reference,
118-35 Queens Blvd., Suite 400,
Forest Hills, NY 11375, USA

Visit us on the World Wide Web at:
www.callistoreference.com

ISBN: 978-1-63239-957-1 (Hardback)

Cataloging-in-Publication Data

Integrated waste management : a sustainable approach / edited by Dave Whittaker.
    p. cm.
Includes bibliographical references and index.
ISBN 978-1-63239-957-1
1. Integrated solid waste management. 2. Refuse and refuse disposal.
3. Recycling (Waste, etc.). I. Whittaker, Dave.
TD794.2 .I58 2018
363.728 5--dc23

Table of Contents

Preface

Over the recent decade, advancements and applications have progressed exponentially. This has led to the increased interest in this field and projects are being conducted to enhance knowledge. The main objective of this book is to present some of the critical challenges and provide insights into possible solutions. This book will answer the varied questions that arise in the field and also provide an increased scope for furthering studies.

Waste management is a primary concern of today's era. Waste has a negative impact on health and our environment. Household waste, municipal waste, sewage sludge, hazardous waste, etc. are some of the most popular forms of waste which need to be disposed and managed on a regular basis. Waste recycling, landfill and biological reprocessing are the most common practices of waste management. Most of the topics introduced in this book cover new techniques of waste management. From theories to research to practical applications, case studies related to all contemporary topics of relevance to this field have been included in this book. It is a vital tool for all researching and studying this field.

I hope that this book, with its visionary approach, will be a valuable addition and will promote interest among readers. Each of the authors has provided their extraordinary competence in their specific fields by providing different perspectives as they come from diverse nations and regions. I thank them for their contributions.

Editor

Heavy metals removal from aqueous environments by electrocoagulation process

Edris Bazrafshan[1], Leili Mohammadi[1*], Alireza Ansari-Moghaddam[1] and Amir Hossein Mahvi[2,3,4]

Abstract

Heavy metals pollution has become a more serious environmental problem in the last several decades as a result releasing toxic materials into the environment. Various techniques such as physical, chemical, biological, advanced oxidation and electrochemical processes were used for the treatment of domestic, industrial and agricultural effluents. The commonly used conventional biological treatments processes are not only time consuming but also need large operational area. Accordingly, it seems that these methods are not cost-effective for effluent containing toxic elements. Advanced oxidation techniques result in high treatment cost and are generally used to obtain high purity grade water. The chemical coagulation technique is slow and generates large amount of sludge. Electrocoagulation is an electrochemical technique with many applications. This process has recently attracted attention as a potential technique for treating industrial wastewater due to its versatility and environmental compatibility. This process has been applied for the treatment of many kinds of wastewater such as landfill leachate, restaurant, carwash, slaughterhouse, textile, laundry, tannery, petroleum refinery wastewater and for removal of bacteria, arsenic, fluoride, pesticides and heavy metals from aqueous environments. The objective of the present manuscript is to review the potential of electrocoagulation process for the treatment of domestic, industrial and agricultural effluents, especially removal of heavy metals from aqueous environments. About 100 published studies (1977–2016) are reviewed in this paper. It is evident from the literature survey articles that electrocoagulation are the most frequently studied for the treatment of heavy metal wastewater.

Keywords: Electrocoagulation, Wastewater treatment, Heavy metals removal

Introduction

Environmental issues, mainly concerning chemical and biological water pollution, represent a key priority for civil society, public authorities and, especially, for the industrial sector. In fact, the use of water, both in urban and industrial contexts, implies its subsequent pollution: any activity, whether domestic, agricultural or industrial, produces effluents containing undesirable, and possibly toxic, pollutants. Thus, a constant effort to protect water resources is being made by the various governments, through the introduction of increasingly strict legislation covering pollutant release. In particular for liquid industrial effluents, recent restrictions impose appropriate treatments of wastewater before its release into the environment [1].

This high pollutant load poses complex and extremely varied problems, related to each particular situation. In addition, the release of organic and inorganic pollutants is not uniform (either in quality or in quantity), but always leads to the same result: toxicity for aquatic ecosystems which creates worries for the population [2].

Industrial wastewaters like electroplating or acid mine wastewaters contain various kinds of toxic substances such as cyanides, alkaline cleaning agents, degreasing solvents, oil, fat and metals [3]. Most of the metals such as copper, nickel, chromium, silver and zinc are harmful when they are discharged without treatment [3]. Heavy metals are elements having atomic weights between 63.5 and 200.6 and a specific gravity greater than 5 [4].

* Correspondence: lailimohamadi@gmail.com
[1]Health Promotion Research Center, Zahedan University of Medical Sciences, Zahedan, Iran
Full list of author information is available at the end of the article

With the rapid development of industries such as metal plating facilities, mining operations, fertilizer industries, tanneries, batteries, paper industries and pesticides, etc., heavy metals wastewaters are directly or indirectly discharged into the environment increasingly, especially in developing countries. Unlike organic contaminants, heavy metals are not biodegradable and tend to accumulate in living organisms and many heavy metal ions are known to be toxic or carcinogenic. Toxic heavy metals of particular concern in treatment of industrial wastewaters include zinc, copper, nickel, mercury, cadmium, lead and chromium. Zinc is a trace element that is essential for human health. It is important for the physiological functions of living tissue and regulates many biochemical processes. However, too much zinc can cause eminent health problems, such as stomach cramps, skin irritations, vomiting, nausea and anemia [5]. Copper does essential work in animal metabolism. But the excessive ingestion of copper brings about serious toxicological concerns, such as vomiting, cramps, convulsions, or even death [6]. Nickel exceeding its critical level might bring about serious lung and kidney problems aside from gastrointestinal distress, pulmonary fibrosis and skin dermatitis [7]. And it is known that nickel is human carcinogen. Mercury is a neurotoxin that can cause damage to the central nervous system. High concentrations of mercury cause impairment of pulmonary and kidney function, chest pain and dyspnea [8]. The classic example of mercury poisoning is Minamata Bay. Cadmium has been classified by U.S. Environmental Protection Agency as a probable human carcinogen. Cadmium exposes human health to severe risks. Chronic exposure of cadmium results in kidney dysfunction and high levels of exposure will result in death. Lead can cause central nervous system damage. Lead can also damage the kidney, liver and reproductive system, basic cellular processes and brain functions. The toxic symptoms are anemia, insomnia, headache, dizziness, and irritability, weakness of muscles, hallucination and renal damages [9]. Chromium exits in the aquatic environment mainly in two states: Cr^{3+} and Cr^{6+}. In general, Cr^{6+} is more toxic than Cr^{3+}. Cr^{6+} affects human physiology, accumulates in the food chain and causes severe health problems ranging from simple skin irritation to lung carcinoma [10]. Various regulatory bodies have set the maximum prescribed limits for the discharge of toxic heavy metals in the aquatic systems. However the metal ions are being added to the water stream at a much higher concentration than the prescribed limits by industrial activities, thus leading to the health hazards and environmental degradation (some of permissible limits and health effects of various toxic heavy metals are presented Table 1).

Heavy metals can be easily absorbed by fishes and vegetables due to their high solubility in the aquatic environments and may accumulate in the human body by means of the food chain. So these toxic heavy metals should be removed from the wastewater to protect the people and the environment. In recent years, a variety of techniques are used for heavy metals removal from water and wastewater which include ion-exchange, adsorption, chemical precipitation, membrane filtration, flocculation, coagulation, flotation and electrochemical methods [3].

Electro-coagulation is an electrochemical approach, which uses an electrical current to remove metals from solution. Electro-coagulation system is also effective in removing suspended solids, dissolved metals, tannins and dyes. The contaminants presents in wastewater are maintained in solution by electrical charges. When these ions and other charged particles are neutralized with ions of opposite electrical charges provided by electrocoagulation system, they become destabilized and precipitate in a stable form. Electrochemical methods are simple, fast, inexpensive, easily operable and eco-friendly in nature. Besides, purified water is potable, clear, colorless and odorless with low sludge production. There is no chance of secondary contamination of water in these techniques.

Electrocoagulation process (EC) has been successfully applied to remove soluble ionic species from solutions and heavy metals by various investigators [11, 12]. The EC process is based on the continuous in situ production of a coagulant in the contaminated water. It had been shown that EC is able to eliminate a variety of pollutants from wastewaters, as for example metals and arsenic [3] strontium and cesium[13], phosphate [14], sulfide, sulfate and sulfite [15], boron [16], fluoride [17], nitrate [18], chromium [19–22], cadmium [23], zinc [24], nickel [25, 26], mercury [27], cobalt [28], clay minerals [29], as well as oil [30], chemical oxygen demand [31], color [32] and organic substances [33].

The most widely used method for the treatment of metal polluted wastewater is precipitation with NaOH and coagulation with $FeSO_4$ or $Al_2(SO_4)_3$ with subsequent time-consuming sedimentation [34]. Other methods include adsorption, ion exchange and reverse osmosis [34]. Although precipitation is shown to be quite efficient in treating industrial effluents, the chemical coagulation may induce secondary pollution caused by added chemical substances [34]. These disadvantages encouraged many studies on the use of electrocoagulation for the treatment of several industrial effluents [34]. This technique does not require supplementary addition of chemicals, reduces the volume of produced sludge [33] and first economic studies indicate also a financial advantage compared to the conventional methods [35].

Table 1 Permissible limits and health effects of various toxic heavy metals

Metal contaminant	Permissible limits for industrial effluent discharge (in mg/l)						Permissible limits by international bodies (μg/l)		Health hazards
	Into inland surface waters Indian Standards: 2490 (1974)[a]	Into inland surface waters, Iranian Standards (2009)[b]	Into public sewers, Indian Standards: 3306 (1974)[a]	Into public sewers, Iranian Standards (2009)[b]	On land for irrigation, Indian Standards: 3307 (1974)[a]	On land for irrigation, Iranian Standards (2009)[b]	WHO[a]	USEPA[a]	
Arsenic	0.20	0.10	0.20	0.10	0.20	0.10	10	50	Carcinogenic, producing liver tumors, skin and gastrointestinal effects
Mercury	0.01	–	0.01	–	–	–	01	02	Corrosive to skin, eyes and muscle membrane, dermatitis, anorexia, kidney damage and severe muscle pain
Cadmium	2.00	0.10	1.00	0.10	–	0.05	03	05	Carcinogenic, cause lung fibrosis, dyspnea and weight loss
Lead	0.10	1.00	1.00	1.00	–	1.00	10	05	Suspected carcinogen, loss of appetite, anemia, muscle and joint pains, diminishing IQ, cause sterility, kidney problem and high blood pressure
Chromium	0.10	Cr^{6+} = 0.50, Cr^{3+} = 2.0	2.00	Cr^{6+} = –, Cr^{3+} = 2.0	–	Cr^{6+} = 1.00, Cr^{3+} = 2.0	50	100	Suspected human Carcinogen, producing lung tumors, allergic dermatitis
Nickel	3.0	2.0	3.0	2.0	–	2.0	–	–	Causes chronic bronchitis, reduced lung function, cancer of lungs and nasal sinus
Zinc	5.00	2.0	15.00	2.0	–	2.0	–	–	Causes short-term illness called "metal fume fever" and restlessness
Copper	3.00	1.0	3.00	1.0	–	0.2	–	1300	Long term exposure causes irritation of nose, mouth, eyes, headache, stomachache, dizziness, diarrhea

In the above Table [a] referred to Reference No. [61] and [b] referred to Reference No. [62]

EC process has the potential to extensively eliminate the disadvantages of the classical treatment techniques to achieve a sustainable and economic treatment of polluted wastewater [33, 36]. Since the turn of the 19th century, EC has been applied for wastewater treatment [37] and many studies attended to optimize the process for specific problems. Typically, empirical studies were done [34, 38]. These studies show the successful treatment of the wastewaters, however, they provide little insight into fundamental chemical and physical mechanisms [39]. Therefore, the mechanisms involved are yet not clearly understood [39]. But exactly these physicochemical mechanisms have to be understood to optimize and control the process, to allow modeling of the method and to improve the design of the system. The main objectives of the present work were to gain insight into some fundamental mechanisms and possible interactions influencing the removal process of heavy metals by electrocoagulation.

Table 2 shows the removal efficiency of heavy metals by various treatment technologies. In addition, removal of some of metals and other pollutants by EC process are presented in Table 3.

Description of electrocoagulation process

Electrocoagulation (EC) is a simple and efficient method and has been used for the treatment of many types of wastewaters such as electroplating wastewater [34], laundry wastewater [40], restaurant wastewater [38] and poultry slaughterhouse wastewater [41]. EC has been successfully used for the removal of pollutants from different industrial wastewaters (Table 4). Many studies have been reported in the literature [20, 21, 24, 42].

EC in combination with other treatment processes is a safe and effective way for the removal of pollutants. EC is an efficient technique because adsorption of hydroxide on mineral surfaces are 100 times greater on 'in situ' rather than on pre-precipitated hydroxides when metal hydroxides are used as coagulant [43]. Since the flocs formed by EC are relatively large which contain less bound water and are more stable, therefore, they can be easily removed by

Table 2 Comparison of various treatment technologies for removal of heavy metals from aqueous environments

Treatment method	Metal	pH of solution	Initial concentration (mg/l)	Efficiency (%)	References
Reverse osmosis	Ni^{2+}	3	26	98	[63]
		7	26	99	[63]
	Cu^{2+}	3	17	98	[63]
		7	17	99	[63]
	Cr	3	167	95	[63]
		7	167	99	[63]
Ultrafiltration	Ni^{2+}	7	50	99	[64]
		7	100	99	[64]
	Cu^{2+}	7	50	98	[64]
		7	100	97	[64]
	Cr	7	50	93	[64]
		7	100	76	[64]
	Ni^{2+}	-	25	100	[65]
Nanofiltration	Cu^{2+}	-	200	96	[66]
Electrocoagulation	Ni^{2+}	3	394	98	[67]
		7	394	99	[67]
	Cu^{2+}	3	45	100	[67]
		7	45	100	[67]
	Cr	3	44.5	100	[67]
		7	44.5	100	[67]
	Ni^{2+}, Zn^{2+}	6	248, 270, 282; 217, 232, 236	100	[68]
Chemical precipitation	Cu^{2+}, Zn^{2+}, Cr^{3+}, Pb^{2+}	7- 11	100 mg/L	99.3-99.6	[69]
	Cu^{2+}, Zn^{2+}, Pb^{2+}	3	0.01, 1.34, 2.3 mM	100, >94, >92	[70]
Adsorption	Pb^{2+}	4	2072	-	[71]
	Pb^{2+}	4	1036	55	[72]
	Cd^{2+}, Cr^{6+}	6	2	$Cd^{2+} = 55$, $Cr^{6+} = 60$	[22]

Table 3 Removal of heavy metals and other pollutants by EC process

References	Metals or other compounds	Concentration (mg/L)	Anode–cathode	Removal efficiency (%)
[55]	Cr^{3+}, Cr^{6+}	887.2, 1495.2	Fe-Fe	100
[67]	Cu^{2+}, Cr, Ni^{2+}	45, 44.5, 394	Al-Fe	100
[49]	Cd^{2+}	20	Al-Al	AC: 97.5, DC: 96.2
[18]	NO_3^-	150	Fe-Fe, Al-Al	90, 89.7
[23]	Pb^{2+}, Zn^{2+}, Cd^{2+}	170, 50, 1.5	Al-SS	95, 68, 66
[50]	As	150	Al-Al, Fe-Fe	93.5, 94
[26]	TOC, Ni^{2+}, Zn^{2+}	173, 248, 232	SS $_{304}$-SS $_{304}$	66, 90, 100
[73]	Humic acid	20	Fe-Fe	92.69

Nomenclature: *Cr* chromium, *Ni* nickel, *Cu* copper, *As* arsenic, *Zn* zinc, *pb* lead, *Cd* Cadmium, *Co* cobalt, *Fe* iron, *Al* aluminum, *St* steel, *SS* stainless steel

filtration. It is cost effective and easily Performance. EC needs simple equipment's and can be designed for any capacity of effluent treatment plant. Since no chemical addition is required in this process, it reduces the possibility of generation of secondary pollutants. It needs low current and therefore, can be operated by green processes, such as, solar, windmills and fuel cells [44]. It is an environment friendly technique since the electron is the main reagent and does not require addition of the reagents/chemicals. This will minimize the sludge generation to a great extent and eventually eliminate some of the harmful chemicals used as coagulants in the conventional effluent treatment methods. EC process can effectively destabilize small colloidal particles and generates lower quantity of sludge compared to other processes. The advantages of EC as compared to chemical coagulation are as follows:

1. EC requires simple equipment and is easy to operate with sufficient operational latitude to handle most problems encountered on running. Wastewater treated by EC gives pleasant/edible palatable, clear, colorless and odorless water.
2. Sludge formed by EC tends to be readily settable and easy to de-water, because its main elements/components are metallic oxides/hydroxides. Above all, it is a low sludge producing technique.

3. Flocs formed by EC are similar to chemical flocs, except that EC flocs tends to be much larger, contains less bound water, is acid-resistant and more stable and therefore, can be separated faster by filtration.
4. EC produces effluent with less total dissolved solids (TDS) content as compared with chemical treatments. If this water is reused, the low TDS level contributes to a lower water recovery cost.
5. The EC process has the advantage of removing the smallest colloidal particles, because the applied electric field sets them in faster motion, thereby facilitating the coagulation. The EC process avoids uses of chemicals and so there is no problem of neutralizing excess chemicals and no possibility of secondary pollution caused by chemical substances added at high concentration as when chemical coagulation of wastewater is used.
6. The gas bubbles produced during electrolysis can carry the pollutant to the top of the solution where it can be more easily concentrated, collected and removed. The electrolytic processes in the EC cell are controlled electrically with no moving parts, thus requiring less maintenance.

The EC technique can be conveniently used in rural areas where electricity is not available, since a solar panel

Table 4 Application of electrocoagulation process for treatment of different types of wastewater

References	Type of wastewater	Current density or current	Time (min)	pH	Anode–cathode	COD removal (%)
[47]	Olive oil mill wastewater	39.06, 78.1 and 117.18 A/m^2	60	5.2	Ti-Fe	96.14
[74]	Real dairy wastewater	5A	60	7.24	Al-Al	98.84
[75]	Slaughterhouse wastewater	5A	15	7	Al-Al	99
[76]	Carwash wastewater	5 A	15	7.65 ± 0.02	Al-Al	COD = 96.8, BOD$_5$ = 94,TSS = 98.4, MBAS = 98.6
[77]	Textile wastewater	5 A	60	7	Al-Al	98.28
[44]	Textile wastewater	-	3	10.6	Fe-Fe	84
[34]	Olive mill effluents	75 mA/cm^2	25	4-6	Al-Al	76
[37]	Industrial effluents	0.01 A/m^2	30	10.8	SS-SS	95

Nomenclature: *MS* mild steel, *SS* Stainless steel, *St* steel, *Ti* titanium, *Fe* iron, *Pt* platinum, *Cu* copper attached to the unit may be sufficient to carry out the process. Potentially recoverable metals and reuse of treated effluent are other advantages of EC. EC is an alternative to chemical precipitation for the removal of dissolved and suspended metals in aqueous solutions (see Chemical Precipitation Technology Overview). The quantity of sludge produced is lower. The floc generated is larger and heavier and settles out better than in conventional chemical precipitation processes. Since a large thickener is not required, capital costs can also be lower. The effluent generated by EC contains no added chemicals and is often of better quality, containing TDS and less colloidal particulates. Reduction of TDS has been reported at 27 %-60 %, and reduction of total suspended solids can be as great as 95 %-99 % [45].

Although EC requires energy input, it requires only low currents and can be operated using green technologies such as solar or wind power. Some of the limitations of the electrochemical coagulation are as follows [43, 46]:

1. The sacrificial anodes need to be replaced periodically.
2. EC requires minimum solution conductivity depending on reactor design, limiting its use with effluent containing low dissolved solids.
3. In case of the removal of organic compounds, from effluent containing chlorides there is a possibility of formation of toxic chlorinated organic compounds.
4. An impermeable oxide film may be formed on the cathode which may provide resistance to the flow of electric current. However, change of polarity and periodical cleaning of the electrodes may reduce this interference.
5. The high cost of electricity can result in an increase in operational cost of EC process [43].

Electrocoagulation process involves the generation of coagulants in situ by dissolving electrically either aluminum or iron ions from aluminum or iron electrodes, respectively. In this process, the metal ions generation takes place at the anode and hydrogen gas is released from the cathode. The hydrogen gas bubbles carry the pollutant to the top of the solution where it can be more easily concentrated, collected and removed. Various reactions take place in the electrocoagulation process, where aluminum is used as the electrode:

At the anode:

$$Al \rightarrow Al^{3+}{}_{(aq)} + 3e \qquad (1)$$

At the cathode:

$$3H_2O + 3e \rightarrow 3/2H_2 + 3OH^- \qquad (2)$$

The cathode may also be chemically attacked by OH^- ions generated during H_2 evolution at high pH:

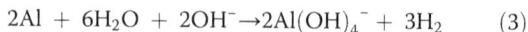

$$2Al + 6H_2O + 2OH^- \rightarrow 2Al(OH)_4{}^- + 3H_2 \qquad (3)$$

$Al^{3+}_{(aq)}$ and OH^- ions generated by electrode reactions (1) and (2) react to form various monomeric species such as $Al(OH)^{2+}$, $Al(OH)_2^+$, $Al_2(OH)_2^{4+}$, $Al(OH)_4^-$, and polymeric species such as $Al_6(OH)_{15}^{3+}$, $Al_7(OH)_{17}^{4+}$, $Al_8(OH)_{20}^{4+}$, $Al_{13}O_4(OH)_{24}^{7+}$, $Al_{13}(OH)_{34}^{5+}$, which transform finally into $Al(OH)_3$ according to complex precipitation kinetics [43].

Freshly formed amorphous $Al(OH)_3$ "sweep flocs" have large surface areas which are beneficial for a rapid adsorption of soluble organic compounds and trapping of colloidal particles. These flocs polymerize as:

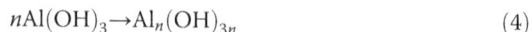

$$nAl(OH)_3 \rightarrow Al_n(OH)_{3n} \qquad (4)$$

and they are easily removed from aqueous environment by sedimentation and by H_2 flotation. Secondary anodic reactions occur also during electrocoagulation process for example, in neutral and acidic chloride solutions, native and free chlorine and hypochlorite are formed which are strong oxidants. On the other hand, the aluminum hydroxide flocs normally act as adsorbents and/or traps for pollutants. Therefore, they would eliminate them from the solution [43].

In addition the main reactions occurring at the iron electrodes are:

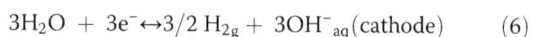

$$Fe\,(s) \leftrightarrow Fe^{+3}{}_{aq} + 3e^- (anode) \qquad (5)$$

$$3H_2O + 3e^- \leftrightarrow 3/2\,H_{2g} + 3OH^-{}_{aq} (cathode) \qquad (6)$$

In addition, Fe^{3+} and OH^- ions generated at electrode surfaces react in the bulk wastewater to form ferric hydroxide:

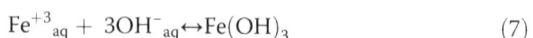

$$Fe^{+3}{}_{aq} + 3OH^-{}_{aq} \leftrightarrow Fe(OH)_3 \qquad (7)$$

The suspended aluminum or iron hydroxides can remove pollutants from the solution by sorption, co-precipitation or electrostatic attraction, followed by coagulation [43].

For a particular electrical current flow in an electrolytic cell, the mass of aluminum or iron theoretically dissolved from the sacrificial anode is quantified by Faraday's law [43]:

$$w = \left[\frac{ItM}{ZF}\right] \qquad (8)$$

where "w" is the amount of anode material dissolved (g), I the current (A), the electrolysis time (t), M the specific molecular weight of electrode (g/mol), Z the number of electrons involved in the reaction and F is the Faraday's constant (96485.34 C/mol). The mass of evolved hydrogen and formed hydroxyl ions can be calculated correspondingly. The amount of coagulant dosed into the solution can be increased by increasing the current and the reaction time. But increasing the current density leads to a decreased current efficiency. Some influencing factors of the EC process are current density (or applied voltage), conductivity and pH of solution, mode of operation, electrolysis time, electrode material and distance between the electrodes [43].

Batch and continuous mode of operation

It can be noticed from the literature that EC has been studied for the removal of a wide range of pollutants using batch and continuous mode of operation. (Diagram of batch and continues flow electrochemical reactor is shown in Figs. 1 and 2). A continuous system operates under steady state conditions, specially a fixed pollutant concentration and effluent flow rate. Comparably, a batch reactor's dynamic nature enables to study the range of operating conditions and is more suited for research work [45]. Continuous systems are better suited to industrial processes for large effluent volumes whereas the batch reactors are suited to laboratory and pilot plant scale applications. The continuous mode of operation is preferred due to its better control than the batch mode of operation.

Batch mode of EC reactors exhibits time dependent behavior as the coagulant is continuously generated in the reactor with the dissolution of anode. The anode material is hydrolyzed, and is capable of aggregating the pollutants. As a result, the concentration of the pollutant, coagulant, and pH keeps on changing with respect to time. A batch reactor has neither inflow nor outflow of effluent during the electrolysis time [45].

Effect of various operating parameters on pollutants removal

The efficiency of the EC process depends on many operational parameters such as conductivity of the solution, arrangement of electrode, electrode shape, type of power supply, pH of the solution, current density, distance between the electrodes, agitation speed, electrolysis time,

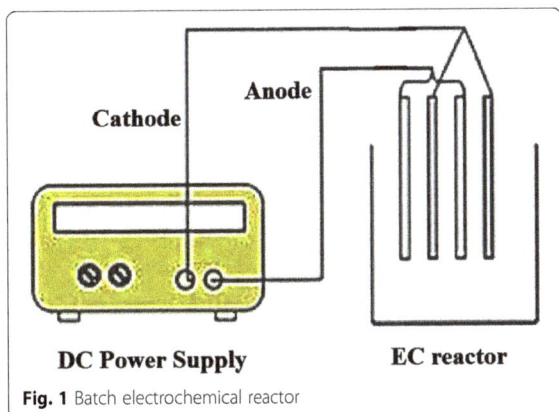

Fig. 1 Batch electrochemical reactor

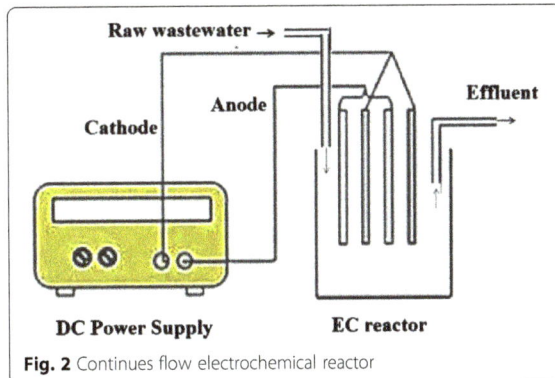

Fig. 2 Continues flow electrochemical reactor

initial pollutant concentration, retention time and passivation of the electrode.

Solution conductivity and type of power supply

Conductivity of the solution is very important parameter in electrolysis process as the removal efficiency of the pollutant and operating cost are directly related to the solution conductivity [45]. The conductivity of an electrolyte solution is a key property. In an electrochemical process, the conductivity determines the cell resistance while the properties of solvent and electrolyte determine their interaction with the electrically active species and thereby influence the electrode reactions [47].

The solution must have some minimum conductivity for the flow of the electric current. The conductivity of the low-conductivity wastewater is adjusted by adding sufficient amount of salts such as sodium chloride or sodium sulphate. There is an increase in the current density with an increase in the conductivity of the solution at constant cell voltage or reduction in the cell voltage at constant current density [48]. The energy consumption is decreased with high performance/approach solution. The energy consumption is decreased with high conductivity solution. In the EC process, there is an in situ generation of metal hydroxide ions by electrolytic oxidation of the sacrificial anode. These metal hydroxide ions act as coagulant and remove the pollutants from the solution by sedimentation. Majority of the studies reported in the literature have used direct current (DC) in the EC process. The use of DC leads to the corrosion formation on the anode due to oxidation. An oxidation layer also form on the cathode reducing the flow of current between the cathode and the anode and thereby lowering the pollutant removal efficiency [49]. These limitations of the DC electrocoagulation process have been decreased to some extent by the addition of parallel plate sacrificial electrodes in the cell configuration. Nevertheless, many have preferred the use of alternating current electrocoagulation (ACE) technology. It is believed that the ac cyclic energization retards the normal mechanisms of electrode attack that are experienced in DC electrocoagulation system, and consequently, ensure reasonable electrode life. In addition to that, since the AC electric fields in an ACE separator do not cause electrophoretic transport of the charged particles due to the frequent change of polarity, it can induce dipole–dipole interactions in a system containing non spherical charged species. Consequently, the AC electric fields may also disrupt the stability of balanced dipolar structures existing in such a system. This is, however, not possible in a DC electrocoagulation separator using DC electric fields [46].

Arrangement of electrodes

The electrode material and the connection mode of the electrodes play a significant role in the cost analysis of the EC process. Kobya et al. [50] studied the treatment of textile wastewater and compared the performances of various electrode connection modes as a function of wastewater pH, current density and operating time. They studied three different modes of electrode connection and areas follow: Monopolar electrodes in parallel connections (MP-P): The anodes and cathodes are connected in parallel due to which the current is divided between all the electrodes to the resistance of individual cells. The parallel connection needs a lower potential difference compared with serial connections [50]. Monopolar electrodes in serial connections (MP-S): In the monopolar configuration, each pair of sacrificial electrodes is internally connected with each other. The addition of the cell voltages leads to a higher potential difference for a given current. Bipolar electrode in serial connections (BP-S): In this connection mode, the outer electrodes are connected to the power supply and there is the no electrical connection between the inner electrodes [50]. Schematic diagram of EC reactor with monopolar and bipolar electrode connections is shown in Figs. 3 and 4.

The shape of the electrodes affects the pollutant removal efficiency in the EC process. It is expected that the

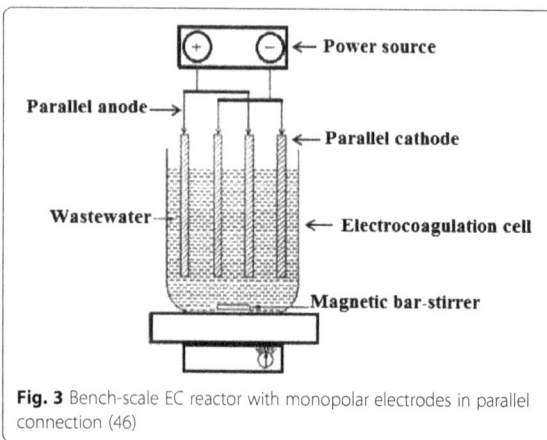

Fig. 4 Bench-scale EC reactor with bipolar electrodes in parallel connection (46)

punched holes type electrodes will result in higher removal efficiency compared to the plane electrodes. Very few studies have been reported in the literature [51] describing the effect of electrode shape on the performance of the electrostatic precipitator. Kuroda et al. [51] performed experiments using metallic electrodes with/without punched holes as a barrier discharge electrode to study the effect of electrode shape of reactor on the collector efficiency in electrostatic precipitator. They have reported higher discharge current for the electrode with punched holes than for plane electrode resulting in higher collection efficiency with punched electrode compared with plane electrode. The electric field intensity at the edge of punched holes type electrodes is higher (1.2 times) than at plane type electrode resulting in an increase in the discharge current at punched type electrode. More studies are needed to establish the effect of the electrode shape (punched hole diameter and pitch of the holes) on the EC process [51].

Current density

Current density is very important parameter in EC as it determines the coagulant dosage rate, bubble production rate, size and growth of the flocs, which can affect the efficiency of the EC. With an increase in the current density, the anode dissolution rate increases. This leads to an increase in the number of metal hydroxide flocs resulting in the increase in pollutant removal efficiency. An increase in current density above the optimum current density does not result in an increase in the pollutant removal efficiency as sufficient numbers of metal hydroxide flocs are available for the sedimentation of the pollutant [52, 53]. Effect of current density or current on removal efficiency of EC process is shown in Table 5.

Distance between the electrodes

Inter-electrode spacing is a vital parameter in the reactor design for the removal of pollutant from effluent. The inter electrode-spacing and effective surface area of electrodes

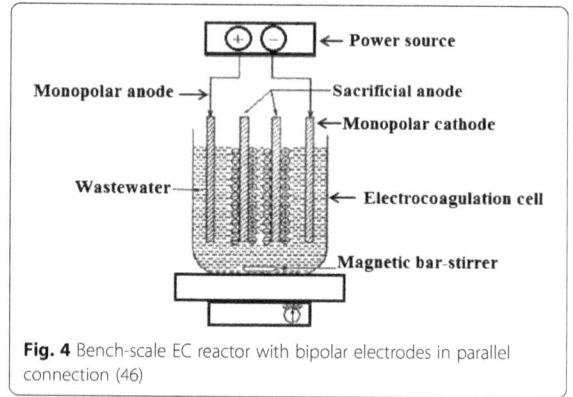

Fig. 3 Bench-scale EC reactor with monopolar electrodes in parallel connection (46)

Table 5 Effect of cell voltage (V), electrode material, electrode connection mode, current or current density, flow rate and pH on removal efficiency of heavy metals in EC process

References	Heavy metals	Current density	Cell voltage (V)	Flow rates	Optimum pH	Electrode materials	Removal efficiency (%)
[78]	Cr^{6+}	$8.33\ A/m^2$	2	$1.2\ m^3/h$	7-8.5	Fe-Fe A_M	70-85
[34]	Cu^{2+}, Zn^{2+}, Cr^{6+}	$4.8A/dm^2$	-	10 ml/min	4	Al-Al	99, 99, 83
[19]	Cr^{6+}	$30\ A/m^2$	-	50 ml/min	5-8	Fe-Fe	80-97
[21]	Cr^{6+}	5A	40	-	10	Al-Al	99
[79]	Cr^{3+}	$A_M = 10.84$, $Bi = 32.52\ mA/cm^2$	-	40 ml/min	$A_M = 5.5$, $Bi = 6$	MS-MS	$A_M = 90.6$, $Bi = 71.4$
[20]	Cr^{6+}	5A	20-40	-	3	Al-Fe	99.9
[80]	Cr^{6+}	0.05, 1 A	30	-	5	Al-Al	100
[81]	Cr^{6+}	$35.7\ mA/\ cm^2$	10-24	22.5 ml/min	5	Al–Al	90.4
[82]	Cr^{6+}	$55.5\ mA/cm^2$	60	12 ml/min	7	Fe-Fe, Pt Ti (platinized titanium)/Fe, Al/Al and Pt Ti/Al	65.3
[12]	Cr^{6+}	5A	40	-	3	Fe-Fe	98
[83]	Cr^{6+}	$153\ A/m^2$	15-25	-	5	Fe-Al	99
[84]	Cr^{6+}	$2-25\ mA/cm^2$	80	-	5.68	Fe-Fe	99
[55]	Cr^{3+}, Cr^{6+}	$50\ mA/cm^2$	-	-	4	Fe-Fe	100
[85]	Fe, Ni^{2+}, Cu^{2+}, Zn^{2+}, Pb^{2+}, Cd^2+	$11.55\ mA/cm^2$	0-30	-	7.6	Al-Al	SS = 86.5, turbidity = 81.56, $BOD_5 = 83$, COD = 68, color > 92.5
[86]	Cd^{2+}, Cu^{2+}	5 A	30	20 L/h	0.64	Ss-ti	$Cd^{2+} = 73.8, Cu^{2+} = 98.8$
[87]	Cd^{2+}	2.2, 3.5 mA/cm^2	6	-	11	Al-Fe	>99.5
[88]	Cd^{2+}	$0.2\ A/dm^2$	AC:270, DC: 25	-	7	Zn-Zn	AC: 97.8, DC: 96.9
[49]	Cd^{2+}	$0.2\ A/dm^2$	DC = 25,AC = 270	-	7	Al-Al	AC: 97.5, DC: 96.2
[89]	Cd^{2+}	$0.04\ A/m^2$	70	5 ml/min	8.9	Al-Al	98.2
[90]	Zn phosphate	$60.0\ A/m^2$	30	400 mL/min	Al-Al = 5, Fe-Fe = 3	Al-Fe	max 97.8
[24]	Zn^{2+}	$15\ mA\ cm^2$	60	-	6	Al-Fe	>99
[42]	Zn^{2+}, Cu^{2+}	5A	40	-	7	Fe-Fe	99.99
[58]	COD, Zn^{2+}	COD = 0.90, Zn^{2+} = 0.45-1.8 A/dm^2	-	-	COD = 3, ZN^{2+} = 10	Fe-graphite	COD = 88, 99.3 Zn^{2+} = 99
[91]	Ni^{2+}, Cu^{2+}	0.3 A	29	-	5	RO-Ti-Ss	$Ni^{2+} = 82$, $Cu^{2+} = 99$
[12]	Zn^{2+}, Cu^{2+}	5 A	40	-	7	Al-Al	$Zn^{2+} = 99.6$, $Cu^{2+} = 99.9$
[67]	Cu^{2+}, Cr, Ni^{2+}	$10\ mA/cm^2$	30	-	3	Fe-Al	100
[17]	Ni^{2+}	5 A	20	-	10	Fe-Fe	99.99
[50]	As	5 A	-	-	Fe = 6.5, Al =7	Al-Al, Fe-Fe	Fe = 93.5, Al = 95.7

Table 5 Effect of cell voltage (V), electrode material, electrode connection mode, current or current density, flow rate and pH on removal efficiency of heavy metals in EC process (Continued)

Ref	Pollutant	Current/current density [Al = 2.5 A/m², Fe =7.5 A/m²]	[Al electrode = 0.8-1.6, Fe electrode =1.5-2.3]	Flow rate [Fe electrode = 60 ml/min, Al electrode = 50 ml/min]	Cell voltage (V)	Electrode	Removal efficiency
[92]	As, Nitrite	-	$NO_3=25$, $As^{5+}=20$	2 L/h	9.5	MS-MS	$NO_3=84$, $As^{5+}=75$
[93]	As	8.86 mA/cm²	17	7 L/h	5 ± 0.2	AL-AL	89
[11]	Oil, grease, heavy metals	0.6 A/cm²	40	1 L/min	2-4	Al-Cs	$Zn^{2+}=99$, Cu^{2+}, $Ni^{2+}=70$, Oil and grease = 99.9, Turbidity = 99.7
[28]	Co	6.25 mA/cm²	30	-	8	Al-Al	99
[94]	Heavy metals	4 mA/cm²	30	600 mL/min	9.56	Cs-Cs	$Cr^{3+}=Cu^{2+}=100$, $Ni^{2+}=99$
[16]	Boron	12.5 mA/cm²	30	-	6.3	Al-Al	99.7
[95]	Ba , Zn^{2+}, Pb^{2+}	350 A/m²	30	-	10	Ss-Ss	97
[96]	Cd^{2+}	3.68 mA/cm²	-	-	Bipolar configuration = 10.90, monopolar configuration = 9.03	AL-AL	100
[97]	Ni^{2+}	7.5 A/m²	-	6 ml/min	6	AL-AL, Fe-Fe	100
[98]	Cr^{6+}	0.55 A	20	-	1	Fe-Fe	100
[99]	Zn^{2+}, Cu^{2+}, Ni^{2+}, Ag^+, $Cr_2O_7^{2-}$	33 A/m²	30	-	9	Al-Al	>50
[100]	Cr^{6+}	50-200 A/m²	-	2.5 cm³/s	7.5	Fe-Fe, Al-Al	40

Nomenclature: MS mild steel, SS stainless steel, St steel, Ti titanium, Fe iron, Pt platinum, Cu copper, CS carbon steel electrodes, RuO ruthenium oxide, A_M Monopolar, B Bipolar

are important variable when an operational costs optimization of a reactor is needed [52]. To decrease the energy consumption (at constant current density) in the treatment of effluent with a relatively high conductivity, larger spacing should be used between electrodes. For effluent with low conductivity, energy consumption can be minimized by decreasing the spacing between the electrodes [53].

The inter electrode distance plays a significant role in the EC as the electrostatic field depends on the distance between the anode and the cathode. The maximum pollutant removal efficiency is obtained by maintaining an optimum distance between the electrodes. At the minimum inter electrode distance; the pollutant removal efficiency is low. This is due to the fact that the generated metal hydroxides which act as the flocs and remove the pollutant by sedimentation get degraded by collision with each other due to high electrostatic attraction [54]. The pollutant removal efficiency increases with an increase in the inter electrode distance from the minimum till the optimum distance between the electrodes. This is due to the fact that by further increasing the distance between the electrodes, there is a decrease in the electrostatic effects resulting in a slower movement of the generated ions. It provides more time for the generated metal hydroxide to agglomerate to form the flocs resulting in an increase in the removal efficiency of the pollutant in the solution. On further increasing the electrode distance more than the optimum electrode distance, there is a reduction in the pollutant removal efficiency. This is due to the fact that the travel time of the ions increases with an increase in the distance between the electrodes. This leads to a decrease in the electrostatic attraction resulting in the less formation of flocs needed to coagulate the pollutant [54]. The pollutant removal efficiency is low at the minimum inter electrode distance. Effect of distance between the electrodes and also type of reactor (batch or continuous) on removal efficiency of EC process are presented in Table 6.

Electrolysis time

The pollutant removal efficiency is also a function of the electrolysis time. The pollutant removal efficiency increases with an increase in the electrolysis time. But beyond the optimum electrolysis time, the pollutant removal efficiency becomes constant and does not increase with an increase in the electrolysis time. The metal hydroxides are formed by the dissolution of the anode. For a fixed current density, the number of generated metal hydroxide increases with an increase in the electrolysis time. For a longer electrolysis time, there is an increase in the generation of flocs resulting in an increase in the pollutant removal efficiency. For an electrolysis time beyond the optimum electrolysis time, the pollutant removal efficiency does not increase as sufficient numbers of flocs are available for the removal of the pollutant [45]. Bazrafshan et al. [20] determined that Cr^{6+} reduction from synthetic chromium solution could be under legal limits as long as treatment was between 20 and 60 minutes. Effect of different electrolysis time on removal efficiency of EC process is shown in Table 6.

Effect of initial pH on the efficiency of Heavy metal removal

It has been established that pH is an important parameter influencing the performance of the electrochemical process [38]. The maximum pollutant removal efficiency is obtained at an optimum solution pH for a particular pollutant. The precipitation of a pollutant begins at a particular pH. The pollutant removal efficiency decreases by either increasing or decreasing the pH of the solution from the optimum pH. Verma et al. [55] studied the removal of hexavalent chromium from synthetic solution using EC and found that the pH of the solution has a significant effect on the Cr^{6+} removal efficiency. They performed the experiments at different pH of the synthetic solution and obtained the maximum chromium removal efficiency at the pH 4. They further reported that the pH of the synthetic solution after the EC process increased with an increase in the electrolysis time due to the generation of OH in the EC process [55].

The pH changed during batch EF, Its evolution depended on the initial pH. EF process exhibits some buffering capacity because of the balance between the production and the consumption of OH [56].The pH has a significant influence on the coagulant species formed during coagulation processes. It also has influence on the superficial charge of the aluminum hydroxide precipitates (caused by the adsorption of ionic species) [57]. During the time-course of coagulation and EC processes, the pH changes in an opposite way and this affects significantly to the coagulant species formed, and hence to the efficiencies obtained in the removal of pollutants [57].

It cannot be said that any process is better than the other for all wastes. Under the same fluid dynamic conditions, doses of aluminum, pH, the efficiencies obtained by coagulation and EC are very similar. The pH of the waste can be a key parameter in the choice of the coagulation technology [57]. Effect of different initial pH on removal efficiency in EC process is shown in Table 5.

Cost analysis

Cost analysis plays an important role in industrial wastewater treatment procedure/method as the wastewater treatment technique should be cost attractive. The costs involved in EC include, the cost of energy consumption,

Table 6 Effect of inter electrodes distance, conductivity of solutions, energy consumption and electrolysis time on heavy metals removal efficiency in EC process

References	Heavy metal	Reactor	Electrolysis time	Inter electrode distance	Conductivity (mS/cm)	Energy consumption	Efficiency (%)
[78]	Cr^{6+}	Continuous	10-12 min	-	-	-	70-85
[34]	Cu^{2+}, Zn^{2+}, Cr^{6+}	Continuous	20 min	5 mm	-	-	99, 99, 83
[19]	Cr^{6+}	Continuous	72 min	4 mm	1.5	1 kWh/m^3	80-97
[21]	Cr^{6+}	Batch	20 min	1.5 cm	1.6	1.92-2.29 kwh/m^3	99
[79]	Cr^{3+}	Continuous	20-25 min	22 mm	5.73 , 7.36	0.1KWh/m^3	A_M = 90.6, Bi = 71.4
[20]	Cr^{6+}	Batch	20, 60 min	1.5 cm	1.6	2.11 kWh/m^3	99.9
[80]	Cr^{6+}	Batch	45 min	5 mm	20	9.0 kWh/m^3	100
[81]	Cr^{6+}	Continuous	24 min	15 mm	2	137.2 KWh/m^3	90.4
[82]	Cr^{6+}	Continuous	75 min	4 cm	2.41, 1.70	-	65.3
[12]	Cr^{6+}	Batch	60 min	1.5 cm	1.6	35.06 kwh/g	98
[83]	Cr^{6+}	Batch	25 min	1.5 cm	0.59- 3.4	16.3 kWh/m^3	99
[84]	Cr^{6+}	Batch	5-10 min	0.3 cm	365	38 kWh/m^3	99
[55]	Cr^{3+}, Cr^{6+}	Batch	15 min	0.5 cm	-	-	100
[85]	Fe, Ni^{2+}, Cu^{2+}, Zn^{2+}, Pb^{2+}, Cd^{2+}	Batch	10 min	1 cm	2.1	-	SS = 86.5, Turbidity = 81.56, BOD_5 = 83, COD = 68, Color > 92.5
[19]	Cd^{2+}	Batch	20 min	1.5 cm	-	9.37 kwh/kg	>99
[86]	Cd^{2+}, Cu^{2+}	Continuous	120 min	1.5 cm	-	10.99 kWh/kg	Cd^{2+} = 73.8,Cu^{2+} = 98.8
[87]	Cd^{2+}	Batch	10 min	-	1.05- 5.22	-	>99.5
[88]	Cd^{2+}	Batch	30 min	5 mm	-	AC:0.6, DC: 1.2 kWh/m^3	AC: 97.8, DC: 96.9
[49]	Cd^{2+}	Batch	AC: 30, DC: 45 min	5 mm	-	AC:0.4, DC:1 kWh/ kg	AC: 97.5, DC: 96.2
[89]	Cd^{2+}	Continuous	200 min	1 cm	1.06	-	98.2
[90]	Zn phosphate	Batch and continuous	15 min = Fe electrode, 25 min = Al electrode	Batch = 11, continuous = 20 mm	Batch = 5.1-5.3, Continuous = 4.8- 4.9	Al electrode =0.18–11.29, Fe electrode = 0.24-8.47 kWh/m^3	Max 97.8
[24]	Zn^{2+}	Batch	10 min	11 mm	3000 µS/cm	3.3 kWh/kg	>99
[42]	Zn^{2+}, Cu^{2+}	Batch	60 min	1.5 cm	1.6	Zn^{2+} = 22.31, Cu^{2+} = 35.63KWh/g	99.99
[58]	COD, Zn^{2+}	Batch	50 min	16 mm	0.49	1.7 kWh/kg	COD = 88, 99.3, Zn^{2+} = 99
[91]	Ni^{2+}, Cu^{2+}	Batch	60 min	1 cm	634 µS/cm	-	Ni^{2+} = 82, Cu^{2+} = 99
[12]	Zn^{2+}, Cu^{2+}	Batch	60 min	1.5 cm	1.6	Zn^{2+} = 19.98, Cu^{2+} = 35.06 kWh	Zn^{2+} = Cu^{2+} = 99.9
[67]	Cu^{2+}, Cr, Ni^{2+}	Batch	20 min	10 mm	2	10.07 kWh/m^3	100
[17]	Ni^{2+}	Batch	20, 40 min	1 cm	1.6	9.37 kWh/kg	99.9
[50]	As	Continuous	Fe electrode =12.5, Al electrode = 15 min	13 mm	1.55	Fe electrode =0.015, Al electrode = 0.032 kWh/m^3	Fe = 93.5, Al = 95.7
[92]	As^{5+}, NO_2^-	Continuous	120 min	7 cm	-	-	Nitrite = 84, As^{5+} = 75

Table 6 Effect of inter electrodes distance, conductivity of solutions, energy consumption and electrolysis time on heavy metals removal efficiency in EC process (Continued)

Ref	Pollutant	Mode	Electrolysis time	Inter electrode distance	Conductivity	Energy consumption	Removal efficiency
[93]	As	Continuous	30 min	1.2 cm	1700 ± 37 μS/cm	3.03 kWh/m^3	89
[11]	Oil, grease, heavy metals	Continuous	105 s	-	-	Cu^{2+}, Ni^{2+} = 0.166, Zn^{2+} = 0.117, Oil and grease = 0.116, Turbidity = 0.117 kwh/m^3	Zn^{2+} = 99, Cu^{2+}, Ni^{2+} = 70, Oil and grease = 99.9, Turbidity = 99.7
[28]	Co	Batch	15 min	-	6.5	3.3 kwh/m^3	99
[94]	Heavy metal	Continuous	45 min	15 mm	8.9 ± 0.2	6.25 kWh/m3	Cr^{3+} = Cu^{2+} = 100, Ni^{2+} = 99
[16]	Boron	Batch	89 min	0.5 cm	30,000 mS/cm	2.4 kWh/m3	99.7
[95]	Ba, Zn^{2+}, Pb^{2+}	Continuous	20 min	10 mm	-	14 kWh/m3	97
[96]	Cd^{2+}	Batch	5 min	0.5 cm	1.176 mS/cm	1.6 kW h m 3	100
[97]	Ni^{2+}	Continuous	20 min	10 mm	1 mS/cm	-	100
[98]	Cr^{6+}	Batch	14 min	0.87 cm	adjusted	0.007 kWh/g	100
[99]	Zn^{2+}, Cu^{2+}, Ni^{2+}, Ag^+, $Cr_2O_7^{2-}$	Batch	30 min	5 mm	20 mS/cm	-	>50
[100]	Cr^{6+}	Continuous	60 min	20 mm	2.4 mS /cm	-	40

cost of the dissolved electrode (electrode consumption) and the cost of addition of any external chemical (for increasing the solution conductivity or varying the pH of the solution).

Electrode consumption can calculate by equation 8 which presented earlier. In addition, electrical energy consumption is a very important economical parameter in the electrocoagulation process and can calculated using the following equation [33]:

$$E = \left[\frac{UIt}{1000\ V} \right] \qquad (9)$$

where E is the energy consumption (kWh/m^3), U is the applied voltage (V), I is the current intensity (A), t is the electrocoagulation time (h), and V is the volume of the treated wastewater (m^3).

The detailed calculation of operating cost for the treatment of fluoride containing drinking water using EC has been reported by Ghosh et al. [58]. Espinoza-Quinones et al., (2009) studied the removal of organic and inorganic pollutants from a wastewater of lather finishing industrial process using EC. They found the EC to be cheaper compared to the conventional method. The operational cost for the EC was found to be US $ 1.7 per cubic meter of the treated tannery effluent as compared to the cost of US $ 3.5 per cubic meter of the treated effluent for conventional methods [59]. Similarly Bayramoglu et al. [60] have been reported that the operating cost of chemical coagulation is 3.2 times as high as that of EC for the treatment of textile wastewater.

Conclusions

The rapid urbanization and industrialization in the developing countries are creating high levels of water pollution due to harmful industrial effects and sewage discharges. The characteristics of industrial effluents in terms of nature of contaminates, their concentrations, treatment technique and required disposal method vary significantly depending on the type of industry. Further, the choice of an effluent treatment technique is governed by various parameters such as contaminants, their concentration, volume to be treated and toxicity to microbes. Electrocoagulation is a treatment process that is capable of being an effective treatment process as conventional methods such as chemical coagulation. Having observed trends over the last years, it has been noted that electrocoagulation is capable of having high removal efficiencies of color, chemical oxygen demand (COD), biochemical oxygen demand (BOD$_5$) and achieving a more efficient treatment processes quicker than traditional coagulation and inexpensive than other methods of treatment such as

ultraviolet (UV) and ozone. Unlike biological treatment which requires specific conditions, therefore limiting the ability to treat many wastewaters with high toxicity, xenobiotic compounds, and pH, electrocoagulation can be used to treat multifaceted wastewaters, including industrial, agricultural, and domestic. Continual research using this technology will not only improve new modeling techniques can be used to predict many factors and develop equations that will predict the effectiveness of treatment.

Electrocoagulation is an attractive method for the treatment of various kinds of wastewater, by virtue of various benefits including environmental capability, versatility, energy efficiency, safety, selectivity and cost effectiveness. The process is characterized by simple equipment, easy operation, less operating time and decreased amount of sludge which sediments rapidly and retain less water. However, further studies needs to be performed to study the effect of shape and geometry of the electrodes (punched hole and pitch of the holes) to possibly improve the pollutant removal efficiency. Efforts should be made to study the phenomena of electrode passivation to reduce the operating cost of the EC process. Most of the studies reported in the literature have been carried out at the laboratory scale using synthetic solutions. Efforts should be made to perform EC experiments at pilot plant scale using real industrial effluent to explore the possibility of using EC for treatment of real industrial effluents.

Competing interests

The authors declare that they have no competing interests.

Authors' contribution

EB and LM were performed design, analysis and writing the study. AA and AM contributed in writing of the manuscript. All authors read and approved the final manuscript.

Author details

[1]Health Promotion Research Center, Zahedan University of Medical Sciences, Zahedan, Iran. [2]Department of Environmental Health Engineering, School of Public Health, Tehran University of Medical Sciences, Tehran, Iran. [3]Center for Solid Waste Research, Institute for Environmental Research, Tehran University of Medical Sciences, Tehran, Iran. [4]National Institute of Health Research, Tehran University of Medical Sciences,, Tehran, Iran.

References

1. Bertrand S, Giuseppe T, Jeremie C, Jean-François M, Sophie G, Pierre-Marie B, et al. Heavy metal removal from industrial effluents by sorption on cross-linked starch: Chemical study and impact on water toxicity. J Environ Manage. 2011;92:765–72.
2. Blais JF, Dufresne S, Mercier G. State of the art of technologies for metal removal from industrial effluents. J Wat Sci. 1999;12/4:687–711.
3. Hunsom M, Pruksathorn K, Damronglered S, Vergnes H, Duverneuil P. Electrochemical treatment of heavy metals (Cu2+, Cr6⁺, Ni2+) from industrial effluent and modeling of copper reduction. Water Res. 2005;39:610–16.

4. Srivastava NK, Majumder CB. Novel biofiltration methods for the treatment of heavy metals from industrial wastewater. J Hazard Mater. 2008;28, 151(1):1–8.

5. Oyaro N, Juddy O, Murago ENM, Gitonga E. The contents of Pb, Cu, Zn and Cd in meat in Nairobi. Kenya Int J Food Agric Environ. 2007;5:119–21.

6. Paulino AT, Minasse FAS, Guilherme MR, Reis AV, Muniz EC, Nozaki J. Novel adsorbent based on silkworm chrysalides for removal of heavy metals from wastewaters. J Colloid Interf Sci. 2006;301:479–87.

7. Borba CE, Guirardello R, Silva EA, Veit MT, Tavares CRG. Removal of nickel^{2+} ions from aqueous solution by biosorption in a fixed bed column: experimental and theoretical breakthrough curves. Bio Chem Eng J. 2006;30:184–91.

8. Namasivayam C, Kadirvelu K. Uptake of mercury^{2+} from wastewater by activated carbon from unwanted agricultural solid by- product: coirpith. Carbon. 1999;37:79–84.

9. Naseem R, Tahir SS. Removal of Pb^{2+} from aqueous solution by using bentonite as an adsorbent. Water Res. 2001;35:3982–6.

10. Khezami L, Capart R. Removal of chromium (VI) from aqueous solution by activated carbons: Kinetic and equilibrium studies. J Hazard Mater. 2005;123:223–31.

11. Rincon GJ, Motta EJL. Simultaneous removal of oil and grease, and heavy metals from artificial bilge water using electrocoagulation/flotation. J Environ Manage. 2014;144:42–50.

12. Nouri J, Mahvi AH, Bazrafshan E. Application of electrocoagulation process in removal of zinc and copper from aqueous solutions by aluminum electrodes. Int J Environ Res. 2010;4:201–8.

13. Kamaraj R, Vasudevan S. Evaluation of electrocoagulation process for the removal of strontium and cesium from aqueous solution. Che Eng Res Des. 2015;93:522–30.

14. Mahvi AH, Ebrahimi SJA, Mesdaghinia A, Gharibi H, Sowlat MH. Performance evaluation of a continuous bipolar electrocoagulation/electrooxidation–electroflotation (ECEO–EF) reactor designed for simultaneous removal of ammonia and phosphate from wastewater effluent. J Hazard Mater. 2011;192(3):1267–74.

15. Apaydin KU, Gonullu MT. An investigation on the treatment of tannery wastewater by electrocoagulation. Global Nest J. 2009;11:546–55.

16. Mohamed-Hasnain I, Ezerie HE, Zubair A, Saleh FM, Shamsul R, Mohamed K. Boron removal by electrocoagulation and recovery. Water Res. 2014;51:113–23.

17. Bazrafshan E, Ownagh K, Mahvi AH. Application of electrocoagulation process using iron and aluminum electrodes for fluoride removal from aqueous environment. E- J Chem. 2012;9(4):2297–308.

18. Malakootian M, Yousefi N, Fatehizadeh A. Survey efficiency of electrocoagulation on nitrate removal from aqueous solution. Int J Environ Sci Technol. 2011;8(1):107–14.

19. Bazrafshan E, Mahvi AH, Nasseri S, Mesdaghinia AR, Vaezi F, Nazmara S. Removal of cadmium from industrial effluents by electrocoagulation process using iron electrodes. Iran J Environ Health Sci Eng. 2006;3(4):261–6.

20. Bazrafshan E, Mahvi AH, Naseri S, Mesdaghinia AR. Performance evaluation of electrocoagulation process for removal of chromium^{6+} from synthetic chromium solutions using iron and aluminum electrodes. Turk J Eng Environ Sci. 2008;32:59–66.

21. Mahvi AH, Bazrafshan E. Removal of cadmium from industrial effluents by electrocoagulation process using aluminum electrodes. World Appl Sci J. 2007;2(1):34–9.

22. Gholami Borujeni F, Mahvi AH, Nejatzadeh-Barandoozi F. Removal of heavy metal ions from aqueous solution by application of low cost materials. Fresen Environ Bull. 2013;22(3):655–8.

23. Pociecha M, Lestan D. Using electrocoagulation for metal and chelant separation from washing solution after EDTA leaching of Pb, Zn and Cd contaminated soil. J Hazard Mater. 2010;174(1–3):670–8.

24. Kobya M, Demirbas E, Sozbir M. Depolarization of aqueous reactive dye Remazol Red 3B by electrocoagulation. Color Technol. 2010;126(5):282–8.

25. Mansoorian HJ, Rajabizadeh A, Bazrafshan E, Mahvi AH. Practical assessment of electrocoagulation process in removing nickel metal from aqueous solutions using iron-rod electrodes. Desalin Water Treat. 2012;44:29–35.

26. Kabdaşlı I, Vardar B, Alaton AI, Tünay O. Effect of dye auxiliaries on color and COD removal from simulated reactive dye bath effluent by electrocoagulation. Chem Eng J. 2009;148:89–96.

27. Chaturvedi SI. Mercury removal using Al-Al electrodes by electrocoagulation. IJMER. 2013;3(1):109–15.

28. Shafaei A, Pajootan E, Nikazar M, Arami M. Removal of Co^{2+} from aqueous solution by electrocoagulation process using aluminum electrodes. Desalination. 2011;279(1–3):121–6.

29. Holt PK, Barton GW, Mitchell CA. Deciphering the science behind electrocoagulation to remove clay particles from water. Water Sci Technol. 2004;50(12):177–84.

30. Pouet MF, Grasmick A. Urban wastewater treatment by electrocoagulation and flotation. Water Sci Technol. 1995;31(3–4):275–83.

31. Jung KW, Hwang MJ, Park DS, Ahn KH. Combining fluidized metal-impregnated granular activated carbon in three-dimensional electrocoagulation system: Feasibility and optimization test of color and COD removal from real cotton textile wastewater. Sep Purif Technol. 2015;146:154–67.

32. Adjeroud N, Dahmoune F, Merzouk B, Leclerc JP, Madani K. Improvement of electrocoagulation–electroflotation treatment of effluent by addition of Opuntia ficus indica pad juice. Sep Purif Technol. 2015;144:168–76.

33. Bazrafshan E, Biglari H, Mahvi AH. Phenol removal by electrocoagulation process from aqueous solutions. Fresen Environ Bull. 2012;21(2):364–71.

34. Adhoum N, Monser L, Bellakhal N, Belgaied JE. Treatment of electroplating wastewater containing Cu^{2+}, Zn^{2+} and Cr^{6+} by electrocoagulation. J Hazard Mater. 2004;B 112(3):207–13.

35. Meunier N, Drogui P, Montane C, Hausler R, Mercier G, Blais JF. Comparison between electrocoagulation and chemical precipitation for metals removal from acidic soil leachate. J Hazard Mater. 2006;137:581–90.

36. Yu MJ, Koo JS, Myung GN, Cho YK, Cho YM. Evaluation of bipolar electrocoagulation applied to biofiltration for phosphorus removal. Water Sci Technol. 2005;51(10):231–9.

37. Matteson MJ, Dobson RL, Glenn J, Robert W, Kukunoor NS, Waits I, et al. Electrocoagulation and separation of aqueous suspensions of ultrafine particles. Colloid Surface A. 1995;104(1):101–9.

38. Chen X, Chen G, Yue PL. Separation of pollutants from restaurant wastewater by electrocoagulation. Sep Purif Technol. 2000;19(1–2):65–76.

39. Chen G. Electrochemical technologies in wastewater treatment. Sep Purif Technol. 2004;38:11–41.

40. Janpoor F, Torabian A, Khatibikamal V. Treatment of laundry wastewater by electrocoagulation. J Chem Technol Biotechnol. 2011;86(8):1113–20.

41. Kobya M, Senturk E, Bayramoglu M. Treatment of poultry slaughterhouse wastewaters by electrocoagulation. J Hazard Mater. 2006;133(1–3):172–6.

42. Bazrafshan E, Mahvi AH, Zazoli MA. Removal of zinc and copper from aqueous solutions by electrocoagulation technology using iron electrodes. Asian J Chem. 2011;23(12):5506–10.

43. Mollah MYA, Morkovsky P, Gomes JAG, Kesmez M, Parga J, Cocke DL. Fundamentals, present and future perspectives of electrocoagulation. J Hazard Mater. 2004;B114(1–3):199–210.

44. Zaroual Z, Azzi M, Saib N, Chainet E. Contribution to the study of electrocoagulation mechanism in basic textile effluent. J Hazard Mater. 2006;B131:73–8.

45. Khandegar V, Saroha AK. Electrocoagulation for the treatment of textile industry effluent- A review. J Environ Manage. 2013;128:949–63.

46. Mollah M, Schennach R, Parga JR, Cocke DL. Electrocoagulation (EC) science and applications. J Hazard Mater. 2001;84:29–41.

47. Yazdanbakhsh AR, Massoudinejad MR, Arman K, Aghayani E. Investigating the Potential of Electrocoagulation- Flotation (ECF) Process for Pollutants Removal from Olive Oil Mill Wastewater. J Appl Environ Biol Sci. 2013;3(3):22–8.

48. Merzouk B, Madani K, Sekki A. Using electrocoagulation- electroflotation technology to treat synthetic solution and textile wastewater.two case studies. Desalination. 2010;250(5):573–7.

49. Vasudevan S, Lakshmi J, Sozhan G. Effects of alternating and direct current in electrocoagulation process on the removal of cadmium from water. J Hazard Mater. 2011;192(1):26–34.

50. Kobya M, Ulu F, Gebologlu U, Demirbas E, Oncel MS. Treatment of potable water containing low concentration of arsenic with electrocoagulation: different connection modes and Fe-Al electrodes. Sep Purif Technol. 2011;77(3):283–93.

51. Kuroda Y, Kawada Y, Takahashi T, Ehara Y, Ito T, Zukeran A, et al. Effect of electrode shape on discharge current and performance with barrier discharge type electrostatic precipitator. J Electrostat. 2003;57(3):407–15.

52. Bukhari AA. Investigation of the electrocoagulation treatment process for the removal of total suspended solids and turbidity from municipal wastewater. Bioresour Technol. 2008;99(5):914–21.

53. Vik EA, Carlson DA, Eikum AS, Gjessing ET. Electrocoagulation of potable water. Water Res. 1984;18(11):1355–60.

54. Aoudj S, Khelifa A, Drouiche N, Belkada R, Miroud D. Simultaneous removal of chromium (VI) and fluoride by electrocoagulation- electroflotation: Application of a hybrid Fe-Al anode. Chem Eng J. 2015;267:153–62.

55. Verma SK, Khandegar V, Saroha AK. Removal of chromium from electroplating industry effluent using electrocoagulation. J Hazard Toxic Radio Waste. 2013;17(2):146–52.

56. Chen G. Electrochemical technologies in wastewater treatment. Sep Purif Technol. 2004;38:11–41.

57. Canizares P, Jiménez C, Martínez F, Rodrigo M, Saez C. The pH as a key parameter in the choice between coagulation and electrocoagulation for the treatment of wastewaters. J Hazard Mater. 2009;163:158–64.

58. Ghosh P, Samanta AN, Ray S. Reduction of COD and removal of Zn^{2+} from rayon industry wastewater by combined electrofenton treatment and chemical precipitation. Desalination. 2011;266:213–17.

59. Espinoza- Quinones FR, Fornari MMT, Modenes AN, Palacio SM, DaSilva FG, Szymanski N, et al. Pollutant removal from tannery effluent by electrocoagulation. Chem Eng J. 2009;151(1–3):59–65.

60. Bayramoglu M, Eyvaz M, Kobya M. Treatment of the textile wastewater by electrocoagulation economical evaluation. Chem Eng J. 2007;128:155–61.

61. Orisakwe OE, Nduka JK, Amadi CN, Dike DO, Bede O. Heavy metals health risk assessment for population via consumption of food crops and fruits in Owerri, South Eastern, Nigeria. Chem Cent J. 2012;6:77.

62. Institute of Standards and Industrial Research of Iran (ISIRI). (2009). Drinking water: Physical and chemical specifications. 5th edition. Standard No. 1053.

63. Ozaki H, Sharmab K, Saktaywirf W. Performance of an ultra- low- pressure reverse osmosis membrane (ULPROM) for separating heavy metal: effects of interference parameters. Desalination. 2002;144:287–94.

64. Barakat MA, Schmidt E. Polymer- enhanced ultrafiltration process for heavy metals removal from industrial wastewater. Desalination. 2010;256:90–3.

65. Kryvoruchko A, Yurlova L, Kornilovich B. Purification of water containing heavy metals by chelating- enhanced ultrafiltration. Desalination. 2002;144:243–8.

66. Qdaisa HA, Moussa H. Removal of heavy metals from wastewater by membrane processes: a comparative study. Desalination. 2004;164:105–10.

67. Akbal F, Camci S. Copper, chromium and nickel removal from metal plating wastewater by electrocoagulation. Desalination. 2011;269:214–22.

68. Kabdasli I, Arslan T, Olmez HT, Alaton AI, Tünay O. Complexing agent and heavy metal removals from metal plating effluent by electrocoagulation with stainless steel electrodes. J Hazard Mater. 2009;15, 165(1–3):838–45.

69. Chen QY, Luo Z, Hills C, Xue G, Tyrer M. Precipitation of heavy metals from wastewater using simulated flue gas: sequent additions of fly ash, lime and carbon dioxide. Water Res. 2009;43:2605–14.

70. Alvarez MT, Crespo C, Mattiasson B. Precipitation of Zn^{2+}, Cu^{2+} and Pb^{2+} at bench- scale using biogenic hydrogen sulfide from the utilization of volatile fatty acids. Chemosphere. 2007;66:1677–83.

71. Inglezakis VJ, Grigoropoulou HP. Modeling of ion exchange of Pb^{2+} in fixed beds of clinoptilolite. Micropor Mesopor Mat. 2003;61:273–82.

72. Inglezakis VJ, Stylianou MA, Gkantzou D, Loizidou MD. Removal of Pb^{2+} from aqueous solutions by using clinoptilolite and bentonite as adsorbents. Desalination. 2007;210:248–56.

73. Bazrafshan E, Biglari H, Mahvi AH. Humic acid removal from aqueous environments by electrocoagulation process using iron electrodes. E J Chem. 2012;9(4):2453–61.

74. Bazrafshan E, Moein H, Kord Mostafapour F, Nakhaie S. Application of electrocoagulation process for dairy wastewater treatment. J Chem. 2013;Article. ID 640139:8.

75. Bazrafshan E, Farzadkia M, Kord Mostafapoor F, Ownagh KA, Mahvi AH. Slaughterhouse wastewater treatment by combined chemical coagulation and electrocoagulation process. PLOS ONE. 2012;7(6):1–8.

76. Bazrafshan E, KordMostafapoor F, Soori MM, Mahvi AH. Application of combined chemical coagulation and electrocoagulation process to carwash wastewater treatment. Fresen Environ Bull. 2012;21(9a):2694–701.

77. Bazrafshan E, Mahvi AH, Zazouli MA. Textile wastewater treatment by electrocoagulation process using aluminum electrodes. Iran J Health Sci. 2014;2(1):16–29.

78. Osipenko VD, Pogorelyi PI. Electrocoagulation neutralization of chromium containing effluent. Metallurgist. 1977;21(9–10):44–5.

79. Golder AK, Samanta AN, Ray S. Removal of trivalent chromium by electrocoagulation. Sep Purif Technol. 2007;53(1):33–41.

80. Heidmann I, Calmano W. Removal of Cr^{6+} from model wastewaters by electrocoagulation with Fe electrodes. Sep Purif Technol. 2008;61:15–21.

81. Bhatti MS, Reddy AS, Thukral AK. Electrocoagulation removal of Cr^{6+} from simulated wastewater using response surface methodology. J Hazard Mater. 2009;172:839–46.

82. Mouedhen G, Feki M, De Petris WM, Ayedi HF. Electrochemical removal of Cr^{6+} from aqueous media using iron and aluminum as electrode materials: towards a better understanding of the involved phenomena. J Hazard Mater. 2009;168(2–3):983–91.

83. Keshmirizadeh E, Yousefi S, Rofouei MK. An investigation on the new operational parameter effective in Cr^{6+} removal efficiency: a study on electrocoagulation by alternating pulse current. J Hazard Mater. 2011;190(1–3):119–24.

84. Bassam AA. Electrocoagulation of chromium containing synthetic wastewater using monopolar iron electrodes. Damascus University J. 2012;28(2):79–89.

85. Merzouk B, Gourich B, Sekki A, Madani K, Chibaned M. Removal turbidity and separation of heavy metals using electrocoagulation-electroflotation technique a case study. J Hazard Mater. 2009;164:215–22.

86. BashaAhmed C, Bhadrinarayana NS, Anantharaman N. Heavy metal removal from copper smelting effluent using electrochemical cylindrical flow reactor. J Hazard Mater. 2008;152:71–8.

87. Chien H, Luke C, Chen-Lu Y. Effect of anions on electrochemical coagulation for cadmium removal. Sep Purif Technol. 2009;65:137–46.

88. Vasudevan S, Lakshmi J. Effects of alternating and direct current in electrocoagulation process on the removal of cadmium from water e a novel approach. Sep Purif Technol. 2011;80(3):643–51.

89. Saber EM, ElSayed MN, Ibrahim HH, Osama AD, Abdykalykova R, Beisebekov M. Removal of Cadmium Pollutants in Drinking Water Using Alternating Current Electrocoagulation Technique. G J E R. 2013;7(3):45–51.

90. Kobya M, Demirbas E, Dedeli A, Sensoy MT. Treatment of rinse water from zinc phosphate coating by batch and continuous electrocoagulation processes. J Hazard Mater. 2009;173(1–3):326–34.

91. Khelifa A, Moulay S, Naceur AW. Treatment of metal finishing effluents by the electroflotation technique. Desalination. 2005;181:27–33.

92. Kumar NS, Goel S. Factors influencing arsenic and nitrate removal from drinking water in a continuous flow electrocoagulation (EC) process. J Hazard Mater. 2010;173(1–3):528–33.

93. Emilijan M, Srdjan R, Jasmina A, Aleksandra T, Milena M, Mile K, et al. Removal of arsenic from groundwater rich in natural organic matter (NOM) by continuous electrocoagulation/flocculation (ECF). Sep Purif Technol. 2014;136:150–6.

94. Al-Shannag M, Al-Qodah Z, Bani-Melhem K, Rasool-Qtaishat M, Alkasrawi M. Heavy metal ions removal from metal plating wastewater using electrocoagulation: Kinetic study and process performance. Chem Eng J. 2015;260:749–56.

95. Mota IO, Castro JA, Casqueira RG, Junior AG. Study of electroflotation method for treatment of wastewater from washing soil contaminated by heavy metals. J Mater Res Technol. 2015;4(2):109–13.

96. Brahmi Kh, Bouguerra W, Hamrouni B, Elaloui E, Loungou M, Tlili Z. Investigation of electrocoagulation reactor design parameters effect on the removal of cadmium from synthetic and phosphate industrial wastewater. Arab J of Chem. 2015; http://dx.doi.org/10.1016/j.arabjc.2014.12.012.

97. Jun L, Yan L, Mengxuan Y, Xiaoyun M, Shengling L. Removing heavy metal ions with continuous aluminum electrocoagulation: A study on back mixing and utilization rate of electro-generated Al ions. Chem Eng J. 2015;267:86–92.

98. El-Taweel Y, Nassef E, Elkheriany I, Sayed D. Removal of Cr(VI) ions from waste water by electrocoagulation using iron electrode. Egypt J Petrol. 2015; http://dx.doi.org/10.1016/j.ejpe.2015.05.011.

99. Heidmann I, Calmano W. Removal of Zn(II), Cu(II), Ni(II), Ag(I) and Cr(VI) present in aqueous solutions by aluminium electrocoagulation. J Hazard Mater. 2008;152:934–41.

100. Zongo I, Leclerc JP, Maiga HA, Wéthé J, Lapicque F. Removal of hexavalent chromium from industrial wastewater by electrocoagulation: A comprehensive comparison of aluminium and iron electrodes. Sep Purif Technol. 2009;66:159–66.

2

Influence of upflow velocity on performance and biofilm characteristics of Anaerobic Fluidized Bed Reactor (AFBR) in treating high-strength wastewater

Jalil Jaafari[1], Alireza Mesdaghinia[1], Ramin Nabizadeh[1], Mohammad Hoseini[2], Hossein kamani[3] and Amir Hossein Mahvi[1,4,5*]

Abstract

One of the key parameters in Fluidized Bed reactors is the control of biofilm thickness and configuration. The effect of upflow velocity on performance and biofilm characteristics of an Anaerobic Fluidized Bed Reactor was studied in treating Currant wastewater at various loading rates. The reactor used this study was made of a plexiglass column being 60 mm diameter, 140 cm height, and a volume of 3.95 L. The results demonstrated that the AFBR system is capable of handling an exceptionally high organic loading rate. At organic loading rates of 9.4 to 24.2 (kg COD m^{-3}) at steady state, reactor performances with upflow velocities of 0.5, 0.75 and 1 (m min^{-1}) were 89.3- 63.4, 96.9 – 79.6 and 95 – 73.4 percent, respectively. The average biomass concentration per unit volume of the AFBR (as gVSSatt L^{-1} expended bed) decreased with the increase of upflow velocity in the range of 0.5–1 m min^{-1} at all applied organic loading rates. The total biomass in the reactor increased with increases in the organic loading rate. The peak biomass concentration per unit volume (as gVSSatt L^{-1} expended bed) was observed at the bottom part of the reactor, then it droped off slowly towards the top. The biofilm thickness increased from the bottom to the top of the reactor representing a stratification of the media in the AFBR. The bed porosity increased from the bottom to the top of the reactor.

Keywords: Biofilm characteristics, Biomass concentration, Anaerobic Fluidized Bed Reactor, Currant wastewater, Upflow velocity

Introduction

In recent years many alternatives have been performed to treatment of high-strength wastewaters [1-4]. Anaerobic Fluidized Bed Reactors (AFBR) were originally a chemical engineering tool used to perform phase transformations, reactions, and diffusions of various chemicals existing in solid, liquid, and vapor phases. With the concept of maximum diffusion and maximum chemical reaction within a minimum volume in mind, AFBRs have been used in biological wastewater treatment and are utilized in several process configurations [5-7]. The results

from recent studies have consistently illustrated the technical advantage of the fluidized bed over most other suspended and attached growth biological systems. Typically, in a similar capacity, efficiency of the AFBR can be more than 10 times of the activated sludge system while the total space occupied by AFBR is about 10 percent of the required space for stirred tank in activated sludge process [8]. This is due to the AFBR ability in maintaining high concentration of biomass compared with conventional activated sludge system (40,000 mg L^{-1} vs. 3000 mg L^{-1}) [9]. Fluidization can overcome operating problems such as bed clogging and high pressure drop, which happen if the media with high surface area used in packed-bed reactor. Another advantage of using media is possibility of elimination of secondary clarifiers [10]. Anaerobic Fluidized bed reactors (AFBR) are high-load wastewater treatment

* Correspondence: ahmahvi@yahoo.com
[1]School of Public Health, Tehran University of Medical Sciences, Tehran, Iran
[4]Center for Solid Waste Research, Institute for Environmental Research, Tehran University of Medical Sciences, Tehran, Iran
Full list of author information is available at the end of the article

systems, which have been studied by numerous authors to treat different industrial wastewaters. For example, this system has been used for treatment of textile wastewater [11], ice-cream wastewater [12], and brewery wastewater [13], winery wastewater from Grape-Red and tropical fruit [14], currant [15] and sanitary landfill leachate [16]. The microbial population is the critical parameter in the performance of biological process that the influenced by operational parameters, physicochemical properties of the carrier material (density, roughness, porosity) on the fixed bed process are critical considerations [17,18]. One of the key operational parameters in attached biofilm reactors is the control of biofilm thickness and configuration, and research on biofilm formation and detachment has developed considerably in the past years, although there is no design rule for the rate of detachment. The prediction of biofilm structure (density, porosity, roughness, shape) and thickness is most important in designing and operation of biofilm processes, because hydrodynamics, mass transfer and conversion in biofilm processes depend on these variables. In attached growth process, biofilm accumulation is a dynamic process that is the net result of growth and the detachment processes. This is affected by several external factors, including composition and concentration of the feed, concentration of particles, particle–particle collisions, and particle–wall collisions, and velocity of the liquid phase (shear stress). This is the most important factor influencing formation, structure and stability of biofilms. In a biofilm system, higher hydrodynamic shear force take a stronger biofilm, and the biofilm tends to become a heterogeneous, porous and weaker structure when the shear force is too weak [19-22].

The main objective of this study was to investigate the influence of different upflow velocity on performance and biofilm characteristics of Anaerobic Fluidized Bed Reactor in treating a real Currant wastewater in various HRT and loading rates.

Materials and methods
Anaerobic Fluidized bed reactor
The reactor was made of a plexiglass column being 60 mm diameter, 140 cm height, and a volume of 3.95 L. The enlarged top section of column was used as a gas–solid separator. The enlarged section had diameter of 100 mm with a volume of 1.48 L (Figure 1). The bottom of the reactor was flat with symmetrically placed four pores through which flow was equally distributed into the reactor. The column has six sampling ports located at 5, 30, 55, 80, 105 and 130 cm above the reactor bottom. The recycled flow was drawn from the top section using a Circulator Pump and then fed upward into the reactor. Reactor temperature was controlled by Aquarium Heater at $35 \pm 2°C$. The reactor was loaded with 1.48 kg

media made of PVC with a mean diameter of 2 mm as a biofilm carrier to a settled depth of 0.6 m. The particles had a specific gravity of 1.45, a porosity of 0.4, and a specific surface area of $1800 \ m^2 \ m^{-3}$. The bed expansion in fluidized bed reactor was 30% during the start-up period. The expansion of the bed should be determined based on consideration of the minimum fluidization velocity. Some studies have been determined the factors affecting minimum fluidization velocity and maximum pressure drop [23-25]. Also, Peng and Fan [26] developed theoretical models for estimating of minimum fluidization velocity and maximum pressure drop, based on the dynamic balance of forces exerted on the particle. Some of the well known correlations available for predicting the minimum fluidization velocity (Umf) and maximum pressure drop (Pmax) for tapered beds are those by Peng and Fan. The influence of superficial velocity on pressure drop in reactor is illustrated and shown in Figure 2. By increasing upflow velocity to 0.75 (m min^{-1}), pressure drop increased, then, with increasing superficial velocity to more than 0.75 (m min^{-1}), pressure drop remained constant. So, to ensure that the fluidization condition is exist, the superficial velocity should not be less than the minimum fluidization velocity which is 0.75 m min^{-1} in our reactor. Also, upflow velocity in start-up period was adjusted 0.75 (m min^{-1}). After the start-up period, the real Currant wastewater was fed to the reactor and upflow velocities in different organic loading rate were adjusted to 0.5, 0.75 and 1 m min^{-1}.

Start-up period
Anaerobic reactor was seeded with 1 L of aerobic active sludge obtained from aerobic digesters of municipal wastewater treatment plant with MLSS and MLVSS of 24.84 and 16.9 g L^{-1}, respectively. Table 1 is a summary of conditions tested during the start-up. The reactor fed with synthetic wastewater contained methanol, glucose, and Currant wastewater. Some macro and micronutrients such as $CaCl_2.2H_2O$ (50 mg L^{-1}), $(NH_4)_2.HPO_4$ (80 mg L^{-1}), $FeCl_2.4H_2O$ (40 mg L^{-1}), NH_4Cl (1200 mg L^{-1}), $Na_2S.9H_2O$ (300 mg L^{-1}), $CuCl_2.2H_2O$ (0.5 mg L^{-1}), $MgSO_4.7H_2O$ (400 mg L^{-1}), H_3BO_3 (0.5 mg L^{-1}), $MnCl_2.4H_2O$ (0.5 mg L^{-1}), $NaWO_4.2H_2O$ (0.5 mg L^{-1}), $AlCl_3.6H_2O$ (0.5 mg L^{-1}), Na_2SeO_3 (0.5 mg L^{-1}), mg/l), KCl (400 mg L^{-1}), $ZnCl_2$ (0.5 mg L^{-1}), $NaHCO_3$ (3000 mg L^{-1}), $NaMoO_4.2H_2O$ (0.5 mg L^{-1}), $CoCl_2.6H_2O$ (10 mg L^{-1}), KI (10 mg L^{-1}), and $NiCl_2.6H_2O$ (0.5 mg L^{-1}), which are needed for optimal biofilm growth were used. In colonization stage, anaerobic fluidized reactor was run in the batch mode for one week. Then, during the start-up period run in the continuous mode and COD concentration in feed was gradually raised. Also methanol, which included 75% of the total influent COD, was used in the beginning to encouraging the growth of methanosarcina

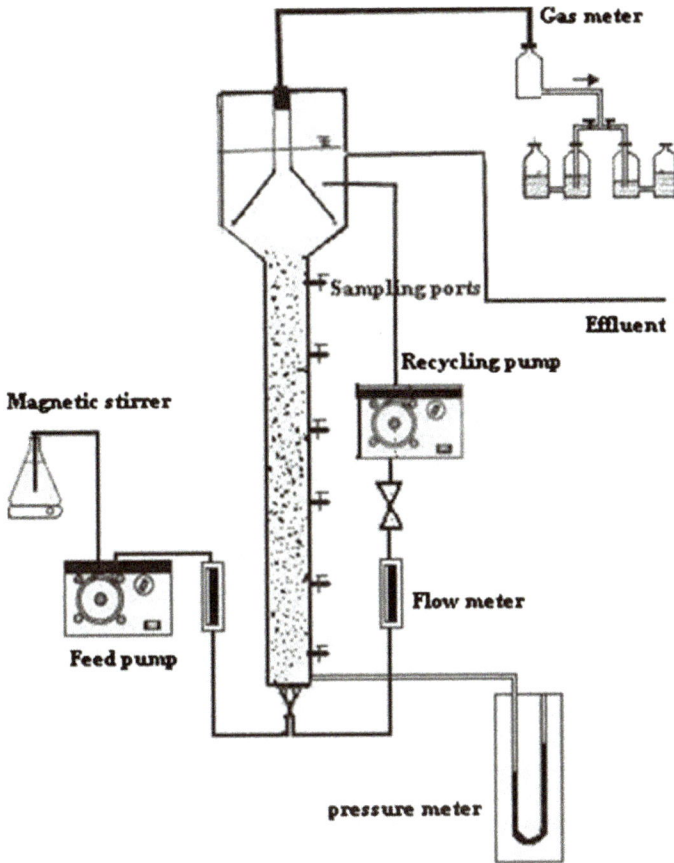

Figure 1 Schematic configuration of Anaerobic Fluidized bed reactor.

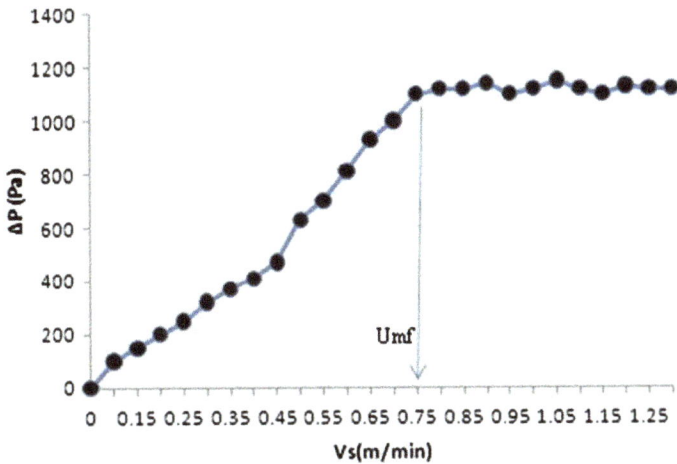

Figure 2 Effect of liquid upflow velocity on pressure drop.

Table 1 Organic loading and characteristics of fed during the start-up

Time (d)	COD loading (kg COD/m^3)	Methanol a	Glucose a	Currant wastewater a	NH$_4$Cl b
0-10	0.5 - 4	75	25	0	50
11-20	4 - 7	50	50	0	75
21-30	7 - 11	25	75	0	100
31-40	11 - 13	0	75	25	100
41-50	13 - 15	0	50	50	100
51-60 ·	13 - 15	0	25	75	100

a- % of total COD b- % of its value at the end of the start -up.

bacteria [27]. Then, the percent of methanol in the influent was gradually decreased to 50%, 25%, and 0% in days 11, 21, and 31, respectively by replacing with glucose and Currant wastewater. Additionally, in the start-up period, NH$_4$Cl concentration was gradually increased to taken high C/N ratios (1200 mg L-1). Part of this N with carbon was used by bacteria for building up the new cell to encourage extra cellular polymer production, which aids bacterial attachment on solid surface [28].

Operation period

In the operational period that lasted 372 d, the Anaerobic Fluidized bed reactor was fed with real Currant wastewater. The real currant wastewater obtained from the factory located in the Safadasht Industrial Zone, Shahriar, Iran, that in Characteristics of Currant wastewater is given in Table 2. The AFBR was operated under five different hydraulic retention times of 48, 40, 32, 24 and 18 h, respectively and each of HRT operated under three upflow velocities of 0.5, 0.75 and 1 m min^{-1}, respectively.

Analytical methods

Samples were analyzed for COD according to standard method [29]. Temperature was measured by a thermometer and pH was measured by a pH-meter (E520 Metrohm Herisau). The biofilm thickness was measured using the method of Schreyer and Coughlin [30], according to the following method. A slurry sample of known volume was smoothly washed to remove the suspended solids and then filtered. The wet bio-particles were carefully removed from the filter into a ceramic dish and weighed to determine its wet mass. After oven-drying for 24 h at 105°C, then cooled in a desiccator and weighed. The dried sample was ignited in a 550°C furnace for 30 min, cooled in a desiccator and then weighed. The difference between two dried weights would yield the weight of immobilized biomass as attached volatile solids (AVS). Also for ensure the results obtained from the Schreyer and Coughlin procedure, the biofilm thickness was measured using a high-resolution microscope equipped with a micrometer [30] method. In comparison of two measurements, the relative error was always less than 10%. The bio-particle density was measured from its settling velocity and diameter of bio-particle [31].

Results and discussion

The start-up period was completed in 60 d. So, that the feed COD increased stepwise and effluent COD of the anaerobic decreased and the COD removal efficiency gradually increased. In the end of the start-up period, in Anaerobic Fluidized Bed Reactor, attached volatile solid (AVS) concentration reached to 0.0185 gvss g^{-1} which is in accordance with ranges 0.074–0.11 reported by Farhan et al., 1997 [32], 0.039 [33], 0.05 [34], 0.0732 [27] and 0.0375–0.429 gvss g^{-1} by [35]. Table 3 shows operational parameters obtained at the end of start-up period.

Table 2 Characteristics of currant wastewater used in the present study

Parameter	Value		
	Range	Average	SD
pH Value	5.2-7.3	6	0.7
COD (mg/L)	17200-19000	18250	447
BOD5 (mg/L)	12500-13000	12748	185
TSS (mg/L)	331-410	365	23.3
COD–BOD ratio	1.45	-	-
Tot-P (mg/L)	12-25	18	3.8
Tot-N (mg/l)	41-86	60	13.1

Table 3 Operational parameters obtained at the end of start-up period for AFBR

Operational parameters	FBR
OLR, kg COD/m3	15
HRT	24
Upflow velocity (m/min)CBU	0.75
Expansion %	30
Volume of expanded bed (cm3)	2210
M support (g)	1480
VSatt (g)	27.5
g VSatt/g support	0.0185
g VSatt/l expanded bed	11.9

Effect of organic loading rate and upflow velocity on COD removal

Figure 3 and Table 4 show the effect of the OLR on the COD removal efficiency (E) and COD effluent in reactor throughout the operation time for the reactor studied. As shown, during stage 1, OLR in AFBR was kept at around 9.4 g COD L.d^{-1} with the feed COD concentration of $18,000 \pm 300$ mg L^{-1} and HRT around 48 h, the reactor performance was investigate for 0.5, 0.75 and 1 m min^{-1} upflow velocities. At steady state, with 0.5, 0.75 and 1 m min^{-1} upflow velocities, the reactor performances in stage 1 were 89.3, 96.6 and 95 percent, respectively. At stage 2, the HRT of reactor was decreased from 48 h to 40 h and OLR in Anaerobic Fluidized Bed Reactor increased from 9.4 to 10.8 g COD L.d^{-1} and feed COD concentration was as same as the stage 1, the reactor performance dropped to 86, 95.2 and 94 percent, respectively. As can be seen, in the stage 1 and 2, in the second set, 0.75 m min^{-1} upflow velocity had more removal efficiency than other upflow velocities. Also In other stages, in the second set, 0.75 m min^{-1} upflow velocity had more removal efficiency than other upflow velocities. In the stage 5, the OLR was further increased to 24.2 g COD L.d^{-1}. By decreasing HRT to 18 h, reactor performance dropped to 63.4, 79.6 and 73.4 percent, respectively. The average COD concentration in the effluent of the three sets at a loading rate of 9.4 g COD L.d^{-1} was 2020, 630 and 940 mg L^{-1}, respectively. Then, the average COD concentration in the effluent of the three sets at a loading rate of 24.2 g COD L.d^{-1} increased gradually to 6515, 3666 and 4835 mg L^{-1}, respectively.

Higher biodegrading rates were generally achieved at relatively lower superficial velocities. However there was a minimum practical velocity (0.5 m min^{-1}) below which would agglomeration of media occur in the reactor and the anaerobic process might disrupt. Also the subsequent decrease of the fluidization percentage in 0.5 m min^{-1} upflow velocity, which is below the minimum fluidization velocity, might have mass transfer limitations caused by accumulation of fatty acids in the reactor [36]. The substrate utilization rate in the biological process, correlated to diffusion resistance, is strongly dependent on reactor design and mixing intensity [37]. In the third set in Vs of 1 m min^{-1}, the reactor performance was lower in compare with the second set with the Vs of 0.75 m min^{-1}, because the biofilm was detached and washed out of the system as a result of the increased shearing force and bed porosity. In the treatment of high-strength distillery wastewater by anaerobic fluidized bed reactor with natural zeolite, COD removals of 80% were achieved at OLR of 20 g COD L.d-1 and HRT of 11 h [38]. In another study with anaerobic fluidized bed reactor for treating ice-cream wastewater, at an organic COD loading rate of 15.6 g L-1. d and HRT of 8 h, COD removal efficiencies of 94.4% was achieved [12]. In the treatment of thin stillage wastewater using an anaerobic fluidized bed with OLR of 29 g COD L.d-1 and HRT of 3.5 h, COD removal efficiencies of 88% was achieved [38]. In the stage 1 and 2, with increasing the upflow velocity, COD removal rate due to appropriate mass balance was improved. But, in the stages 3 to 5, in three set with 0.75 m min^{-1} upflow velocity, Vs was increased due to the decrease in the biomass concentration, which resulted increase in shearing force and increase in bed porosity, while the organic loading rate in the reactor was increasing.

Effect of the upflow velocity and organic loading rate on the biomass concentration

The effect of the upflow velocity on the average biomass concentration in the AFBR is illustrated by Figure 4. As shown, that the average biomass concentration per unit volume of the AFBR decreased with the increase of the upflow velocity at all of organic loading rate. For example, at OLR of 9.4 g COD L.d^{-1} with HRT of 48 h,

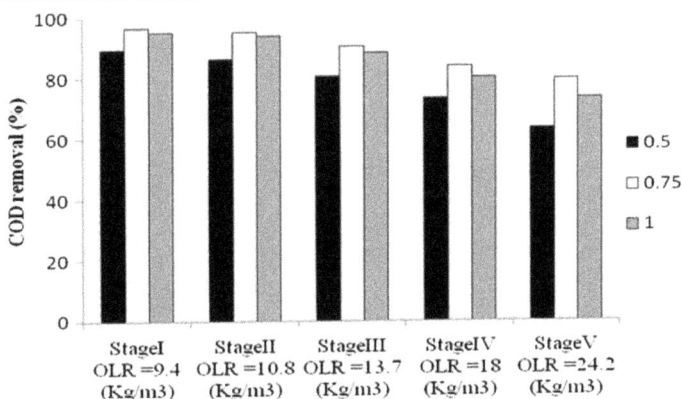

Figure 3 Effect of the upflow velocity and organic loading rate on reactor performance.

Table 4 Summary of the average results of the three sets of experiments at steady state

Stage	Time (d)	Vs (m/min)	Q_{in} (l/d)	HRT (h)	OLR (gCOD/l/d)	CODout (mg/l)	VS_{att} (gvs/l)	Expanded bed (mm)
I	1-18	0.5	2.25	48	9.4 ± 0.2	2020	20.2	690
	19-37	0.75	2.25	48	9.4 ± 0.2	630	15.5	790
	38-61	1	2.25	48	9.4 ± 0.2	940	12.1	905
II	62-78	0.5	2.7	40	10.8 ± 0.2	1540	18.1	780
	79-97	0.75	2.7	40	10.8 ± 0.2	873	13.9	885
	98-118	1	2.7	40	10.8 ± 0.2	1086	11.7	945
III	119-138	0.5	3.375	32	13.7 ± 0.3	3440	17.1	835
	139-163	0.75	3.375	32	13.7 ± 0.3	1544	13.4	930
	164-188	1	3.375	32	13.7 ± 0.3	1780	11.2	1000
IV	189-210	0.5	4.5	24	18 ± 0.3	4815	16.8	870
	211-238	0.75	4.5	24	18 ± 0.3	2970	13.2	955
	239-267	1	4.5	24	18 ± 0.3	3650	10.8	1120
V	268-294	0.5	5.6	18	24.2 ± 0.5	6515	16.7	905
	295-326	0.75	5.6	18	24.2 ± 0.5	3666	13.18	980
	327-372	1	5.6	18	24.2 ± 0.5	4835	10.8	1180

the average biomass concentration decreased from 20.2 to 12.1 g VSS L^{-1}, when Vs was increased from 0.5 to 1 m min^{-1}. The decrease in the average biomass concentration as a result of the increase in upflow velocity is attributed to two main factors. When, Vs was greater than before, the bed porosity increased, which resulted a lower concentration of bio-particles per unit volume of the AFBR and consequently a lower biomass concentration in volume of reactor. Furthermore, shear forces exerted on the biofilm by the fluid increased. This resulted in a denser and thinner biofilms were formed and consequently resulted in a lower biomass concentration. As shown in Figure 5, it was observed that the average biomass concentration in the AFBR generally decreased with increase in the organic loading rate up to stage 3,

wherever the change of biomass concentration as a function of the organic loading rate became insignificant. As at Vs of 0.75 m min^{-1}, the average biomass concentration decreased from 15.5 to 13.2 gvss L^{-1} expanded bed when the organic loading rate was increased from 9.4 to 10.87 g COD L.d^{-1}, respectively. Then, when the loading rate was increased from 10.87 to 24.2 g COD L.d^{-1}, the average biomass concentration accomplished an approximately steady value of 13.18 gvss L^{-1} expanded bed. Also, similar trend of results was obtained from the other sets of experiments. However, at Vs of 0.5 m min^{-1}, the rate of change in the average biomass concentration started to decrease at a higher loading rate (13.72 g COD L.d^{-1}) than that of the other two upflow velocities. The biomass concentration decreases when the organic

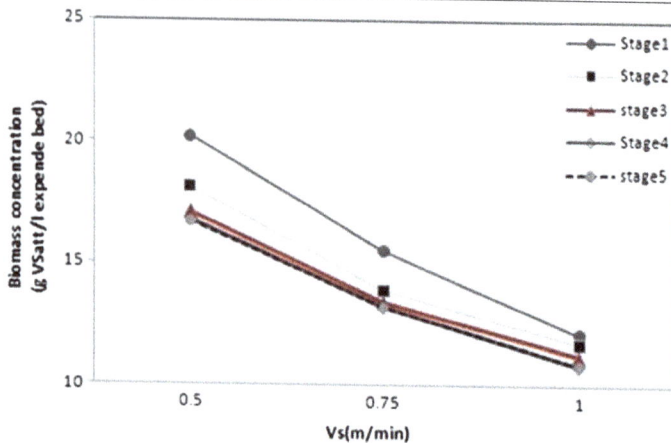

Figure 4 Effect of upflow velocity in biomass concentration.

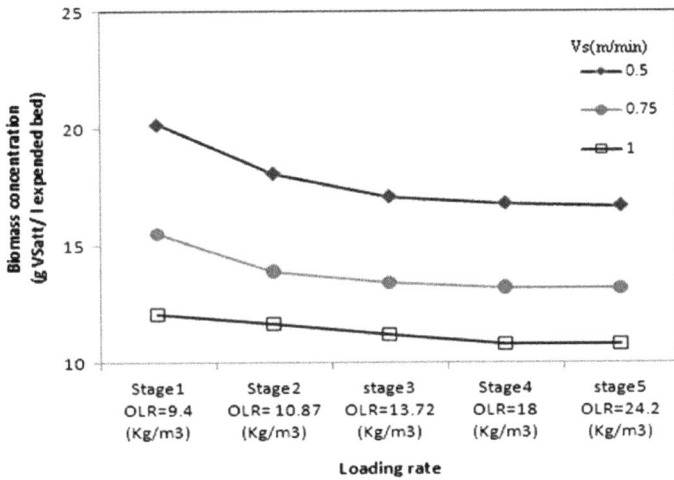

Figure 5 Effect of organic loading rate in biomass concentration.

loading rate increase occurs as a result of the increase in the biofilm thickness due to the increase in the substrate concentration in the bulk liquid [39]. Since, the biofilm thickness increased, the porosity of the AFBR rose, and therefore the average biomass concentration per unit volume of the bed decreased.

Biomass concentration, biofilm thickness and particle density profiles along the AFBR

Figures 6, 7, and 8 show typical profiles of biomass concentration, biofilm thickness and particle density along the AFBR, respectively. As shown in Figure 6, as a result of the increase in the bed porosity along the reactor from the bottom to the top. In this figure, it was observed that biomass concentration as bio-particles per unit volume of

the AFBR decreased along the reactor from the bottom to the top. As shown in Figure 7, biofilm thickness increased from the bottom to the top of the reactor representing a stratification of the bio-particle in the AFBR. Stratification is a consequence of the variability of the bio-particle densities in the reactor. Also, the unequal colonization of the substratum can be one of the causes of the variability of the bio-particle densities in the reactor.

Figure 8 shows the typical pattern of particle density along the AFBR. The biofilm created on the lower levels will be, probably, more dense than that formed in the upper levels as a result of the higher pressure exerted in this zone of the reactor, and this will create denser bio-particles. In the upper part of the bed, a biofilm with a lower density and, proportionally, greater thickness will

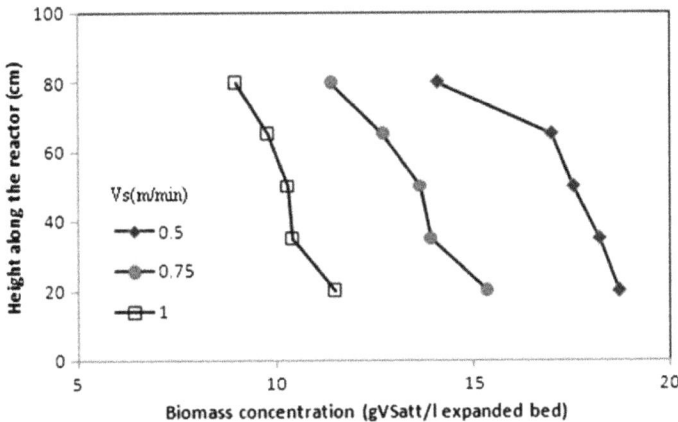

Figure 6 Profiles of biomass concentration at different upflow velocities.

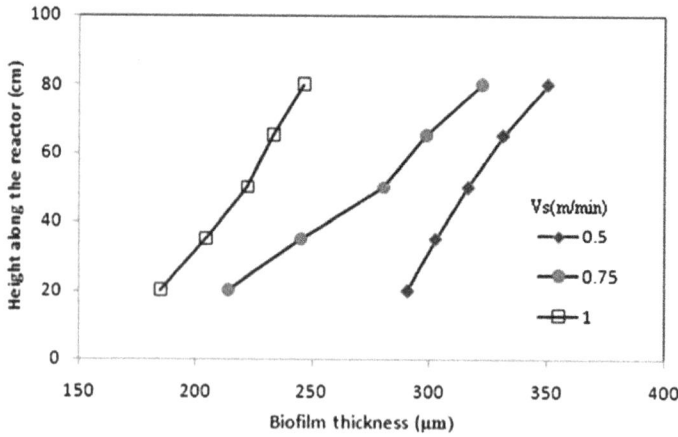

Figure 7 Profiles of biofilm thickness at different upflow velocities.

develop because of the lower pressure presented in this zone. As reported by Zhang and Bishop, the biofilm densities differ with depth within the biofilm layers for the reason that the tops of biofilm are more porous as reported by Zhang and Bishop [40]. The densities in the top layers are usually 5–10 times higher than those in the top layers, and the porosities in the top layers are in the range of 84–93%, while it is in the range of 58–67% in the bottom layers [41]. The lower part of the reactor had denser bio-particles and consequently had lower bed porosity.

Conclusions

Anaerobic fluidized bed reactor with particles made of PVC as the supporting material is highly effective for COD removal for high strength wastewater from currant

wastewater. The results demonstrated that the AFBR system is capable of handling an exceptionally high organic loading rate with very high removal efficiency, up to 96.6%. The average biomass concentration per unit volume of the AFBR (as $gVSSatt \ L^{-1}$ expended bed) decreased with increase in the upflow velocity at all the applied organic loading rates up to some loading rate as a result of the increase in the bed porosity. The total biomass in the reactor increased with increases in the organic loading rate. The peak biomass concentration (as $gVSSatt \ L^{-1}$ expended bed) was observed at the bottom part of the reactor, then it dropped off slowly towards the top. The biofilm thickness increased from the bottom to the top of the reactor representing a stratification of the media in the AFBR. The bed porosity increased from the bottom to the top of the reactor.

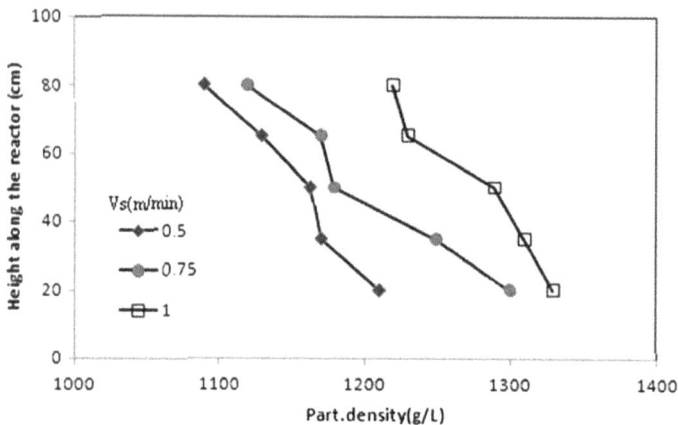

Figure 8 Profiles of particle densiy at different upflow velocities.

Competing interests
The authors declare that they have no competing interests.

Authors' contributions
The overall implementation of this study including experimental design, data analysis and manuscript preparation were done by JJ, AHM, RN, ARM and HK. MH critically reviewed and revised the article. All authors read and approved the final manuscript.

Acknowledgements
The authors are most grateful to the laboratory staff of the Department of Environmental Health Engineering, School of Public Health, Tehran University of Medical Sciences, Iran, for their collaboration in this research.

Author details
[1]School of Public Health, Tehran University of Medical Sciences, Tehran, Iran. [2]Department of Environmental Health Engineering, School of Public Health, Shiraz University of Medical Sciences, Shiraz, Iran. [3]Health Promotion Research Center, Zahedan University of Medical Sciences, Zahedan, Iran. [4]Center for Solid Waste Research, Institute for Environmental Research, Tehran University of Medical Sciences, Tehran, Iran. [5]National Institute of Health Research, Tehran University of Medical Sciences, Tehran, Iran.

References
1. Mahvi A: Application of ultrasonic technology for water and wastewater treatment. *Iranian J Public Health* 2009, **38**:1–17.
2. Karimi B, Ehrampoush MH, Jabary H: Indicator pathogens, organic matter and LAS detergent removal from wastewater by constructed subsurface wetlands. *J Environ Health Sci Eng* 2014, **12**:52.
3. Mahvi A: Sequencing batch reactor: a promising technology in wastewater treatment. *Iranian J Environ Health Sci Eng* 2008, **5**:79–90.
4. Naghizadeh A, Mahvi A, Vaezi F, Naddafi K: Evaluation of hollow fiber membrane bioreactor efficiency for municipal wastewater treatment. *Iranian J Environ Health Sci Eng* 2008, **5**:257–268.
5. Rajasimman M, Karthikeyan C: Aerobic digestion of starch wastewater in a fluidized bed bioreactor with low density biomass support. *J Hazard Mater* 2007, **143**:82–86.
6. Lohi A, Alvarez Cuenca M, Anania G, Upreti S, Wan L: Biodegradation of diesel fuel-contaminated wastewater using a three-phase fluidized bed reactor. *J Hazard Mater* 2008, **154**:105–111.
7. Chen C-L, Wu J-H, Tseng I-C, Liang T-M, Liu W-T: Characterization of active microbes in a full-scale anaerobic fluidized bed reactor treating phenolic wastewater. *Microbes Environments/JSME* 2008, **24**:144–153.
8. Rabah FKJ: Denitrification of high-strength nitrate wastewater using fluidized-bed biofilm reactors. City: Elsevier; 2003:3719–3728.
9. Shieh WK, Sutton P, Kos P: Predicting reactor biomass concentration in a fluidized-bed system. *J Water Pollut Control Fed* 1981, 1574–1584.
10. Fernandez N, Montalvo S, Borja R, Guerrero L, Sĺnchez E, Cortĺes I, Colmenarejo MF, Travieso L, Raposo F: Performance evaluation of an anaerobic fluidized bed reactor with natural zeolite as support material when treating high-strength distillery wastewater. In *Book Performance evaluation of an anaerobic fluidized bed reactor with natural zeolite as support material when treating high-strength distillery wastewater*, Volume 33. City: Elsevier; 2008:2458–2466.
11. Haroun M, Idris A: Treatment of textile wastewater with an anaerobic fluidized bed reactor. In *Book Treatment of textile wastewater with an anaerobic fluidized bed reactor*, Volume 237. City: Elsevier; 2009:357–366.
12. Borja R, Banks CJ: Response of an anaerobic fluidized bed reactor treating ice-cream wastewater to organic, hydraulic, temperature and pH shocks. In *Book Response of an anaerobic fluidized bed reactor treating ice-cream wastewater to organic, hydraulic, temperature and pH shocks*, Volume 39. City: Elsevier; 1995:251–259. 251–259.
13. Alvarado-Lassman A, Rustrian E, Garcia-Alvarado MA, Rodriguez-Jimenez GC, Houbron E: Brewery wastewater treatment using anaerobic inverse fluidized bed reactors. In *Book Brewery wastewater treatment using anaerobic inverse fluidized bed reactors*, Volume 99. City: Elsevier; 2008:3009–3015.
14. Montalvo S, Guerrero L, Borja R, Cortĺos I, Sĺnchez E, Colmenarejo MF: Effect of the influent COD concentration on the anaerobic digestion of

winery wastewaters from grape-red and tropical fruit (guava) wine production in fluidized bed reactors with Chilean natural zeolite for biomass immobilization. *Chem Biochem Eng Q* 2010, **24**:219–226.
15. Jafari J, Mesdaghinia A, Nabizadeh R, Farrokhi M, Mahvi AH: Investigation of Anaerobic Fluidized Bed Reactor/Aerobic Mov-ing Bed Bio Reactor (AFBR/MMBR) System for Treatment of Currant Wastewater. *Iran J Public Health* 2013, **42**:860–867.
16. Turan M, Gulsen H, Celik MS: Treatment of landfill leachate by a combined anaerobic fluidized bed and zeolite column system. *J Environ Eng* 2005, **131**:815–819.
17. Hobson PN, Wheatley A: *Anaerobic digestion: modern theory and practice.* London: Elsevier applied science; 1993.
18. Speece RE: Anaerobic biotechnology for industrial wastewater treatment. In *Book Anaerobic biotechnology for industrial wastewater treatment*, Volume 17. City: ACS Publications; 1983:416A–427A.
19. Kwok WK, Picioreanu C, Ong SL, Van Loosdrecht MCM, Ng WJ, Heijnen JJ: Influence of biomass production and detachment forces on biofilm structures in a biofilm airlift suspension reactor. In *Book Influence of biomass production and detachment forces on biofilm structures in a biofilm airlift suspension reactor*, Volume 58. City: John Wiley & Sons; 1998:400–407.
20. Chang HT, Rittmann BE, Amar D, Heim R, Ehlinger O, Lesty Y: Biofilm detachment mechanisms in a liquid fluidized bed. In *Book Biofilm detachment mechanisms in a liquid fluidized bed*, Volume 38. City: Wiley Online Library; 1991:499–506. 499–506.
21. Van Loosdrecht MCM, Eikelboom D, Gjaltema A, Mulder A, Tijhuis L, Heijnen JJ: Biofilm structures. In *Book Biofilm structures*, Volume 32. City: Elsevier; 1995:35–43.
22. Alves CF, Melo LF, Vieira MJ: Influence of medium composition on the characteristics of a denitrifying biofilm formed by < i > Alcaligenes denitrificans</i > in a fluidised bed reactor. In *Book Influence of medium composition on the characteristics of a denitrifying biofilm formed by < i > Alcaligenes denitrificans</i > in a fluidised bed reactor*, Volume 37. City: Elsevier; 2002:837–845.
23. Koloini T, Farkas E: Fixed bed pressure drop and liquid fluidization in tapered or conical vessels. *Can J Chem Eng* 2009, **51**:499–502.
24. Hsu H: Characteristics of tapered fluidized reactors: two phase systems. In *Book Characteristics of tapered fluidized reactors: two phase systems*. City: Tennessee Univ., Knoxville (USA): Dept. of Chemical, Metallurgical and Polymer Engineering; Oak Ridge National Lab., TN (USA); 1978.
25. Shi YF, Yu Y, Fan L: Incipient fluidization condition for a tapered fluidized bed. *Industrial Eng Chem Fundamental* 1984, **23**:484–489.
26. Peng Y, Fan LT: Hydrodynamic characteristics of fluidization in liquid–solid tapered beds. In *Book Hydrodynamic characteristics of fluidization in liquid–solid tapered beds*, Volume 52. City: Elsevier; 1997:2277–2290.
27. Sen S, Demirer G: Anaerobic treatment of real textile wastewater with a fluidized bed reactor. *Water Res* 2003, **37**:1868–1878.
28. Haroun M, Idris A: Treatment of textile wastewater with an anaerobic fluidized bed reactor. *Desalination* 2009, **237**:357–366.
29. APHA, W. AA: Standard Methods for the Examination of Water and Wastewater. In *Book Standard Methods for the Examination of Water and Wastewater*. City: Mc Graw Hill; 2005.
30. Schreyer HB, Coughlin RW: Effects of stratification in a fluidized bed bioreactor during treatment of metalworking wastewater. In *Book Effects of stratification in a fluidized bed bioreactor during treatment of metalworking wastewater*, Volume 63. City: Wiley Online Library; 1999:129–140.
31. Hidalgo M: Start-up, microbial adhesion in anaerobic fluidized bed bioreactors (Estudio de la puesta en marcha y adhesiĺon de microorganismos en biorreactores anaerobios de lecho fluidizado). Spain: University of Valladolid; 1999.
32. Farhan MH, Chin Hong PH, Keenan JD, Shieh WK: Performance of Anaerobic Reactors during Pseudo Steady State Operation. *J Chem Tech Biotechnol* 1997, **69**:45–57.
33. Koloini T, Farkas EJ: Fixed bed pressure drop and liquid fluidization in tapered or conical vessels. In *Book Fixed bed pressure drop and liquid fluidization in tapered or conical vessels*, Volume 51. City: Wiley Online Library; 1973:499–502.
34. García-Bernet D, Buffiere P, Elmaleh S, Moletta R: Application of the down-flow fluidized bed to the anaerobic treatment of wine distillery wastewater. *Water Sci Technol* 1998, **38**:393–399.
35. Perez M, Romero L, Sales D: Comparative performance of high rate anaerobic thermophilic technologies treating industrial wastewater. *Water Res* 1998, **32**:559–564.

36. Rozzi A: **Operational and control parameters in anaerobic processes.** In
    Book Operational and control parameters in anaerobic processes. City; 1986.
37. Kato MT, Field JA, Kleerebezem R, Lettinga G: **reatment of low strength
    soluble wastewaters in UASB reactors.** In *Book Treatment of low strength
    soluble wastewaters in UASB reactors.* 77th edition. City: Elsevier; 1994:679.
38. Andalib M, Hafez H, Elbeshbishy E, Nakhla G, Zhu J: **Treatment of thin
    stillage in a high-rate anaerobic fluidized bed bioreactor (AFBR).**
    Bioresour Technol 2012, **121**:411–418.
39. Rabah FKJ, Dahab MF: **Biofilm and biomass characteristics in high-
    performance fluidized-bed biofilm reactors.** In *Book Biofilm and biomass
    characteristics in high-performance fluidized-bed biofilm reactors.* 38th edition.
    City: Elsevier; 2004:4262–4270.
40. Zhang TC, Bishop PL: **Density, porosity, and pore structure of biofilms.**
    Water Research 1994, **28**:2267–2277.
41. Zhang TC, Bishop PL: **Structure, activity and composition of biofilms.**
    In *Book Structure, activity and composition of biofilms.* 29th edition. City;
    1994:335–344.

3

Outdoor occupational environments and heat stress in IRAN

Hamidreza Heidari[1], Farideh Golbabaei[1*], Aliakbar Shamsipour[2], Abbas Rahimi Forushani[3] and Abbasali Gaeini[4]

Abstract

Background: The present study aimed at demonstrating the heat stress situation (distribution and intensity) based on a standard and common heat stress index, Wet Bulb Globe Temperature (WBGT), during hot seasons and interpret the obtained results considering global warming and rising temperature in different parts of the country based on climate changes studied in Iran.

Methods: Heat stress assessment was done using WBGT index. Environmental parameters were measured simultaneously in the early, middle and end of shift work. The personal parameters including cloth thermal insulation and metabolic rate of 242 participants from 9 climatic categories were recorded for estimating effective WBGT (measured WBGT plus cloth adjustment factor as well as metabolic rate effect). The values of the indicator were categorized in the statistical software media and then linked to the climatic zoning of the data in the GIS information layers, in which, WBGT values relating to selected stations were given generalization to similar climatic regionalization.

Results: The obtained results showed that in the summer about 60 % and more than 75 % of the measurements relating to 12 pm and 3 pm, respectively, were in heat stress situations (*i.e.* the average amount of heat stress index was higher than 28 °C). These values were found to be about 20–25 % in the spring. Moreover, only in the early hours of shift work in spring could safe conditions be seen throughout the country. This situation gradually decreased in the middle of the day hours and was replaced by the warning status and stress. And finally, in the final hours of shift work thermal stresses reached their peaks. These conditions for the summer were worse.

Conclusions: Regarding several studies related to climate change in Iran and the results of present study, heat stress, especially in the central and southern parts of Iran, can be exacerbated in the decades to come if climate change and rising temperature occurs. Therefore, paying attention to this critical issue and adopting macro-management policies and programs in the field of workplace health is essential.

Keywords: Outdoor worker, Heat stress, Climate change

Background

The Islamic Republic of Iran lies in western Asia. The population now estimated at 72.0 million. The area coverage of different types of climate in Iran is 35.5 % hyper-arid, 29.2 % arid, 20.1 % semi-arid, 5 % Mediterranean and 10 % wet (of the cold mountainous type). Thus more than 82 % of Iran's territory is located in the arid and semi-arid zone of the world [1]. Weather and climate play a significant role in people's health. Changes in climate affect the average weather conditions to which

we are accustomed. Warmer average temperatures will likely lead to hotter days and more frequent and longer heat waves. This, in turn, could increase the number of heat-related illnesses and deaths [2].

In 2009, the U.S. Global Change Research Program presented a report to congress that summarizes current and future impacts of climate change on the U.S., including how it directly affects humans. As extreme heat waves become more common, there is an increased risk of heat-related illness and death. The elderly, the very young and diabetics face the highest threat [3].

The impacts of climate change on health will depend on many factors. These factors include the effectiveness

* Correspondence: Fgolbabaei@sina.tums.ac.ir
[1]Department Occupational Health, School of Public Health, Tehran University of Medical Sciences, Tehran, Iran
Full list of author information is available at the end of the article

of a community's public health and safety systems to address or prepare for the risk and the behavior, age, gender, and economic status of individuals affected. Impacts will likely vary by region, the sensitivity of populations, the extent and length of exposure to climate change impacts as well as society's ability to adapt to change. If high temperatures, especially when combined with high relative humidity, persist for several days (heat waves), and if nighttime temperatures do not drop, extreme heat can be a killer. Of all climate-related projections by scientists, rising temperatures are the most robust [4]. Climate change can affect problems that are already huge, largely concentrated in the developing world, and difficult to combat. WHO and its partners are devising a research agenda to get better estimates of the scale and nature of health vulnerability and to identify strategies and tools for health protection [5]. Researches done in the Iran show that similar the other parts of the worlds, human activities either directly or indirectly cause an increase in most greenhouse gas concentration [6, 7]. They predict this rise may increase greenhouse effect and create a warmer planet. Based on general circulation models (GCM) and scenarios used by Intergovernmental Panel on Climate Change (IPCC) [8], it is anticipated that global temperature in the 2100 s will be warmer than that in the 1900s between 1.1 and 6.4 °C [9]. New climate models also predict that average global temperatures over the range of 1.4–5.8 °C will increase in this century [10]. In Iran some researchers have investigated changes and trends in temperature and precipitation in different regions. Almost all of them find that an increase in air temperature in the decades to come on the other hands, this increase will be more intense in some areas of the south and central regions than the northern areas [6, 7]. The rate of increase in the temperature of 2025 onwards shows higher figures. In the case of reduction of precipitation the results obtained have been roughly the same. Although most studies conducted in different parts of the world are about climate change and the environmental impact of this phenomenon [11–14], in terms of occupational health and the effects of global warming on workers' health, especially workers in outdoor environments has received less attention [15]. Global warming can produce a lot of health problems due to occupational heat exposure [16], which the most of them can be creating heat stress and diseases for all people especially those who work in hot environments. Workers at the risk of heat stress include outdoor workers (who define as any worker who spends time outside) and workers in hot environments such as fire fighters, bakery workers, farmers, construction workers, miners, boiler room workers, factory workers, and others. Workers at the greater risk of heat stress include those who are 65 years of age or older, are

overweight, have heart disease or high blood pressure, or take medications that may be affected by extreme heat. Outdoor workers are exposed to many types of hazards depending on their type of work, geographic region, season, and duration of time they work. Extreme heat conditions can cause heat stroke, heat cramps, heat exhaustion, heat rash, and other problems such as loss of productivity and ability to work [17]. The incidence of heat stroke is unclear; it is due to the lack of accurate statistics on referring patients to medical centers and home treatment. Heat stroke incidence in women and men has been reported alike, but in the case of men in our country -because of their involvement in a position which contributes to heat stroke- the incidence rate has been higher. Tropical diseases in warm seasons are regarded as common diseases in the world. They are frequently seen in our country, especially in hot, dry, warm and moist areas. Therefore awareness of the current status of thermal stress in different parts of the country as well as climate change trends in the future in each area is necessary for taking preventive measures and improving the occupational environments. In spite of the importance of global warming and its adverse health effects on human beings, unfortunately there are a few studies from the viewpoint of occupational health of workers in outdoor workplaces. The Impact of global warming is a growing concern and there is a need for regional-based studies to quantify the degree of stress due to thermal environment. For over a century attempts have been made to construct an index to describe heat stress satisfactorily so that between the years 1905 and 2005 approximately 40 heat stress indices have been provided [18]. Some of them are widely used and general with simple use which can produce appropriate estimations for heat stress. Because of larger number of workers occupied in outdoor environments and existence of different climatic regions in Iran ranging from very hot and dry to very hot and humid regions, it is important to perform a comprehensive investigation to reveal low, moderate and high risk areas considering heat stress. This is important for taking preventive and control measures of regulatory health agencies. As such, the aim of the present study is to demonstrate the heat stress situation (distribution and intensity) based on a standard and common heat stress index, Wet Bulb Globe Temperature (WBGT), during hot seasons and then interpret the obtained results considering rising temperature in different parts of the country based on climate changes studied in Iran.

Methods

In order to reach the aim of the study, that is, to evaluate heat stress in outdoor environments, DeMartonne climate classification was used after a few modifications for Iran. The classification is determined by the 14 climatic zones, including hot and dry, hot and humid, temperate, cold and dry, cold and moist, with different intensities determined. Due to the nature of research that involves the heat stresses, only 9 of 14 above-mentioned categories were selected (Table 1). Cold regions were not included in these 9 climate categories.242 acclimatized subjects from nine climatic regions who worked in outdoor occupations, including railroad workers, farmers, shipbuilding, steel, cement and asphalt industry, *etc.* have been participated in this investigation and their individual heat exposure were assessed. All of them were considered acclimatized and worked at least 6–7 h per day (75 % work plus 25 % rest each hour). To assess heat stress we used WBGT index and corrected it for cloth thermal insulation and metabolic rate of each subject. It means that not only measured WBGT (using environmental parameters) was used in this study, but also effective WBGT was estimated by correction for clothing adjustment factor (CAF), as well as activity level or metabolic rate, two important personal parameters, which should be considered when individual heat stress exposure assessment is necessary. The clothing adjustment factor and metabolic rate were estimated according to ISO-9920 and ISO-8996, respectively for all of 242 participants and then corrections were made to achieve effective WBGT, if necessary.

The WBGT index is by far the most widely used heat stress index throughout the world. The heat effects are dependent upon not only air temperature, but also humidity, air movement, radiated heat, clothing, individual ability to sweat and the workers' physical activity level (metabolic rate). A commonly used "heat stress index" that combines the influence of the four environmental factors into one number is the WBGT [17]. It was developed in a US Navy investigation to heat casualties during training [19] and adopted by ISO-7243 [20] as an approximation to the more cumbersome corrected

effective temperature (CET), modified to account for the solar absorptive of green military clothing. It is given for conditions with solar radiation, outdoor areas, as follows: [21]:

$$WBGT = 0.7tnw + 0.2tg + 0.1ta,$$

Where:

tnw
    the natural wet bulb temperature
tg
    the temperature of a 150 mm diameter black globe
ta
    is the air temperature.

The index has been adapted by the World Health Organization (WHO). In 1982, it was approved by the ISO organization as an international standard for heat load assessment. In 1986, the National Institute for Occupational Safety and Health (NIOSH, 1986) set it as the criterion for occupational exposure to a hot environment. Later on, work-rest regime regulations were made based on this index [22]. In this research WBGT index was determined using the Heat Stress Monitor (Casella Microtherm WBGT, UK). Environmental parameters including air temperature, globe temperature, natural wet temperature, air velocity and relative humidity were measured simultaneously in three interval periods of shift work at 9 am, 12 pm and 3 pm corresponding to early, middle and end of shift work, respectively. In order to have a wide range of weather conditions, all measurements were evaluated twice a year, once in spring and once in summer. To generalize the studied stations to similar climatic regions in country, at first, the measured values of WBGT index for each station were linked to similar climatic stations. The values of the indicator were categorized in the statistical software

Table 1 Description of studied climate categories

Nominal category[a]	Climate category	Synoptic stations
Arid, cool and warm to very warm region	Site 1	Tehran, Semnan, Qazvin, Arak, Qom
Semi arid, moderate and very warm region	Site 2	Ilam, Ahvaz
Semi arid, cool and warm region	Site 3	Bojnourd, Zanjan, Shiraz, Mashhad
Arid, cool and warm region	Site 4	Birjand, Kerman
Arid, cool and very warm region	Site 5	Esfahan, Zahedan, Yazd
Arid, moderate and very warm region	Site 6	Bushehr, Chabhar, Bandarabbas
Humid, cool and warm region	Site 7	Noshahr, Ramsar, Sari
Semi Humid, cool and warm region	Site 8	Gorgan, Yasouj, Ghasre-Shirin
Post Humid, cool and warm region	Site 9	BandarAnzali,rasht

[a]Cool in the nominal category is one of the properties of winter weather conditions, not necessarily spring and summer weather conditions which is emphasized on in this study

media and then linked to the climatic zoning of the data in the GIS information layers, in which, WBGT values relating to selected stations were generalized to similar climatic regionalization. Therefore, in this process the spot information were converted to regional information.

For generalization of the data to the other similar climatic regionalization, the long-term data of each weather station (between 1965 and 2009) including daily minimum and maximum temperature, precipitation rate and sun shine were utilized. Finally, according to the ISO-7243 and ACGIH [23] reference values (Tables 2 and 3), three ranges including safe, caution and stress were assumed for less than 25, 25–28 and more than 28° of centigrade, respectively. In the selection of these values, all workers were supposed to be acclimatized and have moderate work load. The results of heat stress situations in spring and summer as well as interval periods of a shift work were presented graphically on several maps of Iran using Arc/GIS 10.2. Finally according to the obtained results, some recommendations and preventive measures of heat stress were offered.

Results

As mentioned above, after WBGT index was measured, it was corrected for thermal insulation and work load to reach effective WBGT. The thermal clothing insulation of the subjects was ranging from 0.45 to 1.52 Clo (Mean ± SD equal to 0.79 ± 0.16 Clo) and for the case of metabolic rate, it was ranging from 115 to 230 W.m^{-2} (Mean ± SD equal to 179.49 ± 23.36 W.m^{-2}). These amounts of work load indicate moderate to heavy works. Environmental parameters measured for work stations of 242 outdoor workers during spring and summer at 9 am, 12 pm and 3 pm are presented in Table 4.

As noted in Table 4 we can see the range of environmental parameters measured is very broad, so, for example, the dry air temperature parameter has a range from 14.6 to 46° and relative humidity has values between about 21 to the 94 %. In the case of other environmental parameters, a similar situation can be seen. These values, which have been obtained from different

Table 3 WBGT threshold limit values (°C)

Work- rest regimes	Work load		
	Light	Moderate	Heavy
Continuous work	30.0	26.7	25.0
75 % work + 25 % rest; each hour	30.6	28.0	25.9
50 % work + 50 % rest	31.4	29.4	27.9
25 % work + 75 % rest	32.2	31.1	30.0

From ACGIH (American Conference of Governmental Industrial Hygienists, 2013)

parts of the country in spring and summer, confirm the existence of very diverse climatic conditions in the country.

The analysis of covariance in Generalized Linear Model showed that the effect of climatic category, season and time as well as the interaction of these factors is significant in less than 0.001.

As shown in Table 5, based on reference value of the WBGT (28 °C for moderate work load and acclimatized workers) a number of measured stations have a higher average of the maximum limit, in which these conditions are much more abundant in the summer than in the spring. On the other hand in the sites 5 and 6 corresponding to the areas of central and south of the Iran, inappropriate thermal situation was observed in both the spring and summer. The diagrams of Fig. 1 show thermal stress condition in the spring and summer, and in three thermal situations including safe, caution and stress on the basis of what was described in the methods section. These charts which are also provided for different work shift hours show that between 12 and 3 pm (the middle to end hours of work shift) heat stress state is notable, so that in the summer about 60 % and more than 75 % of the measurements which are related to 12 pm and 3 pm, respectively, were in heat stress situations. These values were found to be between about 20 and 25 % in the spring.

In order to clearly show thermal stress situation throughout the country, after the data connection of the measuring stations to the stations with a similar climate and generalizing the results throughout the country

Table 2 ISO 7243: WBGT reference values

Metabolic rate (W.m^{-2})	WBGT reference values	
	Acclimatized (°C)	Not acclimatized (°C)
Resting M < 65	33	32
65 < M < 130	30	29
130 < M < 200	28	26
200 < M < 260	25 (26)[a]	22 (23)[a]
M > 260	23 (25)[a]	18 (20)[a]

[a]Figures in brackets refer to sensible air movement

Table 4 Range of environmental parameters in this study

Environmental parameters	N[a]	Min	Max	M ± SD
Air pressure (mmHg)	1452	594.9	775.6	705.98 ± 57.19
Relative humidity (%)	1452	20.9	93.8	51.78 ± 16.86
Dry air temperature (°C)	1452	14.6	46	31.63 ± 6.18
Globe bulb temperature (°C)	1452	16	49.3	35.09 ± 6.78
Natural wet temperature (°C)	1452	13.2	32.5	22.98 ± 4.29
Air velocity (m/s)	1452	0.02	7.25	1.27 ± 1.35

[a]This number of measurements resulted in 242 work stations by repetitions of 3 times in a day and twice in a year (Spring and summer at 2013)

Table 5 Mean and standard deviation of WBGT$_{effc}$ in spring and summer in different sites and times of shift work

Climate category	Time	WBGT$_{effc}$ (°C)	
		Spring M ± SD	Summer M ± SD
1	9 AM	18.94 ± 2.98	24.30 ± 0.38
	12 PM	21.52 ± 1.50	27.20 ± .54
	3 PM	22.64 ± 1.18	29 ± 0.51
2	9 AM	24.01 ± .97	27.05 ± 1.12
	12 PM	26.20 ± 2.40	31.20 ± 1.35
	3 PM	27.17 ± 2.57	32.97 ± 1.98
3	9 AM	21.72 ± 2.14	25.48 ± 0.63
	12 PM	26.03 ± 1.27	27.38 ± 2.46
	3 PM	25.44 ± 1.13	29.38 ± 1.18
4	9 AM	18.60 ± 1.76	23.55 ± 2.26
	12 PM	19.91 ± 1.62	25.18 ± 0.29
	3 PM	19.84 ± 1.33	25.78 ± 0.99
5	9 AM	27.12 ± 0.93	24.56 ± 0.48
	12 PM	29.85 ± 0.77	28.12 ± 0.93
	3 PM	30.68 ± 2.05	32.38 ± 0.66
6	9 AM	29.40 ± 0.42	32.55 ± 0.37
	12 PM	31.12 ± 0.15	34.68 ± 1
	3 PM	33.53 ± 1.18	35.10 ± 0.55
7	9 AM	17.50 ± 0.99	25.35 ± 0.70
	12 PM	17.85 ± 2.37	28.86 ± 0.45
	3 PM	18.17 ± 2.45	29.77 ± 0.38
8	9 AM	22.90 ± 0.67	27.79 ± 1.34
	12 PM	23.78 ± 0.23	30.84 ± 1.68
	3 PM	23.32 ± 1.18	31.59 ± 0.48
9	9 AM	22.30 ± 0.50	26.04 ± 1.04
	12 PM	25.69 ± 1.20	27.60 ± 2.20
	3 PM	25.18 ± 2.29	28.16 ± 1.95

using Arc/GIS 10.2. program, thermal condition in both spring and summer, as well as at different hours of the day [9, 12, 15] and on the three thermal regions including safe, caution and stress areas were shown in six graphic maps separately (Fig. 2). As is clearing the maps, only in the early hours of shift work in the spring can safe conditions be seen throughout the country. This situation gradually lessens in the middle of the day hours and is replaced by the warning status and stress. And finally, in the final hours of work shift thermal stresses reach their peaks. These conditions, as specified in the maps for the summer, are even worse. So in the final hours of the summer almost neither safe nor warning conditions could be seen while working across the country along with extreme thermal stress. Another noteworthy point is that-as the maps elucidate - it is only in the southern areas of the country and to some extent its central parts corresponding to very hot and humid

and very hot and very dry regions, respectively, that heat stress condition can be seen in both spring and summer and at all periods of shift work (from the early hours to the final hours of shift work).

Discussion

This study aimed to assess the status of the thermal stress in the warm seasons of the year in Iran and demonstrate how thermal stress situations may be changed in different climates regarding predicted rising temperature? It focused on temperature rising in different regions of Iran based on local investigations done so far. This assessment was performed based on the very common and standard index, WBGT, indicating thermal stress conditions in three levels of safe, alert and stress areas. Mention can be made of similar studies worldwide. For example, in a study performed from the years 1995 to 2012 over Dum Dum (22°34'N/88° 22'E) in the district of North 24 Parganas, West Bengal which falls under suburban climate, the distribution of stressful days of summer months (March, April and May) was determined. In that study thermo-hygrometric index, THI, was used to assess heat stress by estimation from the daily weather data. It has been observed that 72.99 % days of summer months have discomfort scores ranging from 3 to 5. The number of stress events is high in the month of May (59.37 %). Estimated values of WBGT (Wet bulb globe temperature) and RSI (Relative strain index) have shown that both the indices are significantly correlated with THI having R^2 values 0.95 and 0.91, respectively. Regional based scales of WBGT and RSI are proposed based on the recognition of discomfort scores of THI [24]. The results of this study, as shown in maps of Fig. 2, specify that the maximum thermal stress can be seen in the Centre and South of the country, so that between 12 and 3 pm in the summer, almost any point of the country is not a safe condition, but conditions have to be alert and stressful. This situation in the Center and South of the country has been observed to be higher than that in other areas (sites 2, 5 and 6 in Table 1). In one comprehensive investigation performed in Iran, the whole country was assessed to create climatic classification based on Tourism Climate Index (TCI). The results showed that with the exception of the Northwest Territories (the areas which are not included in our research; because it was assumed these areas were never involved with heat stress and probability of their placement in heat stress region is almost negligible) and in parts of the Northeast, which has favorable status in the summer, almost in the entire country inappropriate conditions can be seen [25]. Therefore, in non- occupational outdoor environments such as pleasure places, sports fields and so on, the heat stress is also the issue which should be paid more attention regarding rising temperature in the future to take control measures. These results are in tandem with the results of the present study, so that more intense heat

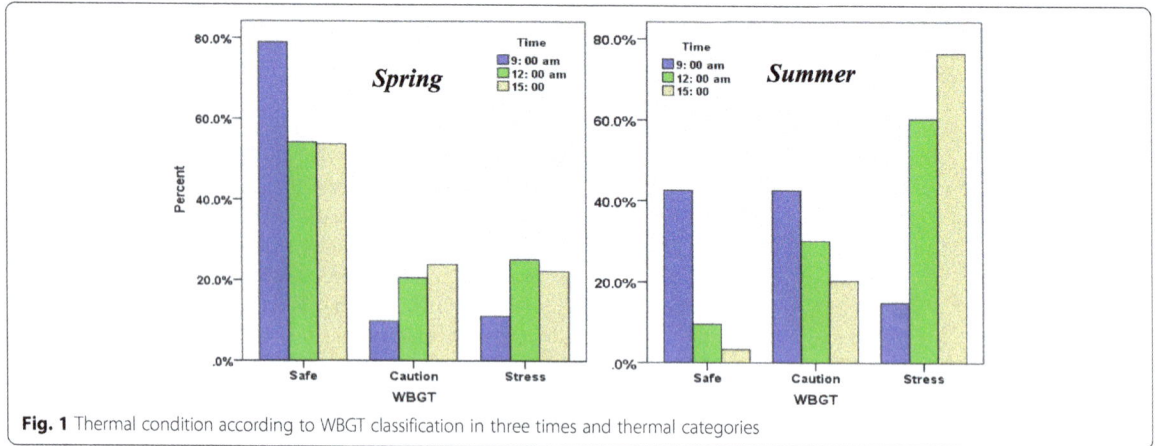

Fig. 1 Thermal condition according to WBGT classification in three times and thermal categories

stress can be seen in the South (the cities of Bandarabbas, Ahvaz, Bushehr) and Centre (the cities of Yazd and Qom) of the country. In Southern regions the WBGT index values were determined during middle and end of work shift higher than 31 °C and 34 °C, for the spring and summer, respectively. In central regions such as Yazd, Esfehan and Qom, the heat stress situation especially for the summer were assessed also higher threshold limit values. The results of the study by Mohammadi et.al on prediction of global warming effects of different areas of Iran yield similar results, which made it clear that in decades to come the country will face an increase in average temperature and that this increase in some areas of the South and Central regions of the country is more noticeable than that of other regions [6]. In another study, Abbasi and Asmari [26] modeled the climate of Iran and changes in perspiration and temperature using General Circulation Models. The prediction of the used models (ECHAM$_4$ and HadCM$_2$) showed that on the average, 3–3.6 °C increases in maximum temperature until decade 2100. Considering rising temperature in the future over the entire of the country especially southern and central regions of it, and WBGT index values obtained in this study, it can be concluded that in spite of huge problems related to heat stress can be see already in several parts of the country, the situation will be worse in future decades regarding to rising temperature and climate change in Iran. There are desert and very hot areas with long hours of sun's radiation in the center of Iran. Moisture is low in these areas and in some sections it is so low that–as is shown in Table 4- moisture level is as low as 20–30 % and the radiant temperatures of about 50° Celsius are the characteristics of these areas. The southern areas of the country which have a desert climate and are very hot, due to the vicinity of the Oman sea and the Persian Gulf, not only experience very high air temperature (36.93 ± 3.00 °C), but also have high

moisture conditions (61.19 ± 8.88 %). The combined effect of these two environmental factors plays a vital role in heat stress incidence in these areas. As the geographical maps show, throughout these areas heat stress exposure exists throughout shift hours in hot months, so that the least amount of WBGT average related to 9 a.m. in the spring and its maximum average is related to the 3 pm in the summer. These values were obtained to be 29.40 ± 0.42 and 35.10 ± 0.55 for minimum and maximum average, respectively. It means that the obtained results of mean WBGT index for regions with minimum and maximum heat stress are, respectively, 1.4 ± 0.42 and 7.1 ± 0.55 °C higher than those of the threshold limit value presented by ACGIH. In another study done in the Persian Gulf weather conditions during hot seasons, the mean (SD) of WBGT values was 33.1 (2.7). The WBGT values exceed those of American Conference of Governmental Industrial Hygienists (ACGIH) standard in 96 % of work stations [27].

In the northern regions of the country (sites 7 and 8) due to the vicinity to the Caspian Sea on the one hand, and the existence of the Alborz Mountains with a little distance from the sea on the other hand, there are highly humid conditions (70.36 ± 11.22). But fortunately, in the spring due to the relatively low average air temperature (25.38 ± 6.59), heat stress usually cannot be seen even in the middle and final hours of shift work. The situation will change in the summer when warming air is combined with high humidity, so, as shown in Table 5, in the middle and final hours of shift work WBGT index is higher than the limit allowed. This is the case which has been well illustrated in maps of Fig. 2. This is compatible with the results of [25]. The remarkable thing found in the current study is that the basis of assessing the situation of thermal stress is the amount of 28 °C reference value, excess of which may bring about heat stress condition. This amount for action limit

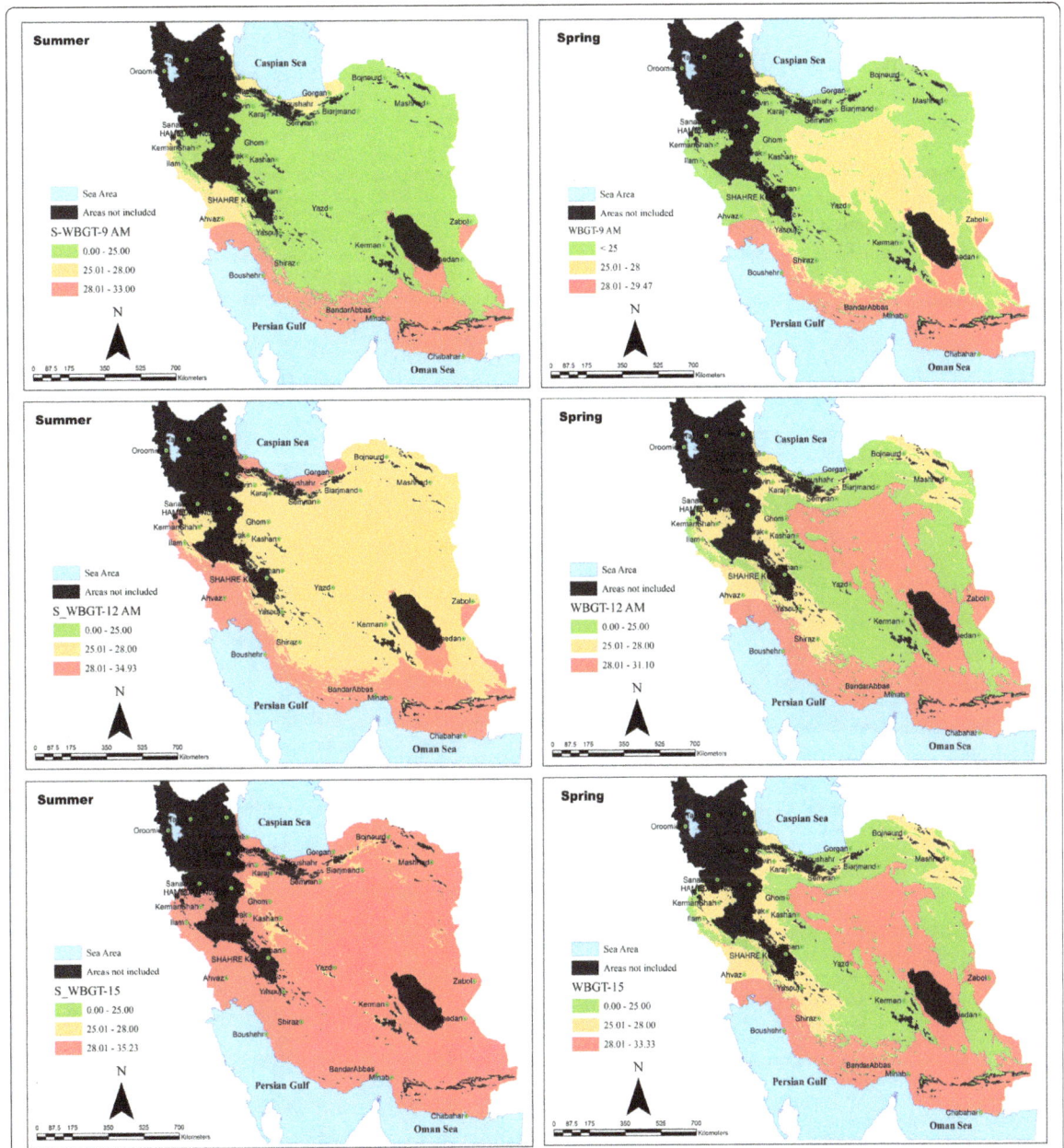

Fig. 2 Assessment of thermal situation based on changes of mean WBGT index in spring and summer during a shift work

has been assumed to be between 25 and 28 °C. Clothing type and amount of thermal insulation, different metabolic rates and work- rest regimens, as well as having protective devices can influence the WBGT reference value. Also, many of workers, particularly workers with low working experience may not be acclimatized. Therefore, with regard to these considerations and the fact that outdoor works in Iran are less organized and health supervision such as provision of plenty of cool and healthy water, regular work

and rest cycles, proper shelter to rest may not be sufficient for them, thermal stress status can be more troublesome and will require special attention.

Conclusion

Heat exposure based on WBGT index exceeded threshold limit value in entire the country in the summer especially between 12 and 3 pm. The south, west south, east south and large areas of central regions of Iran experienced very hot situation, not only in the summer, but also in the spring.

With regard to the status of the existing and projected thermal stress in relation to air temperature increase, especially in the central and southern parts of Iran, workers especially who work outdoor will be encountered with more heat stress in the decades to come.

There is a great need for paying due attention to this critical issue and adopting macro-management policies and programs in the field of occupational health.

Competing interests
The authors declare that they have no competing interests.

Authors' contributions
HH participated in the acquisition, analysis, and interpretation of data and helped to draft the manuscript. FG supervised the study in all steps (acquisition, analysis, and interpretation of data). AS provided geographical maps and has been consulted by AS, ARF and AG. All authors read and approved the final manuscript.

Acknowledgements
This research has been sponsored by Tehran University of Medical Sciences and Institute for Environmental Research (GrantNo.92-02-27-22837). The authors would like to highly appreciate the university cooperation and the Environmental and Occupational Health Center (EOHC) as well as Iran meteorological organization for the provision of required data. The authors declare that there is no conflict of interests.

Author details
[1]Department Occupational Health, School of Public Health, Tehran University of Medical Sciences, Tehran, Iran. [2]Department Physical Geography, School of Geography, University of Tehran, Tehran, Iran. [3]Department Epidemiology and Biostatistics, School of Public Health, Tehran University of Medical Sciences, Tehran, Iran. [4]Department Sport Physiology, School of Physical Education and Sport Science, University of Tehran, Tehran, Iran.

References
1. Amiri M, Eslamian S. Investigation of climate change in Iran. J Environ Sci Technol. 2010;3(4):208–16.
2. EPA. US. assessment of the impacts of global change on regional U.S. air quality: a synthesis of climate change impacts on ground-level ozone (an interim report of the U.S. epa global change research program). Washington, DC: U.S. Environmental Protection Agency; 2009. EPA/600/R-07/094F.
3. Karl TR, Melillo JM, Peterson TC. Global climate change impacts in the United States. Cambridge University Press; 2009.
4. Union of Concerned Scientists. Climate hot map global warming effects around the world. Cambridge University, 2011.
5. Dick T, Rainford J. The impact of climate change on human health, News Team Leader. Geneva: WHO; 2008.
6. Mohammadi H, Taghavi F. Trend of extreme climate indices of temperature and precipitation in Tehran. Geogr Res. 2005;53(37):151–72.
7. Masuodian SA. Investigation of Iran temperature at past half- century. Geogr Res. 2005;54:29–45.
8. Pachauri RK, Reisinger A. IPCC Fourth Assessment Report: Climate Change "Contribution of Working Groups I, II and III to the Fourth Assessment Report of the Intergovernmental Panel on Climate Change". Core Writing Team. IPCC, Geneva, Switzerland: IPCC Fourth Assessment Report: Climate Change; 2007.
9. Ficklin DL, Luo Y, Luedeling E, Zhang M. Climate change sensitivity assessment of a highly agricultural watershed using SWAT. J Hydrol. 2009;374(1):16–29.
10. Pandey M. Global warming and climate change: dominant publishers and distributors. 2005.
11. Chaloupka M, Kamezaki N, Limpus C. Is climate change affecting the population dynamics of the endangered Pacific loggerhead sea turtle? J Exp Mar Biol Ecol. 2008;356(1):136–43.
12. Kojiri T, Hamaguchi T, Ode M. Assessment of global warming impacts on water resources and ecology of a river basinin Japan. Glob Environ Chang. 2008;18:330–40.
13. Mathews JA. Carbon-negative biofuels. Energy Policy. 2008;36(3):940–5.
14. Stafford JM, Wendler G, Curtis J. Temperature and precipitationof Alaska:50 year trend analysis. Theor Appl Climatol. 2000;67:33–44.
15. Dash S, Kjellstrom T. Workplace heat stress in the context of rising temperature in India. Curr Sci (00113891). 2011;101(4).
16. Crowe J, de Joode BW, Wesseling C. A pilot field evaluation on heat stress in sugarcane workers in Costa Rica: What to do next? Global Health Action. 2009; 2: doi:http://dx.doi.org/10.3402/gha.v2i0.2062.
17. Kjellstrom T, Lemke B. Loss of worker productivity due to projected climate change. IOP conf. Series: Earth and Environmental science; 2009. p. 6.
18. Epstein Y, Moran DS. Thermal comfort and the heat stress indices. Ind Health. 2006;44(3):388–98.
19. Yaglon CP, Minard D. Control of heat casuiilties at military training centers. Arch Ind Health. 1957;16:302–5.
20. ISO-7243. Estimation of the heat stress on working man, based on the WBGT-index (wet bulb globe temperature). 1989.
21. Parsons K. Human thermal environments: the effects of hot, moderate, and cold environments on human health, comfort and performance. Taylor and Francis Group. London and New York: Crc Press; 2002.
22. Moran DS, Pandolf KB, Shapiro Y, Heled Y, Shani Y, Mathew W, et al. An environmental stress index (ESI) as a substitute for the wet bulb globe temperature (WBGT). J Therm Biol. 2001;26(4):427–31.
23. ACGIH. TLVs® and BEIs®: threshold limit values for chemical substances and physical agents and biological exposure indices. Cincinnati, OH: American Conference of Governmental Industrial Hygienists; 2013.
24. Bhattacharya R, Pal S, Biswas G, Karmakar S, Banik R. An estimation of heat stressin tropics. Int J Eng Sci Technol. 2012;4(10):4302–7.
25. Farajzadeh M, Ahmadabadi A. Assessment and Zoning of Tourism Climate of Iran Using Tourism Climate Index (TCI). Phys Geogr Res Quart. 2010;71(31):4.
26. Abbasi F, Asmari M. Forecasting and assessment of climate change over IRAN during future decades by using MAGICC-SCENGEN model. J Water Soil. 2011;25(1):70–83.
27. Dehghan H, Mortazavi S, Jafari M, Maracy M, Jahangiri M. The evaluation of heat stress through monitoring environmental factors and physiological responses in melting and casting industries workers. Int J Environ Health Engr. 2012;1(1):21.

4

Removal of amoxicillin from wastewater by self-made Polyethersulfone membrane using nanofiltration

Ahmad Moarefian[1*], Hossein Alizadeh Golestani[2] and Hooman Bahmanpour[3]

Abstract

The present study investigated the performance of a self-made nanofiltration (NF) membrane for the removal of antibiotics from wastewater under changing operating conditions such as pH, initial feed concentration, operating pressure, and temperature. Amoxicillin (AMX) was used as one of the commonly prescribed antibiotics. A self-made NF membrane containing Polyethersulfone (PES), and Polyvinylpyrrolidone (PVP) was modified with Polyethylene glycol hexadecyl ether (Brij®58) surfactant. The self-made membrane was characterized by water contact angle, zeta potential, ATR-FTIR spectroscopy, and scanning electronic microscope (SEM). The obtained results showed that the AMX rejection and permeation flux by the self-made membrane varied from 56.49% to 99.09% and from 15.14 L/m^{2}h to 110.29 L/m^{2}h, respectively. The AMX rejection decreased at a higher level of initial feed concentration while other operating parameters such as pH, operating pressure, and temperature had a negligible effect on the removal of AMX from wastewater by the self-made NF membrane. The highest removal rate was achieved under conditions of pH 9.0, a temperature of 298 K, an operating pressure of 2 MPa, and an initial feed concentration of 20 ppm. According to the research findings, the self-made NF membrane is recommended for the removal of AMX to a considerable extent at low initial feed concentrations.

Keywords: Nanofiltration, Polyethersulfone membrane, Amoxicillin removal, Wastewater treatment

Introduction

The occurrence of antibiotics as emerging containment substances in aquatic environments has always been a cause of concern due to the destructive potential on ecosystems. Prolonged exposure to trace levels of antibiotics leads to the selective proliferation of antibiotic-resistant bacteria and long-term adverse effects on ecosystems and human health [1-7]. However, antibiotic residues have been detected in surface water, ground water, and the final effluent of wastewater treatment plants (WWTPs), even at low concentrations in the range of nanograms/micrograms per liter [8-10].

Although Amoxicillin (AMX) is used to treat a number of infections, however, it is suspected of direct toxic effects on certain aquatic organisms [11-15]. Moreover, amoxicillin-resistant pathogens, such as *Klebsiella pneumoniae*, and

Bacteroides spp [16] in aquatic environments are a potential health threat and make water aesthetically pleasing. According to above-mentioned details, antibiotic residues such as AMX should be removed from aqueous matrices. Common physical and chemical treatment methods are not suitable for the removal of large quantities (mg/L) of pharmaceutically active compounds (PhACs) from wastewater [17,18]. Accordingly, these methods are not efficient for treatment of PhACs [19].

Efficient removal of the polar PhACs can only be ensured using more advanced methods such as ozonation, advanced oxidation processes (AOP), activated carbon, or membrane filtration [19]. However, high cost of equipment and maintenance, as well as energy supply are of disadvantages of the ozonation technique [6]. Capital intensive required to quenching of excess peroxide for some applications [20] rejects the use of AOP. Although granular activated carbons (GACs) are applicable adsorbents, however, they are costly and their regeneration is difficult [6]. The removal efficiency by activated carbon is low for PhACs having low log

* Correspondence: a_moarefian@yahoo.com
[1]Department of Chemical Engineering, College of Science, Shahrood Branch, Islamic Azad University, Shahrood, Iran
Full list of author information is available at the end of the article

K_{ow} values (e.g. AMX by log K_{ow} = 0.87 [21]) and low electrical charges [19].

Membrane technologies such as nanofiltration (NF) and reverse osmosis (RO) have extensively been used for municipal wastewater treatment purposes to ensure the quality of municipal effluents and wastewater reuse [18]. Pressure-driven membrane processes, particularly NF and RO have been the centre of attention in the past few years especially for treatment of drinking water [22,23].

NF is a relatively recent membrane filtration process being used up rapidly [24]. NF membranes are mainly used for the removal of dissolved PhACs from water matrices [22,25]. Compliance of molecular mass (MW) of PhACs ranges between 200 and 1200 Da, with the molecular mass cutoff (MWCO) of NF membranes [24], NF seems to be an efficient technique for the removal of antibiotics from contaminated water [26].

Most NF membranes are charged by the dissociation of surface functional groups such as carboxylic or sulfonic [6,25]. Negatively charged NF membranes are widely used because they can selectively pass or reject the ions from feed solution through the electrostatic interaction between ions and membrane surface [6,25].

Moreover, different rejection mechanisms have been proposed to NF process, which include, molecular sieving (steric hindrance), Donnan exclusion (electrostatic interaction between charged solutes and membrane-attached charges), and dielectric exclusion (interaction between ions and the polarized charge) [25]. In spite of many methods on characterization of NF membranes, the transport mechanisms of solutes through membranes are not completely understood [25].

This research was conducted to investigate the performance of a self-made membrane for removal of AMX from synthetic wastewater under varying operating conditions.

Materials and methods
Materials
Polyethersulfone (Mw = 58000 g/mol) provided by BASF Co. was used to prepare membrane casting solution. Pure AMX (Mw = 365.40 g/mol) was prepared from Dana Pharmacy Co. (Tabriz, Iran). Polyvinylpyrrolidone (Mw = 40000 g/mol), 1-Methyl-2-pyrrolidinone (NMP) as a polymer solvent, and Polyethylene glycol hexadecyl ether (Brij®58) as a non-ionic surfactant (HLB = 15.7) were purchased from Sigma Aldrich Co. (USA). N,N-dimethyl-p-phenylenediamine, potassium hexacyanoferrate (iii), NH_3, and NaOH were bought from Merck Co. (Germany) to determine AMX content in feed and permeate flow. Molecular structure of PES, Brij®58, PVP, NMP, and AMX is shown in Figure 1.

Membrane preparation
The self-made membrane was prepared by phase inversion method using optimal amounts of PES, PVP, and Brij®58

surfactant. The Brij®58 was used as a non-ionic surfactant to modify the PES membrane. The casting solution was prepared by dissolving 1.143 g of Brij®58 in 13.155 mL of NMP, then 4.000 g of PES and 0.381 g of PVP was gently added to residue materials as a pore former and stirred overnight at 300 rpm. After preparing a homogeneous solution, the dope was kept at room temperature for about 24 h until the air bubbles were removed. The homogenous solution was casted onto a glass substrate hovering at a height of 200 μm using a film applicator at room temperature without evaporation. Then, the solution was transferred to a deionized water bath for immersion precipitation at 273 K and allowed it to stand for 5 minutes.

The prepared membrane was stored in distilled water for 24 h to allow the water soluble components to be leached out. Finally, the membrane was dried by two filter papers under very low uniform pressure.

Synthetic wastewater
The synthetic wastewater with initial feed concentrations of 20 and 400 ppm was prepared by dissolving 0.2 and 4 g of AMX in 10 L distilled water. In order to investigate the effect of pH on the performance of the self-made membrane for the removal of AMX, all the experiments were conducted at pH values of 5, 7.0, and 9.0. The pH of feed solution was about 5. 0.1 M (0.1 N) sodium hydroxide (NaOH) was added to the synthetic wastewater in order to adjust the pH at 7.0 and 9.0.

Membrane characterization
Contact angle
In order to evaluate membrane hydrophilicity, static contact angle between water and the membrane was measured directly using an OCA 15 Plus (Data Physics Instruments, Germany). Deionized water was used as a probe liquid in all measurements. All contact angle measurements were made using deionized water drops of 4 μl. To minimize experimental errors, for each sample, the contact angle was measured at 4 random locations and then, the average value was considered.

Zeta potential
To determine electrical charge over the membrane surface, zeta potential was determined by streaming potential measurements using Electro Kinetic Analyzer (EKA 1.00, Anton-Paar, Swiss) equipped with a plated sample cell. Zeta potential was measured in a 0.001 M KCl solution. The measurements were carried out at 298 K in KCl solution (0.001 M) with polymethyl methacrylate (PMMA) plate. Zeta potential was measured at pH values of 5.0, 7.0, and 9.0.

Morphologic study
The morphology of the self-made membrane was studied by an electronic microscope (EM 3200, KYKY, China). For

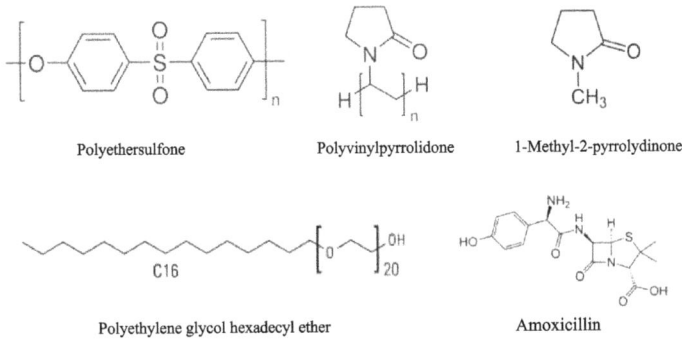

Figure 1 Chemical structure of PES, PVP 40, NMP, Brij®58 and AMX.

this, a sample of the membrane was frozen in liquid nitrogen and then fractured. After gold sputtering, it was examined by an electron microscopy at 20 kV.

ATR-FTIR spectroscopy

In order to ensure the presence of Brij®58 surfactant in the structure of synthetic membrane, the ATR-FTIR spectroscopy was used. The Fourier transform infrared (FTIR) spectrum of the membrane was recorded in the range between 400 and 4000 cm^{-1} using attenuated total reflection (ATR) technique by Nicolet IR 100 FTIR spectrometer (Thermo, USA).

Experimental set-up

NF experiments were conducted in a consecutive lab-scale filtration equipped with a cross-flow permeation cell with an effective filtration area of 6.936×10^{-3} m^2 supported by a porous stainless steel disc. Details of the permeation cell and NF set-up are depicted in Figure 2. The temperature of feed solution was maintained at 298, 308, and 318 K by a shell and tube heat exchanger. The operating pressure

can be varied from 0.5 to 2 MPa. Pressure was adjusted by backpressure and bypass valves. Retentate and permeate streams were directed back into the feed tank in a closed cycle, which makes the feed concentration approximately constant.

Membrane performance

The membrane was cut into appropriate size required to fix up the NF membrane cell and then pressurized with distilled water at 2 MPa for 1 h. After compression, the pressure was dropped to the operating pressure level of 0.5, 1, and 1.5 MPa. Subsequently, the pure water flux (PWF) was collected for 1 h and calculated using the Eq. (1).

$$PWF = Q/A \, \Delta t \qquad (1)$$

Where; PWF = pure water flux (L/m^{2}h), Q = quantity of permeate (L), A = active membrane area (m^2), and Δt = sampling time (h).

After filtration of the pure water, the feed tank was emptied and refilled with the synthetic wastewater. The

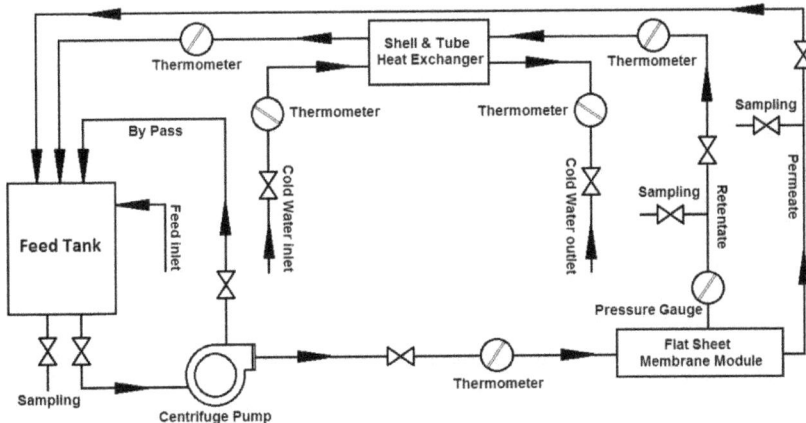

Figure 2 Schematic diagram of the NF set-up.

membrane performance was studied in terms of either permeation flux or AMX rejection. The solute rejection was calculated using Eq. (2).

$$R\ (\%) \ = \ [(1- C_P/C_F)] \times 100 \qquad (2)$$

Where; C_P and C_F are concentrations of the solute in permeate and initial feed solutions, respectively. Concentration of AMX in the permeate stream was determined using N,N-dimethyl-p-phenylenediamine, potassium hexacyanoferrate (iii), and NH_4OH. The absorbance of samples was measured by T60 UV–Vis spectrophotometer (PG Instruments, England) at the maximum wavelength. The measurement technique relies on the Beer-Lambert Law [27].

Results and discussion

Membrane morphology

Separation of neutral solutes by porous membranes is mainly a function of molecular and pore size distribution [28]. Accordingly, membranes with a higher pore density, surface porosity, and porous sub-layer have a higher permeation flux [29]. In other word, higher porosity provides more pore channels for diffusion that leads to a higher flux. However, the morphologic study can help to predict rejection rate of neutral solutes as well as permeation rate of flux. The SEM images of the self-made membrane demonstrate an asymmetric structure consisting of a dense top-layer and a porous sub-layer (Figure 3). The morphology of the asymmetric synthetic membrane demonstrated finger-like macrovoids developed underneath the dense layer. Higher rate of permeation flux and AMX rejection is expected due to the high-porous sub-layer of the self-made membrane (each pore size ranges between 0.5 and 5 μm) and a dense selective top-layer consisting of nanopores.

Pure water flux

Figure 4 shows that the PWF tends to increase with increasing TMP. According to which, increased pressure is an effective way to increase permeation flux. Furthermore, there is a linear relation ($R^2 = 0.994$) between PWF and TMP. As regards, NF is a pressure-driven procedure, transmembrane pressure (TMP) is a very crucial factor in the separation performance. Furthermore, as a constant-pressure procedure, TMP determines membrane

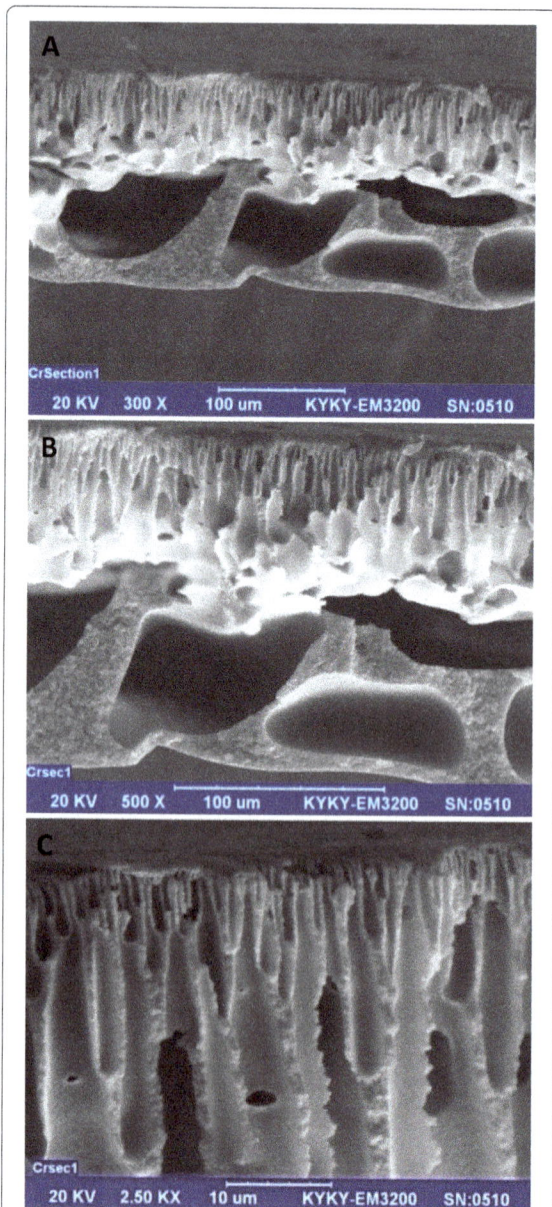

Figure 3 SEM cross-section of the self-made membrane at magnifications of A) 300×, B) 500×, and C) 2500×.

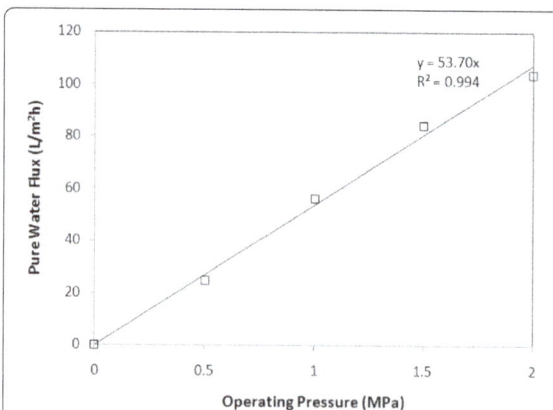

Figure 4 PWF vs. operating pressure at T = 298 K for PES/PVP/ Brij®58 Membrane.

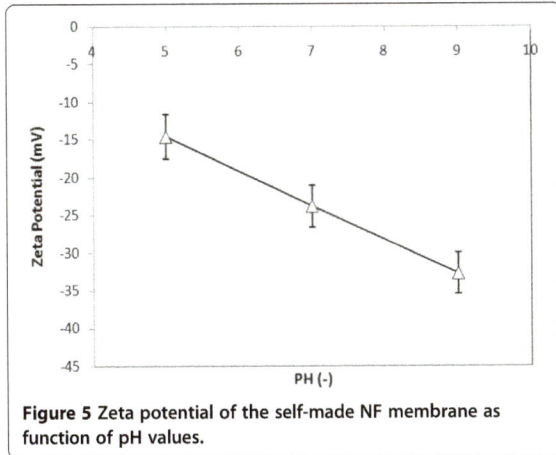

Figure 5 Zeta potential of the self-made NF membrane as function of pH values.

permeability [30]. Accordingly the permeate flux is proportional to the TMP.

Contact angle measurements

As the obtained results show, the highest water contact angle belongs to the unmodified PES membrane, which indicates the lowest hydrophilicity. A low contact angle with water indicates that the surface is hydrophilic. According the obtained result, the water contact angle on the PES membrane decreased from 74.7° to 28.3° once an optimal amount of Brij®58 surfactant was added to the casting solution. This may be attributed to the Brij®58 surfactant and morphology of the top and bottom surfaces of the membrane. As mentioned earlier, the PES membrane was modified by adding an optimal of Brij®58 to the casting solution. Therefore, changes in contact angle

may relate to membrane modification. According to literature [31,32], the contact angle values depend on chemistry, roughness, and heterogeneity of surface as well as membrane parameters. This is consistence with the findings of the current study.

Membrane surface charge

Zeta potential values presented in Figure 5 reveal negative charge of the NF membrane surface as well as a decrease in the absolute value of zeta potential at acidic pH. Zeta potential refers to surface charge that occurs in the presence of an aqueous solution when functional groups dissociate on surface or ions adsorb onto surface from the solution [33,34]. Since PES has no dissociated functional groups [35], specific ionic adsorption is the only possible process for the formation of surface charge.

ATR-FTIR spectroscopy

ATR-FTIR spectroscopy was used to ensure the presence of Brij®58 surfactant in the structure of the synthetic membrane during preparation process, especially in the coagulation stage. Figure 6 exhibits the surface ATR–FTIR spectra of the synthetic membrane. The functionalized self-made membrane illustrates four main peaks. The PES contains repeated ether and sulfone linkage alternating between aromatic rings. Accordingly, the bands at 1151 and 1241 cm^{-1} can be attributed to the stretching vibrations of S = O symmetric and S = O asymmetric, respectively. Besides, the bands at 1663, 3300–3600 and 2919 cm^{-1} depict the amide group of PVP, (–OH), and C–H groups of Brij®58 surfactant, respectively. However, the spectrum of PES/PVP/Brij®58 membrane shows that Brij®58 surfactant was retained in the membrane structure.

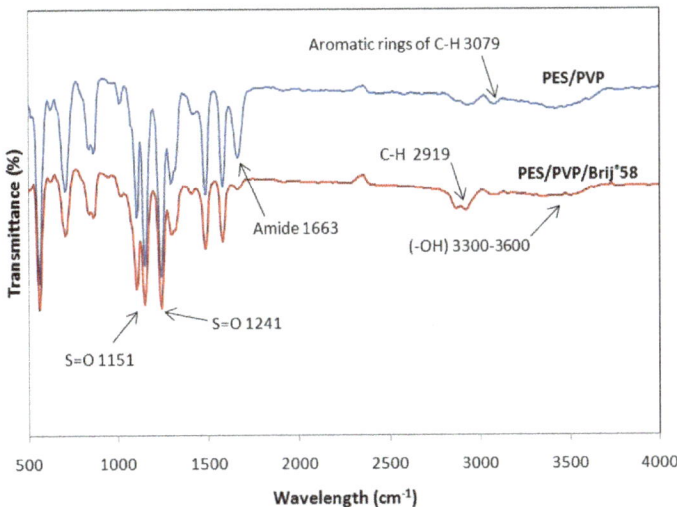

Figure 6 ATR-FTIR spectra of the PES/PVP and PES/PVP/Brij®58 membranes.

Antibiotic rejection and permeation flux

The performance of the self-made membrane for the removal of AMX from wastewater is depicted in Figures 7 and 8. As the figures show, rejection and permeation flux by the self-made membrane varied from 56.49% to 99.09% and from 15.14 L/m^2h to 110.29 L/m^2h, respectively. As shown by Figure 7, the effect of initial feed concentration on AMX rejection was much more than that of

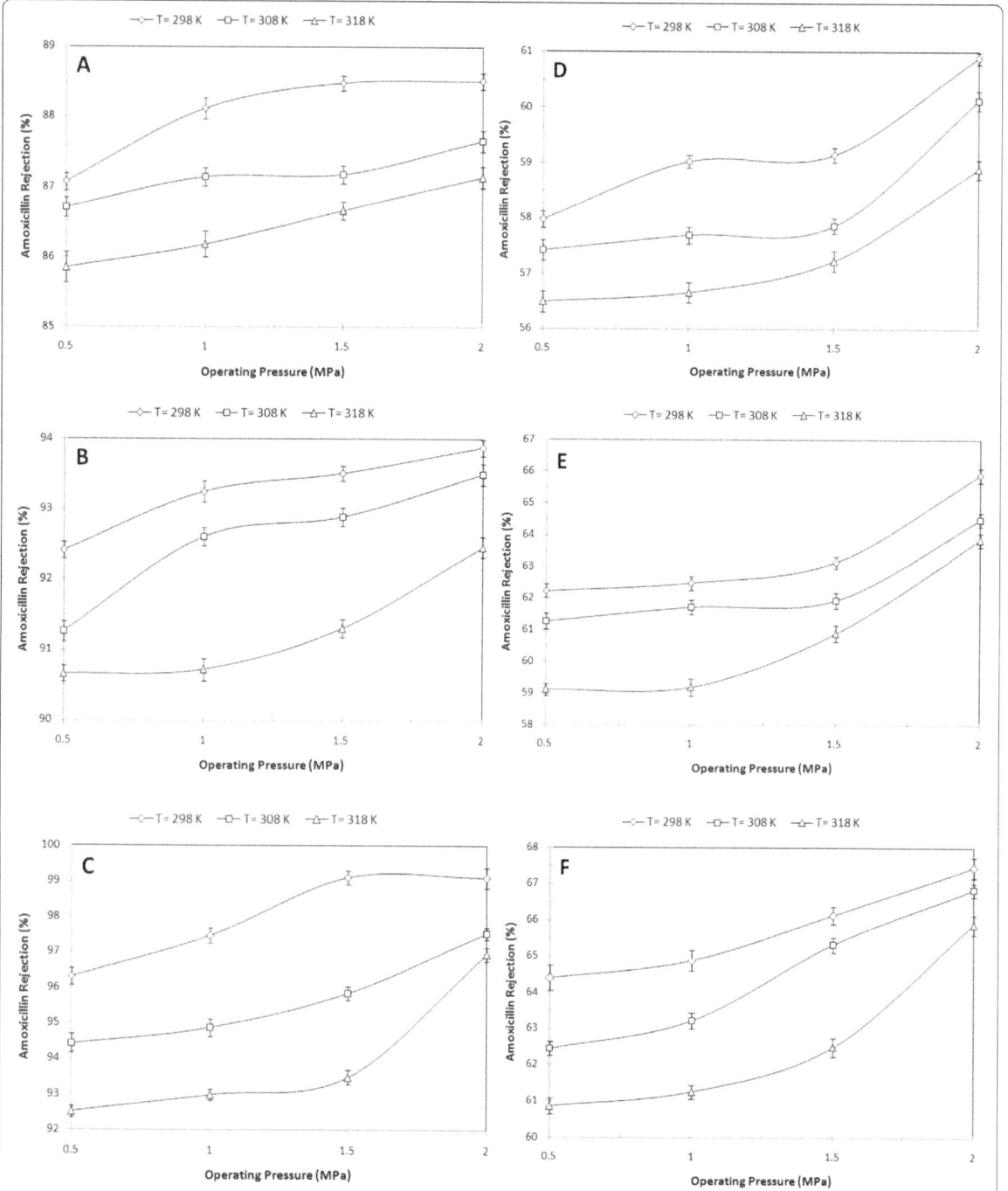

Figure 7 Rejection of AMX vs. operating pressure at varying temperature and initial feed concentration. **A)** 20 ppm/pH = 5.27, **B)** 20 ppm/pH = 7.0, **C)** 20 ppm/pH = 9.0, **D)** 400 ppm/pH = 5.01, E) 400 ppm/pH = 7.0, **F)** 400 ppm/pH = 9.0.

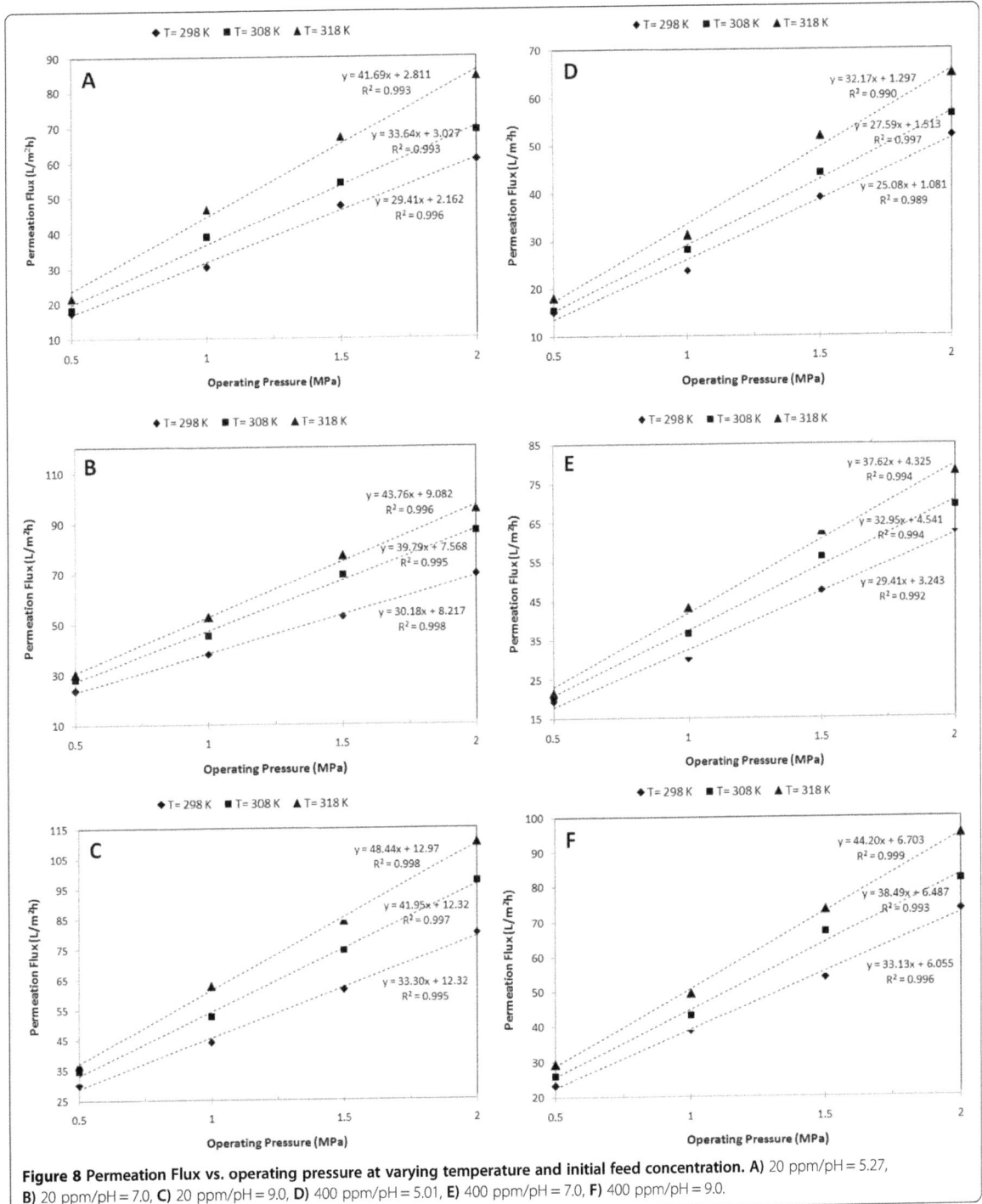

Figure 8 Permeation Flux vs. operating pressure at varying temperature and initial feed concentration. A) 20 ppm/pH = 5.27, **B)** 20 ppm/pH = 7.0, **C)** 20 ppm/pH = 9.0, **D)** 400 ppm/pH = 5.01, **E)** 400 ppm/pH = 7.0, **F)** 400 ppm/pH = 9.0.

pH. Furthermore, the effect of operating pressure and temperature on the AMX rejection was noticeable.

According to Figure 8, the operating pressure, pH, and temperature have remarkable effects on the permeation flux.

The Pearson Correlation test results on AMX rejection and permeation flux are shown in Table 1. According to the results, the feed concentration has a significant effect on AMX rejection (P < 0.001) while it has no significant

Table 1 Pearson Correlation test for AMX rejection and permeation flux

Correlations

		pH	Concentration	Pressure	Temperature	Rejection	Permeation flux
pH	Pearson Correlation	1					
	Sig. (2-tailed)						
	R Square						
	N	72					
Concentration	Pearson Correlation	.000	1				
	Sig. (2-tailed)	1.000					
	R Square						
	N	72	72				
Pressure	Pearson Correlation	.000	.000	1			
	Sig. (2-tailed)	1.000	1.000				
	R Square						
	N	72	72	72			
Temperature	Pearson Correlation	.000	.000	.000	1		
	Sig. (2-tailed)	1.000	1.000	1.000			
	R Square						
	N	72	72	72	72		
Rejection	Pearson Correlation	.192	-.975**	.067	-.067	1	
	Sig. (2-tailed)	.105	.000	.575	.576		
	R Square	.037	.950	.005	.004		
	N	72	72	72	72	72	
Permeation flux	Pearson Correlation	.318**	-.211	.868**	.261*	.308**	1
	Sig. (2-tailed)	.007	.076	.000	.027	.008	
	R Square	.101	.044	.754	.068		
	N	72	72	72	72	72	72

**Correlation is significant at the 0.01 level (2-tailed).
*Correlation is significant at the 0.05 level (2-tailed).

effect on the permeation flux. As such, AMX rejection decreased at high initial feed concentration of 400 ppm (Figure 7). This may be due to increased corresponding ionic strength that tends to neutralize the negative charges of the membrane and consequently, decrease electrostatic repulsion. Finally, a large number of ions pass through the membrane pores that result in a reduction in AMX rejection.

Although, based on Figure 7, rejection rate of the AMX changed by pH changes, however, the impact of these changes is not noticeable according to Table 1. The results showed that increasing the pH from 5 to 9.0 increased rejection efficiency of AMX by 7%. AMX is an amphoteric substance with $pKa_1 = 2.4$, $pKa_2 = 7.4$, and $pka_3 = 9.6$ [36]. AMX is zwitterions at medium pHs, a cation at pH = 2, and an anion at pHs above 7.4. Therefore, at higher pHs, AMX converts anionic forms. Surface charge of the self-made membrane becomes more negative with increasing pH (Figure 5). At a lower pH (5.27), molecular sieve mechanism dominates and results in medium rejection

while at higher pH (9.0), Donnan repulsion mechanism involving electrostatic charge interactions between solute and membrane surface occurs. This leads to electrostatic repulsion between the AMX and self-membrane and higher membrane permeability (P < 0.001).

Significant effect of pH on permeability and AMX rejection by NF membrane has been confirmed by many researchers worldwide. As such, Derakhsheshpoor et al. investigated the effect of pH on the AMX rejection by high permeability polysulfone NF membrane. They observed that increasing feed pH from 6.3 to 8.3, improved 30% of AMX recovery [36].

Since NF is a pressure-driven process, increased pressure leads to an increase in permeate flux. According to Figure 8, the permeation flux increases with increasing operating pressure. This is due to the solution-diffusion model. According to Table 1, operating pressure has a significant effect on permeation flux (P < 0.01).

As Figure 7 suggests, increasing operating pressure from 0.5 to 2 MPa leads to an increase of approximately

3% in the AMX rejection. In other words, at high pressure, water permeability increases rapidly compared to the AMX by which greater number of water molecules can pass through the membrane. Based on Pearson Correlation test results (Table 1), the effect of operating pressure on AMX rejection is negligible.

Increased temperature leads to decreased AMX rejection (Figure 7) and increased permeation flux (Figure 8). This is due to the fact that increased temperature expedites thermal motion of molecules within the membrane, which contributes to the increase in the diffusion coefficient. Thus, transport of components is mainly controlled by the diffusion process in the membrane. In addition, the average pore size of the active separation layer increases slightly when the operating temperature goes up. It is in favor of an increase in permeation flux (P < 0.05).

Considering to Table 1, temperature change has no significant effect on the diffusion coefficient and rejection of AMX because the huge molar volume of AMX blocks the movement of AMX in the membrane.

Conclusion

In this study, the self-made membrane was modified by a non-ionic surfactant (Brij®58). The self-made membrane was characterized by measuring zeta potential and contact angle. Modification of the membrane by Brij®58 surfactant was detected through ATR-FTIR spectroscopy. Moreover, the effect of operating conditions such as pH, feed concentration, operating pressure, and temperature on performance of the self-made membrane for removal of AMX from synthetic wastewater.

The results showed that adding an optimal amount of Brij®58 into the casting solution leads to a decrease in water contact angle. In other words, low water contact angle indicates that the membrane is hydrophilic. As the obtained results revealed, a increase in pH from 5 to 9.0 increases the permeation flux of approximately 18 L/m^{2}h on average. Besides, an increase in pressure and temperature leads to increased permeation flux of the self-made membrane. The analysis results of the Pearson Correlation (Table 1) confirm that pH (P < 0.01), operating pressure (P < 0.01), and temperature (P < 0.05) have a significant effect on permeation flux, while operating pressure is not an effective parameter to improve the rejection efficiency. According to which, an increase in the operating pressure from 0.5 to 2 MPa increased the AMX rejection by approximately 3%. Moreover, pH and temperature have no significant effect on AMX rejection by the self-made membrane, as well.

Increased concentration adversely affects the efficiency of AMX rejection. In overall, maximum AMX rejection of 99.09% was achieved at operating temperature of 298 K, operating pressure of 2 MPa, initial feed concentration of 20 ppm, and pH of 9.0.

However, retention of organic pollutants in membrane separation process depends on the operating condition and characteristics of both membrane and pollutants. Generally, feed concentration is the only parameter affects adversely the rejection of AMX by self-made membrane. Considering AMX rejection efficiency of 99.09% by the self-made membrane at the initial feed concentration of 20 ppm, it can be concluded that this modified membrane is well suited for the removal of AMX from aquatic matrices containing low concentrations of this kind of pollutant. Other operating parameters such as temperature and pressure as well as the qualitative parameters of wastewater such as pH have negligible effect on AMX rejection by the self-made membrane. As a result, the self-made membrane is highly recommended under the above mention conditions.

Competing interests
The authors declare that they have no competing interest.

Authors' contributions
AM participated in data gathering and design of the study. He also coordinated research activities, and revised the manuscript. HAG carried out technical analysis of data, participated in design of the study, and drafted the manuscript. HB participated in the design of the study, final revision of the manuscript, and data analysis. All authors read and approved the finalized manuscript.

Acknowledgements
The authors wish to thank for technical assistance provided by Maryam Omidvar and Dana Pharmacy Company for amoxicillin supply.

Author details
[1]Department of Chemical Engineering, College of Science, Shahrood Branch, Islamic Azad University, Shahrood, Iran. [2]Department of Chemical Engineering, Quchan Branch, Islamic Azad University, Quchan, Iran. [3]Department of Environment, College of Science, Shahrood Branch, Islamic Azad University, Shahrood, Iran.

References
1. Batt AL, Bruce IB, Aga DS: Evaluating the vulnerability of surface waters to antibiotic contamination from varying wastewater treatment plant discharges. *Environ Pollut* 2006, **142**:295–302.
2. Huang JJ, Hu HY, Lu SQ, Li Y, Tang F, Lu Y, Wei B: Monitoring and evaluation of antibiotic-resistant bacteria at a municipal wastewater treatment plant in China. *Environ Int* 2012, **42**:31–36.
3. Zhang Y, Marrs CF, Simon C, Xi C: Wastewater treatment contributes to selective increase of antibiotic resistance among Acinetobacter spp. *Sci Total Environ* 2009, **407**:3702–3706.
4. Tennstedt T, Szczepanowski R, Braun S, Puhler A, Schluter A: Occurrence of integron-associated resistance gene cassettes located on antibiotic resistance plasmids isolated from a wastewater treatment plant. *FEMS Microbiol Ecol* 2003, **45**:239–252.
5. Rizzo L, Manaia C, Merlin C, Schwartz T, Dagot C, Ploy MC, Michael I, Fatta-Kassinos D: Urban wastewater treatment plants as hotspots for antibiotic resistant bacteria and genes spread into the environment: a review. *Sci Total Environ* 2013, **447**:345–360.
6. Homem V, Santos L: Degradation and removal methods of antibiotics from aqueous matrices–a review. *J Environ Manage* 2011, **92**:2304–2347.
7. Watkinson AJ, Murby EJ, Kolpin DW, Costanzo SD: The occurrence of antibiotics in an urban watershed: from wastewater to drinking water. *Sci Total Environ* 2009, **407**:2711–2723.
8. Carrithers S, Sessoms H, Crone T: *Pharmeceutical Screening of Surface Water, Ground Water, and Stream Bed Sediment in Gallatin Valley, Montana.* Montana: Water Quality Research Associates; 2007:1–31.

9.   Brown KD: *Pharmaceutically Active Compounds in Residential and Hospital Effluent, Municipal Wastewater, and the Rio Grande in Albuquerque, New Mexico.* Albuquerque, New Mexico: Water Resources Program, The University of New Mexico; 2004:1–56.

10. Verlicchi P, Al Aukidy M, Zambello E: Occurrence of pharmaceutical compounds in urban wastewater: removal, mass load and environmental risk after a secondary treatment–a review. *Sci Total Environ* 2012, **429:**123–155.

11. Pan X, Deng C, Zhang D, Wang J, Mu G, Chen Y: Toxic effects of amoxicillin on the photosystem II of Synechocystis sp. characterized by a variety of in vivo chlorophyll fluorescence tests. *Aquat Toxicol* 2008, **89:**207–213.

12. Sun J, Li W, Zheng P, Zhu J: Toxicity evaluation of antibiotics in piggery wastewater by luminescent bacteria. *Pol J Environ* 2012, **21:**741–747.

13. Holten Lu¨tzhøft HC, Halling-Sørensen B, Jørgensen SE: Algal toxicity of antibacterial agents applied in Danish fish farming. *Arch Environ Contam Toxicol* 1999, **36:**1–6.

14. Liu Y, Gao B, Yue Q, Guan Y, Wang Y, Huang L: Influences of two antibiotic contaminants on the production, release and toxicity of microcystins. *Ecotoxicol Environ Saf* 2012, **77:**79–87.

15. Gonzalez-Pleiter M, Gonzalo S, Rodea-Palomares I, Leganes F, Rosal R, Boltes K, Marco E, Fernandez-Pinas F: Toxicity of five antibiotics and their mixtures towards photosynthetic aquatic organisms: implications for environmental risk assessment. *Water Res* 2013, **47:**2050–2064.

16. Ball P: The clinical development and launch of amoxicillin/clavulanate for the treatment of a range of community-acquired infections. *Int J Antimicrob Agents* 2007, **30**(Suppl 2):S113–S117.

17. Li S-Z, Li X-Y, Wang D-Z: Membrane (RO-UF) filtration for antibiotic wastewater treatment and recovery of antibiotics. *Sep Purif Technol* 2004, **34:**109–114.

18. Zhang J, Giorno L, Drioli E: Study of a hybrid process combining PACs and membrane operations for antibiotic wastewater treatment. *Desalination* 2006, **194:**101–107.

19. Vergili I: Application of nanofiltration for the removal of carbamazepine, diclofenac and ibuprofen from drinking water sources. *J Environ Manage* 2013, **127:**177–187.

20. Sharma S, Ruparelia JP, Patel ML: A general review on Advanced Oxidation Processes for waste water treatment. In *Nirma University International Conference, Ahmedabad, Gujarat*. 2011.

21. Surampalli RY, Adams CD, Champagne P, Ong SK, Tyagi RD, Zhang TC, Bhandari A: *Contaminants of Emerging Environmental Concern.* Reston, Virginia: American Society of Civil Engineers; 2009:483.

22. Radjenovic J, Petrovic M, Ventura F, Barcelo D: Rejection of pharmaceuticals in nanofiltration and reverse osmosis membrane drinking water treatment. *Water Res* 2008, **42:**3601–3610.

23. Gholami M, Mirzaei R, Rezaei Kalantary R, Sabzali A, Gatei F: Performance evaluation of reverse osmosis technology for selected antibiotics removal from synthetic pharmaceutical wastewater. *J Environ Health Sci Eng* 2012, **9:**1–19.

24. Zhang W, He G, Gao P, Chen G: Development and characterization of composite nanofiltration membranes and their application in concentration of antibiotics. *Sep Purif Technol* 2003, **30:**27–35.

25. Wang KY, Chung T-S: The characterization of flat composite nanofiltration membranes and their applications in the separation of Cephalexin. *J Membr Sci* 2005, **247:**37–50.

26. Omidvar M, Mousavi SM, Soltanieh M, Safekordi AA: Preparation and characterization of poly (ethersulfone) nanofiltration membranes for amoxicillin removal from contaminated water. *J Environ Health Sci Eng* 2014, **12:**1–18.

27. Al-Abachi MQ, Haddi H, Al-Abachi AM: Spectrophotometric determination of amoxicillin by reaction with N, N-dimethyl-p-phenylenediamine and potassium hexacyanoferrate(III). *Anal Chim Acta* 2005, **55:**184–189.

28. Baker RW: *Membrane Technology and Applications.* 2nd edition. Membrane Technology and Research, Inc., Menlo Park, California: John Wiley & Sons; 2004.

29. Sharmiza A, Manh H, Huanting W, Zongli X: Commercial PTFE membranes for membrane distillation application: effect of microstructure and support material. *Desalination* 2012, **284:**297–308.

30. Lin C-J, Rao P, Shirazi S: Effect of operating parameters on permeate flux decline caused by cake formation — a model study. *Desalination* 2005, **171:**95–105.

31. Reddy AVR, Patel HR: Chemically treated polyethersulfone/polyacrylonitrile blend ultrafiltration membranes for better fouling resistance. *Desalination* 2008, **221:**318–323.

32. Qiu Y, Rahman N, Matsuyama H: Preparation of hydrophilic poly(vinyl butyral)/Pluronic F127 blend hollow fiber membrane via thermally induced phase separation. *Sep Purif Technol* 2008, **61:**1–8.

33. Zazouli MA, Susanto H, Nasseri S, Ulbricht M: Influences of solution chemistry and polymeric natural organic matter on the removal of aquatic pharmaceutical residuals by nanofiltration. *Water Res* 2009, **43:**3270–3280.

34. Salgın S, Salgın U, Soyer N: Streaming potential measurements of polyethersulfone ultrafiltration membranes to determine salt effects on membrane zeta potential. *Int J Electrochem Sci* 2013, **8:**4073–4084.

35. Arkhangelsky E, Kuzmenko D, Gitis V: Impact of chemical cleaning on properties and functioning of polyethersulfone membranes. *J Membr Sci* 2007, **305:**176–184.

36. Derakhsheshpoor R, Homayoonfal M, Akbari A, Mehrnia MR: Amoxicillin separation from pharmaceutical wastewater by high permeability polysulfone nanofiltration membrane. *J Environ Health Sci Eng* 2013, **11:**1–9.

5

Application of concrete surfaces as novel substrate for immobilization of TiO$_2$ nano powder in photocatalytic treatment of phenolic water

Mohammad Delnavaz[1,2], Bita Ayati[1*], Hossein Ganjidoust[1] and Sohrab Sanjabi[3]

Abstract

Background: In this study, concrete application as a substrate for TiO$_2$ nano powder immobilization in heterogeneous photocatalytic process was evaluated. TiO$_2$ immobilization on the pervious concrete surface was done by different procedures containing slurry method (SM), cement mixed method (CMM) and different concrete sealer formulations. Irradiation of TiO$_2$ was prepared by UV-A and UV-C lamps. Phenolic wastewater was selected as a pollutant and efficiency of the process was determined in various operation conditions including influent phenol concentration, pH, TiO$_2$ concentration, immobilization method and UV lamp intensity.

Findings: The removal efficiency of photocatalytic process in 4 h irradiation time and phenol concentration ranges of 25–500 mg/L was more than 80 %. Intermediates were identified by GC/Mass and spectrophotometric analysis.

Conclusions: According to the results, photocatalytic reactions followed the pseudo-first-order kinetics and can effectively treat phenol under optimal conditions.

Keywords: Concrete, Immobilization, Intermediates, Water cleaning technology

Background

The ability of Advanced Oxidation Processes (AOP$_s$) in treating a wide range of hazardous wastes has brought this technology to the forefront of research over the last decade. Among AOP$_s$, application of heterogeneous photo-catalysis by using semiconductors has been proved to be real interest as an efficient tool for degrading both aquatic and atmospheric organic contaminants [1]. Semiconductors are photo-reactive metal oxides for contaminants eradication that refer to photo-catalysts [2]. Titanium dioxide (TiO$_2$) is an established photocatalyst utilized in the photo-oxidation process. When TiO$_2$ is exposed to the appropriate wavelength of ultra-violet light (UV-A), electrons in the low-energy valence band will absorb the photon's energy and move into the high-energy conduction band. The result of this electron excitation is a hole, or positive charge, in the valence band (h$^+$)

and an electron in the conduction band (e$^-$) [3]. Reaction yield and photocatalytic activity will be increased when the diameter of TiO$_2$ particles becomes smaller especially below 100A° [1].

Photocatalytic reactors for water and wastewater treatment can be classified to slurry and photocatalytic ones. In the slurry reactors, the catalyst particles are freely dispersed in the fluid phase (water) and consequently the photo-catalyst is fully integrated in the liquid mobile phase [4]. Whereas the immobilized catalyst reactor design features a catalyst anchored to a fixed support and dispersed on the stationary phase. Slurry systems require the separation of the fine sub micron particles TiO$_2$ from the treated milk-like water suspension. Separation steps cause to complicate the treatment process and decrease the economical viability of the slurry reactor approach [5]. Therefore, these difficulties have led many researchers to study reactors with thin immobilized films of catalyst bonded to a solid substrate such as activated carbon [6], fiber optic cables [7] fiberglass [8], glass beads [9], quartz sand [10], silica gel [11] and stainless steel [12]. Dip coating from suspension, spray coating, sputtering, sol–gel related

* Correspondence: ayati_bi@modares.ac.ir
[1]Civil and Environmental Engineering Faculty, Tarbiat Modares University, Tehran, Iran
Full list of author information is available at the end of the article

methods, and electrophoretic deposition [4, 13] are techniques developed for immobilizing TiO$_2$ catalysts on these substrates. Although these techniques and substrates had a suitable performance for treating different types of wastewater in laboratory scale experiments, but their application in wastewater treatment plants and pilot scale studies are questionable.

Most applications of concrete modified by TiO$_2$ were done for air pollutant removal and self-cleaning surfaces [14]. Although many studies in the field of air pollution have been done by concrete-TiO$_2$ photocatalyst process [15, 16], but the number of wastewater treatment researches is very low. Application of concrete as TiO$_2$ substrate has unique characteristics such as porosity, natural abundance, absence of toxicity, and low price. This construction material is used in all of the water and wastewater treatment plants (WWTP) all over the world. Therefore, the photocatalytic process by TiO$_2$ photocatalyst that immobilized on concrete surfaces can be used in large scale WWTP.

Phenols and its compounds are widely found in paint, leather and textile, disinfectants, medicine, oil refinery and lubricant production wastewater industries in Iran [17–19]. Phenol is rapidly absorbed through the skin and can cause skin and eye burns upon contact. Comas, convulsions, cyanosis and death can be resulted from its overexposure [20, 21]. Therefore Phenol-containing wastewater may not be conducted in open water without treatment because of its toxicity.

Several investigations using different physical, chemical and biological systems and their combination for phenol elimination from wastewater have been reported [22–27]. Biological processes are preferred to other conventional methods due to their ability to effectively destroy the pollutants in an environmentally benign and cost effective way [19]. Sensitive of process to organic shocking load and need to high control for microbial acclimation for hard degradable compound such as phenol are limiting factor in biological treatment. Photocatalytic process didn't have these limitations and can be used separately or in joint with biological process. Different researches were done related study to photocatytic process in wastewater treatment. Chiou et al. (2008) have studied degradation of phenol and m-nitrophenol using a photocatalytic process in aqueous solution by commercial TiO$_2$ powders (Degussa P-25) under UV irradiation [28]. The optimal solution pH of single phenol and m-NP was at around 7.4 and 8.9, respectively. Immobilization of TiO$_2$ on perlit granules for photocatalytic degradation of phenol was done by Hosseini et al. (2009) [13]. The Results showed uniform coating on perlit and good photocatalytic activity for the catalysts. Pumic stone was applied as TiO$_2$ substrate for degradation of dyes and dye industry pollutants [29]. Real

wastewaters collected before biological treatment treated in photocatalytic reactor and color disappeared after 4-h.

The main objective of this research was to study the feasibility of using concrete as substrate for TiO$_2$ nano powder immobilization to treat phenolic wastewater. For this reason, four methods were applied for immobilization of nano TiO$_2$ on concrete. Kinetic of the reactions, long term use of the process and intermediates formed during the photo degradation were other objectives of this research.

Material and methods
Chemicals
Cement in the concrete was an ordinary Portland type V that is usually applied in WWTP construction in Iran. Selection of concrete mix proportion was done according to standard practice for selecting proportions of normal concrete (ACI 211.1 :1996) [30]. Light expanded clay aggregate (LECA) was used as light coarse aggregate to lead specific gravity of concrete to 1200 kg/m^3. LECA application increased surface porosity of concrete and promoted the specific surface area as an important immobilization parameter. Concrete surface was fabricated in wooden moulds with an internal dimension of $500 \times 250 \times 50$ mm. TiO$_2$ Degussa P25, anatase/rutile ca. 70/30, with an average particle size of 20 nm and a BET surface area 55 ± 15 m^2.g^{-1} was used as photocatalyst. UV radiation was provided by different powers of UV-A Philips medium pressure lamps. The spectral irradiance of UV-A lamp ranges from 300 to 400 nm. The primary wavelength distribution of the UV-A lamp was 365 nm and the light intensity was among 4.42-8.9 mW.cm^{-2}. The incident UV-A light intensity was measured by a UV power meter (UVA-365- Lutron Taiwan). Also UV illumination was provided by UV-C lamps (20 Watts, maximal light intensity at 256 nm).

Phenol (purity above 99 %) as a contaminant, NaOH, HCl for pH adjustment, NH$_4$OH, K$_2$HPO$_4$, KH$_2$PO$_4$, K$_3$Fe(CN)$_6$ and 4-aminoantiprine as phenol concentration reagents and other chemicals all offered by Merck Co., as an analytical reagent grade. Wastewater produced by adding deionized water and phenol in various concentrations (25–500 mg/L).

Immobilization methods
Immobilization of TiO$_2$ nano powder on concrete surfaces was carried out by four different procedures. Adhesion properties of cement and concrete sealers were applied to fix photocatalyst on pervious concrete surfaces.

Slurry method (SM)
The details of this technique are given as follows:

- 20 g/L slurry TiO_2 was prepared using 25 % (v/v) methanol in deionized water. Methanol would help with the attachment of TiO_2 to the surface.
- The solution was stirred vigorously for 10 min at 20 °C.
- The slurry was sonicated for 5 min to separate the flocculated TiO_2 and to obtain more uniform slurry.
- One half of slurry sprayed on fresh concrete to use adhesive properties of cement.
- The coated surface was placed in oven at 100 °C for 2 h to remove the moisture content.
- The rest of the slurry sprayed on hardened concrete and annealed at 450 °C for 2 h to remove any organics from the surface.
- The support was let to dry and was washed with pure water to eliminate the excess of the catalyst.

Cement mixed method (CMM)

In this method, 20 g/L TiO_2 was mixed by cement grout (40 g cement + 20 mL water) and distributed on hardened concrete base by an ordinary brush. After 8 h the concrete surface was placed in oven at 100 °C for 2 h to be dried. The support was washed with pure water to eliminate the excess of the catalyst.

Epoxy sealer method (ESM) & Waterproof sealer method (WSM)

In these procedures, the effect of epoxy concrete sealer (Nitofix- Fars Iran Company) and waterproof concrete sealer (Nitotile- Fars Iran Company) for adhesion of TiO_2 was examined. The details of this technique are as follows:

- 100 mL of the selected concrete sealer mixed with 1000 mL of pure water and stirred vigorously for 10 min at 20 °C.
- Prepared emulsion was distributed on the hardened concrete surface with a trowel and let to dry for 20 min.
- After that time, 20 gr/L of TiO_2 poured on concrete surface for adhesion on sealer.
- The support was let to dry and was washed with pure water to eliminate the excess of the catalyst.

Photocatalytic reactor setup

The immobilized concrete surfaces were placed in the pilot scale photocatalytic reactor (Fig. 1). A dosing pump was used to feed the phenol-laden wastewater into photocatalytic zone and distributed by a pipe with 1 cm mesh in its length. Reactor hydraulics parameters were controlled by water depth limited to 4 mm. Three concrete surfaces and circulation mode was used to prepare efficient contact time between photocatalyst and wastewater. Dissolved oxygen (DO) and pH was monitored in all the experimental period time by Crison-Oxi45 and Metrohm690, respectively. UV radiation was provided by UV-A and UV-C Philips medium pressure lamps. Samples were taken periodically form the sample ports for phenol concentration analysis. Prior to measurement, the liquid samples were centrifuged by Sigma101 at about 4000 rpm for 10 min to remove detached TiO_2 particles. The phenol concentrations were measured by colorimetric 4-aminoantipyrine procedure using a Perkin Elmer-Lambda EZ 150 UV/vis spectrophotometer as described in standard methods (2005). The structure and

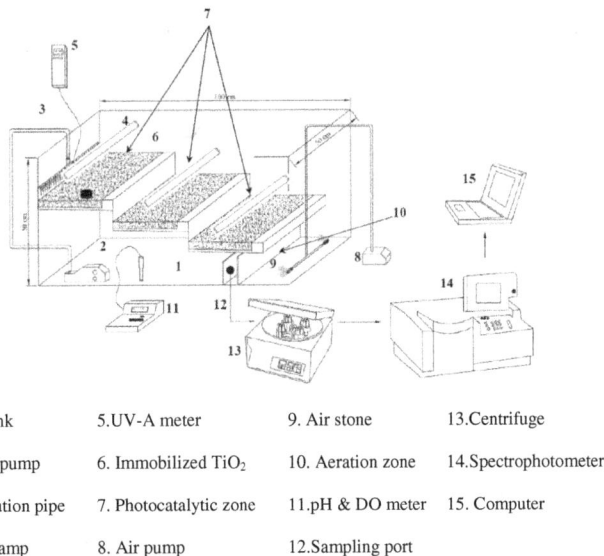

1.Feed tank	5.UV-A meter	9. Air stone	13.Centrifuge
2.Dosing pump	6. Immobilized TiO_2	10. Aeration zone	14.Spectrophotometer
3.Distribution pipe	7. Photocatalytic zone	11.pH & DO meter	15. Computer
4.UV-A lamp	8. Air pump	12.Sampling port	

Fig. 1 Schematic design of photo-catalyst reactor

morphology of prepared catalysts and concrete surface were determined using Philips XL30 scanning electron microscope (SEM) followed by AU-coated by sputtering method using a coater sputter (Bal-Tec, Switzerland). All other parameters were determined according to the standard methods (2005). The intermediates determined by GC/Mass (column: chrompack CP-Sil 8 CB 50 m × 250 μm × 0.12 μm). The gas vector used was helium and the detector was a FID. The program of temperature was: detector temperature = 240 °C, injector temperature = 230 °C, oven initial and final temperatures = 105 °C & 190 °C, oven rise 10 °C min^{-1}, initial and final time = 2 & 15 min). All other parameters were determined according to the Standard Methods [31].

Results and discussion
Characteristics of immobilized concrete surfaces
In all immobilization procedures, SEM and energy dispersive X-ray microanalysis (EDX) were done to confirm the presence of TiO_2 on concrete surfaces (Fig. 2). Images of non-immobilized surface proved high surface porosity of concrete as a good support for TiO_2. SEM analysis showed a uniform appearance of TiO_2 catalyst in ESM, WSM and SM but dispersed coating in CMM.

Mixing TiO_2 by cement as an adhesive agent caused the minimum level of active surface catalyst in CMM (Fig. 2c). In other methods, catalyst was poured on concrete surface that covered by adhesive agent and consequently uniform TiO_2 cover was prepared. EDX analysis

of TiO_2 coatings showed no significant levels of noticeable impurities in all immobilizations cases.

Effect of influent phenol concentration
Effect of different initial concentrations (C_0 = 25–500 ppm) in removal efficiency is shown in Fig. 3.

At first, in all experiments the removal efficiency was measured in pH = 7 without UV lamps for 60 min to determine the pollutants adsorption to concrete and phenol volatility. Removal efficiency in these conditions was less than 5 % and stripping of phenol by aeration was negligible because of its very low Henry's constant [32]. Photocatalytic process was then provided by turning on UV lamps and removal efficiency was measured in different retention times. In SM method after 4 h, 95 and 69 % of phenol was degraded in 25 and 500 ppm initial concentration, respectively. Equal degradation was achieved in ESM when C_0 was 25 ppm but the removal efficiency in WSM was 75 % at the same conditions. The removal efficiency of the CMM was lower than that of other methods. So that after 4 h, 28 % phenol with initial concentration of 500 ppm was degraded. At the same conditions removal efficiency for ESM and WSM were 56 and 43 %, respectively. The results of this study are comparable with similar researches in recent years. For example Chiou et al. (2008) in slurry photo-reactor treated 60 % of 0.52 mM (~49 mg/L) phenol after 3 h and 400 W UV lamp. In other research, Hosseini et al. (2007) reached to 88.3 % phenol degradation after 4 h in initial concentration = 1 mM (~94 mg/L) and UV lamp intensity = 120 W [13].

Fig. 2 SEM picture and EDX spectrum of TiO_2 film coated on concrete surface: **a** SM, **b** CMM, **c** ESM, **d** WSM

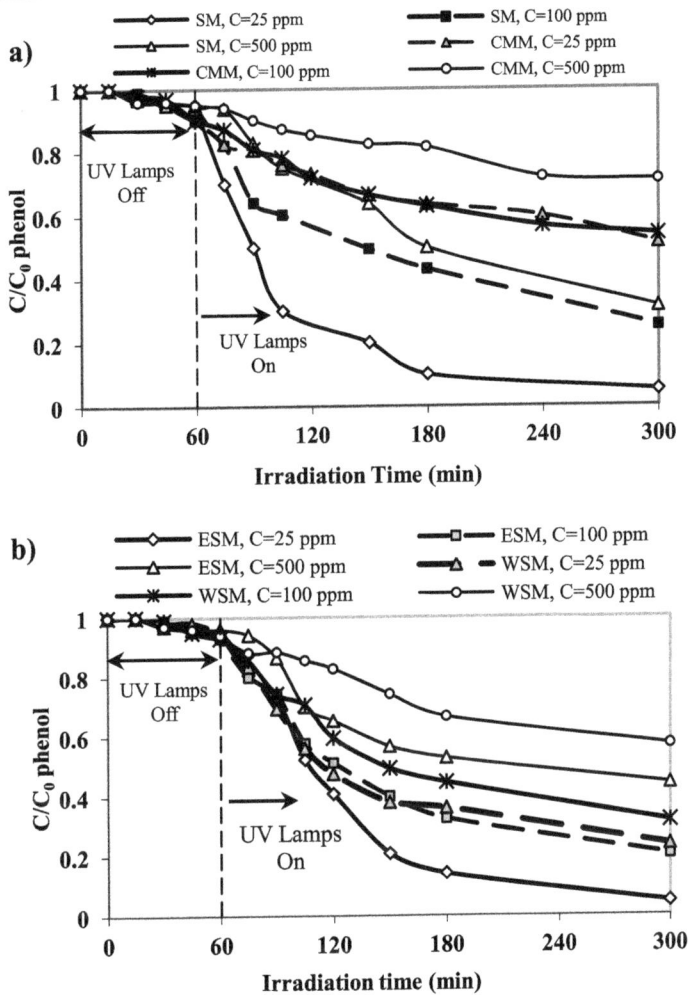

Fig. 3 Effect of phenol initial concentration in removal efficiency, **a** SM & CMM, **b** WSM & ESM

Effect of pH solution

Electrostatic interaction between semiconductor surface, solvent molecules, substrate and charged radicals formed during photocatalytic oxidation is strongly dependent on the pH of the solution [28]. Solution pH dominates photo-degradation process due to the strong pH dependence of many related properties such as the semiconductor's surface charge state, flat band potential, and the solution dissociation. In alkaline pH, phenol was converted to phenoxide group (remove of H^+ and creation of negative charge on hydroxyl group) that was more reactive than phenol in a solution [33]. In the other hand, the ionization state of the surface of the photocatalyst can also be protonated and deprotonated under acidic and alkaline conditions. The pH_{ZPC} of Degussa P-25 TiO_2 used here is 6.5,

and phenol is 9.89, respectively [28]. While under acidic conditions, the positive charge of the TiO_2 surface increases as the pH decreases ($TiOH_2^+$); above pH 6.5 the negative charge at the surface of the TiO_2 increases with increasing pH (TiO^-). The optimum pH can be obtained between pH_{ZPC} of photocatalyst and contaminant. Effect of pH in the range of 4–12 on phenol removal efficiency in ESM was evaluated (Fig. 4).

The highest efficiency was observed at pH of 9–12. The difference between phenol removal efficiency at pH of 12 and 4 was determined 39 % in ESM (52 and 86 % removal efficiency) when initial concentration was 100 mg/L and UV-A lamp intensity was equal to 5.33 mW.cm^{-2}. The rate of removal efficiency in different pH was similar in different immobilization methods and determined between pH_{ZPC} of phenol

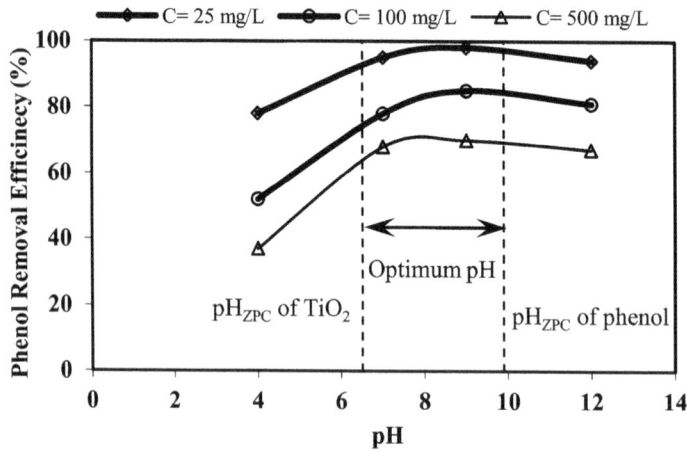

Fig. 4 Effect of pH in phenol removal efficiency in ESM

and TiO_2. Several researchers have observed different results about the effect of the pH on the TiO_2 photocatalytic decomposition of phenol. This discrepancy on the optimum pH may be a function of the various operating conditions considered.

Effect of immobilization methods

The photocatalytic activities for all combinations of coating methods are summarized in Fig. 5. Influent phenol concentration was 50 ppm, pH = 7, UV intensity = 5.33 mW.cm^{-2} and UV lamp distance to concrete surface = 10 cm in these experiments. Results showed ESM had the highest photocatalytic activities so that the removal efficiency of this method was more than 90 % after 4 h. The photocatalytic efficiency of the

coating methods was in the following order: ESM > SM > WSM > CMM. The removal efficiency of CMM method after 4 h was 42 % that was the minimum efficiency between other immobilization techniques. The most important reason that proved by SEM-EDX analysis was reduction levels of active surface of nano particles that mixed with cement. While in other methods, TiO_2 nano particles attached to a concrete top surface, in the CMM nano TiO_2 particles mixed with cement as cohesive agent and a lot of percents of photoctalyst disappeared. Comparison between ESM and WSM showed that ESM had better performance due to hydrophobic properties of waterproof sealers that decreased connection between contaminants and immobilized pohotocatalyst.

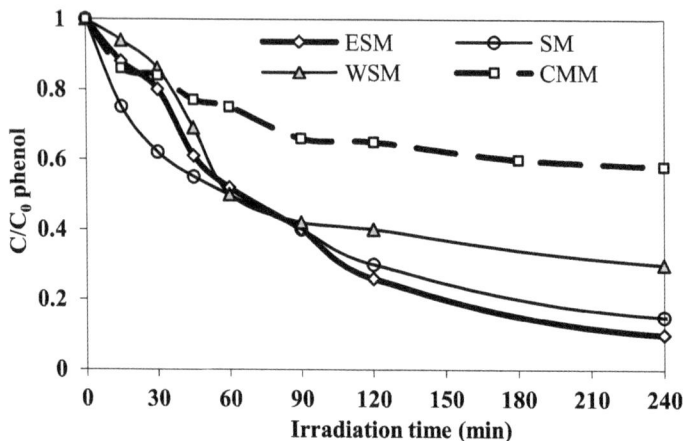

Fig. 5 Comparison between phenol removal efficiency by different immobilization methods

Effect of UV lamp Intensity

The effect of UV lamp intensity on the phenol photo degradation at constant initial concentration (100 ppm) and pH = 7 is presented in Fig. 6. The results showed that phenol removal efficiency increased when the UV lamps intensity promoted. In other words, photocatalytic efficiency in all immobilization methods improved about 50 % when UV-A lamp intensity increased from 4.42 to 8.9 mW.cm^{-2}. This is reasonable because the stronger the irradiating UV, the more the UV penetrating. A few differences between phenol degradation since 8.9 and 8.1 UV intensity was confirmed that further increase in UV intensity couldn't increase the amount of phenol destroyed.

Application of UV-C lamp promoted phenol removal efficiency more than 18 % compared to UV-A lamps at a constant irradiation time. Capability of UV-C lamp in degradation of 50 % of phenol without TiO$_2$ photocatalyst showed proper spectral irradiance of UV-A lamp that degraded only 7 % of phenol when photocatalyst was deleted. On the other hand, UV-A lamps had appreciated feature in the photocatalytic process compare to UV-C lamps.

Effect of long-term use

One of the most important parameters in photocatalytic reactors with immobilized photocatalyst is a reduction in removal efficiency because of TiO$_2$ particles detachment and catalyst surface fouling by formation of by-products during the degradation process. This limitation inhibited the application of immobilized procedure as a long term process for wastewater treatment. The influence of long term use in the degradation efficiency has been examined at a constant initial concentration (100 ppm), pH = 7 and UV lamp intensity of 5.33 mW. cm^{-2}. Proper connection between TiO$_2$ nano particles and concrete surface led to 2 and 3.5 % reduction in

phenol removal efficiency for ESM and WSM, respectively. In SM that TiO$_2$ poured on concrete surfaces and cement was used as a cohesive agent instead of concrete sealers, the reduction in removal efficiency after 45 times iterations was more than 21 %. Mixing cement by TiO$_2$ caused 4 % removal efficiency reduction after 45 times in CMM. In other research, the elimination of some TiO$_2$ from the pumice stone surface showed significant decrease of photocatalytic efficiency in long term use of process when SM was used [29].

Kinetics of the reactions

Langmuir-Hinshelwood kinetic has been used to characterize the destruction of many contaminants in different structure of catalyst [34]. The final form of this model can be expressed as Eq. 1:

$$r = -\frac{dC}{dt} = \frac{K_r K_s C_0}{1 + K_s C_0} \quad (1)$$

Where (r) is the rate of the photocatalytic degradation, (C$_0$) is the organic concentration, (K$_r$) the reaction rate constant, (K$_s$) the apparent adsorption constant and (t) is the time of reaction. The term K$_s$C$_0$ is often negligible when the concentration is low, and the reaction rate can be expressed as pseudo-first-order model as shown in Eq. 2:

$$-\frac{dC}{dt} = K_r K_s C_0 = K_{app} C_0 \quad (2)$$

Where K$_{app}$ is called apparent first order reaction rate. Integration of the equation yields to Eq. 3:

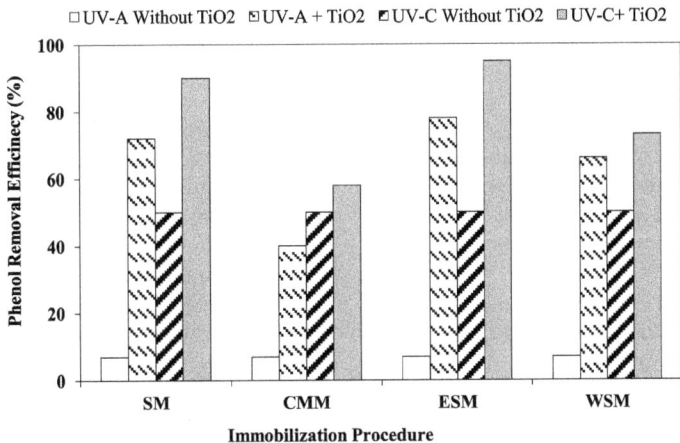

Fig. 6 Effect of UV lamp intensity in phenol removal efficiency

$$\ln\left(\frac{C}{C_0}\right) = -K_{app}t \qquad (3)$$

A plot of $-\ln(C/C_0)$ against (t) and slope of linear regression analysis is equal to the value of K_{app}. The calculated result indicated that photocatalytic degradation of phenol at the reactions conditions follows a pseudo-first-order kinetics that compared with other researches [13, 35].

Regression coefficients (R^2) and K_{app} for the photo degradation of phenol at different immobilization method, pH values and initial concentration are shown in Table 1.

The results showed that K_{app} increased in all immobilization methods and different initial concentration when pH value changes from 4 to 12. Comparison among different immobilization methods approved that ESM had the highest removal efficiency compared to the other methods that had the most K_{app}.

Phenol photo degradation intermediates

The intermediates of photo degradation of 100 mg/L phenol at pH = 9 and intensity = 8.1 mW.cm^{-2} were obtained by GC/Mass analysis. The results showed phenol hydroxylated via hydroxyl radical attack to different extent forming a variety of intermediates from the very beginning of the reaction. Firstly, aromatic ring (hydroquinone, catechol and benzoquinone) produced and after 60 min, intermediates convert to linear compounds such as oxalic acids and formic acids. Lastly, phenol is completely mineralized to CO_2 and H_2O. This can be explained given the fact that adsorbed phenol molecules on TiO_2 are subject to variable interaction with hydroxyl radicals.

Changes of the UV spectrum with respect to destruction of aromatic ring and phenol and variation of the absorbance at 268 nm in different irradiation time are shown in Fig. 7. It can be seen that when the irradiation time increased, maximum peak of the absorption spectrum decreased and finally a spectrum without noticeable peak was observed. At the end, a residual absorbance around 200 nm corresponding to the organic acid formation as final degradation products could be seen.

Conclusions

In this study, the performances of various coating methods alternatives on concrete surfaces for treating phenolic wastewater were compared. Based on the findings, the following conclusions can be drawn:

1. SEM determination has showed well uniformity of coating processes and EDX spectrum approved nano TiO_2 on concrete surface.
2. The photocatalytic tests of phenol degradation showed a good photocatalytic activity for TiO_2.
3. The ESM coating method was found to be the best technique, with the highest photocatalytic activity

Table 1 Regression coefficients (R^2) and K_{app} for phenol photo degradation

C_0 (ppm)	pH		SM	CMM	ESM	WSM
25	4	R^2	0.94	0.9	0.96	0.95
		K_{app}	0.0044	0.002	0.012	0.003
	7	R^2	0.99	0.81	0.97	0.93
		K_{app}	0.0054	0.0029	0.0135	0.01
	9	R^2	0.97	0.88	0.95	0.93
		K_{app}	0.0061	0.0035	0.0161	0.0055
	12	R^2	0.97	0.87	0.99	0.96
		K_{app}	0.007	0.0041	0.0162	0.006
100	4	R^2	0.94	0.9	0.99	0.98
		K_{app}	0.0044	0.0025	0.0058	0.0032
	7	R^2	0.94	0.91	0.91	0.9
		K_{app}	0.0057	0.0025	0.0073	0.0052
	9	R^2	0.97	0.89	0.92	0.95
		K_{app}	0.0061	0.0033	0.01	0.0058
	12	R^2	0.97	0.88	0.97	0.95
		K_{app}	0.007	0.005	0.012	0.0065
500	4	R^2	0.86	0.92	0.98	0.99
		K_{app}	0.0023	0.001	0.0042	0.002
	7	R^2	0.99	0.95	0.98	0.95
		K_{app}	0.0047	0.0013	0.006	0.0023
	9	R^2	0.93	0.92	0.99	0.98
		K_{app}	0.0057	0.0023	0.0075	0.004
	12	R^2	0.96	0.88	0.98	0.93
		K_{app}	0.0058	0.004	0.0089	0.0052

and adhesion. The removal efficiency of this method was more than 90 % after 4 h.

4. Effect of pH showed that the removal efficiency of phenol increased as pH value increased from 4 to 9 while phenol was converted to phenoxide group.
5. Photocatalytic efficiency in all immobilization methods improved about 50 % when UV lamp intensity increased 2 times.
6. Process kinetics by Langmuir-Hinshelwood model was approved pseudo-first-order reaction for phenol photo degradation.
7. Reduction of phenol removal efficiency was done as a reverse method for evaluation of nano TiO_2 detachment from concrete surfaces.
8. Results showed that application of concrete sealers had better performance in comparison with SM and CMM that cement was used as a cohesive agent. In WSM and ESM, reduction of removal efficiency was less than 2 and 3.5 % after 45 times iteration of process, respectively.

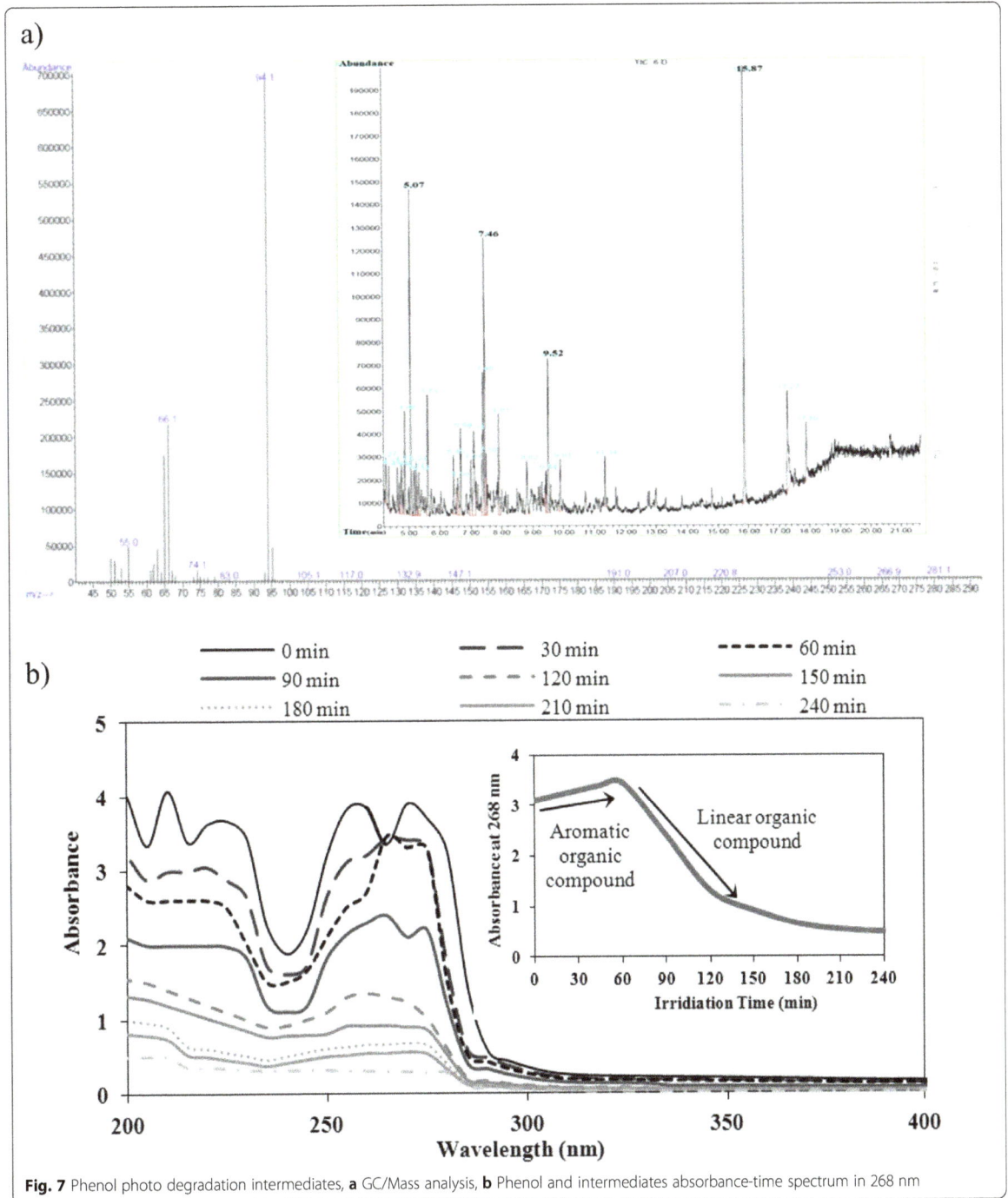

Fig. 7 Phenol photo degradation intermediates, **a** GC/Mass analysis, **b** Phenol and intermediates absorbance-time spectrum in 268 nm

9. Consequently, concrete due to its special characteristics and large consumption in WWTP as a construction material seems to be an ideal support for the immobilization of TiO_2 photo-catalysts. Also concrete sealers showed good capability for attachment of TiO_2 nano particles to concrete surfaces.

Competing interests
The authors declare that they have no competing interests.

Authors' contributions
All authors participated in conception and design, generation of data, analysis of data, interpretation of data and revision of the manuscript. All authors read and approved the final manuscript.

Acknowledgements

The authors would like to acknowledge Iranian Nano Technology Initiative Council and Vice-Chancellor for Research Affairs of Tarbiat Modares University for their partial financial support in this research. The valuable supports of Environmental Engineering Laboratory of Civil and Environmental Engineering Faculty of Tarbiat Modares University are also appreciated.

Author details

[1]Civil and Environmental Engineering Faculty, Tarbiat Modares University, Tehran, Iran. [2]Civil Engineering Department, Faculty of Engineering, Kharazmi University, Tehran, Iran. [3]Material Engineering Department, Nano Materials Division, Tarbiat Modares University, Tehran, Iran.

References

1. Gaya UI, Abdullah AH. Heterogeneous photocatalytic degradation of organic contaminants over titanium dioxide: a review of fundamentals. Progress and problems, J Photoch Photobio C. 2008;9:1–12.

2. Li Y, Li S, Li Y, Guo Y, Wang J. Visible-light driven photocatalyst (Er^{3}+:YAlO$_3$/Pt–NaTaO$_3$) for hydrogen production from water splitting. Int J Hydrogen Energ. 2014;39:17608–16.

3. Rizzo L, Koch J, Belgiorno V, Anderson MA. Removal of methylene blue in a photocatalytic reactor using polymethylmethacrylate supported TiO$_2$ nanofilm. Desalination. 2007;211:1–9.

4. De Lasa H, Errano B, Salaices M. Photocatalytic reaction engineering. USA: Springer Science Pub; 2005.

5. Bickley RI, Slater MJ, Wang WJ. Engineering development of a photocatalytic reactor for wastewater treatment. Process Saf Environ. 2005;83:205–16.

6. Shi J, Zheng J, Wu P, Ji X. Immobilization of TiO$_2$ films on activated carbon fiber and their photocatalytic degradation properties for dye compounds with different molecular size. Catal Commun. 2008;9:1846–50.

7. Lim LLP, Lynch RJ, In SI. Comparison of simple and economical photocatalyst immobilisation procedures. Appl Catal A-Gen. 2009;365:214–21.

8. Horikoshi S, Watanabe N, Onishi H, Hidaka H, Serpone N. Photodecomposition of a nonylphenol polyethoxylate surfactant in a cylindrical photoreactor with TiO$_2$ immobilized fiberglass cloth. Appl Catal B-Environ. 2002;37:117–29.

9. Sakthivel S, Shankar MV, Palanichamy M, Arabindoo B, Murugesan V. Photocatalytic decomposition of leather dye: comparative study of TiO$_2$ supported on alumina and glass beads. J Photoch Photobio A. 2002;148:153–9.

10. Choy CC, Wazne M, Meng X. Application of an empirical transport model to simulate retention of nanocrystalline titanium dioxide in sand columns. Chemosphere. 2008;7:1794–01.

11. Zainudin NF, Abdullah AZ, Mohamed AR. Characteristics of supported nano-TiO$_2$/ZSM-5/silica gel (SNTZS): photocatalytic degradation of phenol. J Hazard Mater. 2010;174:299–06.

12. Chen Y, Dionysiou DD. TiO$_2$ photocatalytic films on stainless steel: the role of Degussa P-25 in modified sol–gel methods. Appl Catal B-Environ. 2006;62:255–64.

13. Hosseini SN, Borghei SM, Vossoughi M, Taghavinia N. Immobilization of TiO$_2$ on perlite granules for photocatalytic degradation of phenol. Appl Catal B-Environ. 2007;74:53–62.

14. Chen J, Poon CS. Photocatalytic activity of titanium dioxide modified concrete materials – influence of utilizing recycled glass cullets as aggregates. J Environ Manage. 2009;90:3436–42.

15. Poon CS, Cheung E. NO removal efficiency of photocatalytic paving blocks prepared with recycled materials. Constr Build Mater. 2007;21:1746–53.

16. Husken G, Hunger M, Brouwers HJH. Experimental study of photocatalytic concrete products for air purification. Build Environ. 2009;44:2463–74.

17. Mortazavi SB, Sabzali A, Rezaee A. Sequence-Fenton Reaction for decreasing phenol formation during benzene chemical conversion in aqueous solutions. Iranian J Environ Health Sci Eng. 2005;2:62–71.

18. Kidak R, Ince NH. Ultrasonic destruction of phenol and substituted phenols: a review of current research. Ultrason Sonochem. 2006;13:195–9.

19. Moussavi G, Mahmoudi M, Barikbin B. Biological removal of phenol from strong wastewaters using a novel MSBR. Water Res. 2009;43:1095–302.

20. Busca G, Berardinelli S, Resini C, Arrighi L. Technologies for the removal of phenol from fluid streams: a short review of recent developments. J Hazard Mater. 2008;160:265–88.

21. Gholizadeh A, Kermani M, Gholami M, Farzadkia M. Kinetic and isotherm studies of adsorption and biosorption processes in the removal of phenolic compounds from aqueous solutions: comparative study. J Environ Health Sci Eng. 2013;11:29.

22. Asadgol Z, Forootanfar H, Rezaei S, Mahvi AH, Faramarzi MA. Removal of phenol and bisphenol-a catalyzed by laccase in aqueous solution. J Environ Health Sci Eng. 2014;12:93.

23. Hemmati Borji S, Nasseri S, Mahvi A, Nabizadeh R, Javadi A. Investigation of photocatalytic degradation of phenol by Fe(III)-doped TiO$_2$ and TiO$_2$ nanoparticles. J Environ Health Sci Eng. 2014;12:101.

24. Mahvi AH, Maleki A, Alimohamadi M, Ghasri A. Photo-oxidation of phenol in aqueous solution: toxicity of intermediates. Korean J Chem Eng. 2007;24:79–82.

25. Mahvi AH, Maleki A. Photosonochemical degradation of phenol in water. Deaslin Water Treat. 2010;20:197–02.

26. Maleki A, Mahvi AH, Mesdaghinia A, Naddafi K. Degradation and toxicity reduction of phenol by ultrasound wavesSoc. Bull Chem Ethiop. 2007;21:33–8.

27. Shahamat YD, Farzadkia M, Nasseri S, Mahvi A, Gholami M, Esrafili A. Magnetic heterogeneous catalytic ozonation: a new removal method for phenol in industrial wastewater. J Environ Health Sci Eng. 2014;12:50.

28. Chiou CH, Wu CY, Juang RS. Photocatalytic degradation of phenol and m-nitrophenol using irradiated TiO$_2$ in aqueous solutions. Sep Purif Technol. 2008;62:559–64.

29. Venkata Subba Rao K. Immobilization of TiO$_2$ on pumice stone for the photo-catalytic degradation of dyes and dye industry pollutants. Appl Catal B-Environ. 2003;46:77–85.

30. ACI 211. Manual of concrete practice. Farmington Hill: ACI; 1996.

31. APHA, AWWA, WEF. Standard methods for the examination of water and wastewater. 21st ed. Washington: American Public Health Association; 2005.

32. Feigenbrugel V, Le Calve S, Mirabel P, Louis F. Henry's law constant measurements for phenol, o-, m-, and p-cresol as a function of temperature. Atmos Environ. 2004;38:5577–88.

33. Morrison RT, Boyd RN. Organic chemistry. 6th ed. USA: McGraw-Hill; 2000.

34. Zheng S, Cai Y, O'Shea KE. TiO$_2$ photocatalytic degradation of phenylarsonic acid. J Photoch Photobio A. 2010;210:61–8.

35. Royaee SJ, Sohrabi M. Application of photo-impinging streams reactor in degradation of phenol in aqueous phase. Desalination. 2010;253:57–61.

6

Hematological changes of silver carp (*hypophthalmichthys molitrix*) in response to Diazinon pesticide

Aliakbar Hedayati[*] and Elaheh Hassan Nataj Niazie

Abstract

Diazinon is widely consumed for plague control in the agricultural farms and in domestic and aquaculture aspects. The present research purposed to evaluate the effect of half-lethal concentration (LC50) of diazinon on some biochemical and hematological parameters of silver carp, *hypophthalmichthys molitrix*, after 0, 24, 48 and 96 h. The results showed that the values of leukocytes (WBC), haematocrit (Ht), hemoglobin (Hb), MCHC, lymphocyte, cortisol and glucose were significantly increased ($P < 0.05$). The mount of MCV and MCH were significantly increased at 48 h and then decreased at 96 h ($P < 0.05$). Moreover, there was a significant increasing in neutrophils count at 48 h, and then a significant decreasing at 96 h ($P < 0.05$). There were no significant differences in RBC, monocyte and eosinophile counts among treatment groups at different sampling intervals. Thereupon, low-term (96 h) exposure to diazinon at half-lethal concentration (LC50) caused biochemical and hematological changes in silver carp, and offers a simply implement to appraise toxicity-derivatived changes.

Keywords: Blood indices, Diazinon, Fish, Pollution

Introduction

Diazinon is a vast spectrum OP, often utilized in residential areas, while its major utilization is with regard to agricultural activities. It has been categorized as very toxic and its use has been prohibited in many developed countries, however, it is commercialized in expansing ones [1]. Diazinon is partly water-soluble, and simply comes into contiguity with aquatic organisms. Afterwards, inward the cells of these organisms, diazinon is metabolized to diazoxon, which is the toxic composition of this insecticide. In addition, diazinon affects abroad range of non-target organisms, such as invertebrates, birds, mammals and fishes, chiefly, those are habiting aquatic environment [2]. Kalender et al. [3] showed that diazinon causes alters in liver enzymes and biochemical indicators. Diazinon is widely utilized in the rice paddy farms of Mazandaran, Gillan, Golestan and other areas in Iran [4]. Based on reports of Bulletin of Agriculture Ministry, the annual utilization of diazinon in iran is appraised to be 3775 t [5]. Several studies related that

some of the surface waters and the surrounding environments in Iran were infected with organophosphate pesticides like diazinon and its derivates [6].

Hematological parameters like MCH, MCV and MCHC and Hematological indices such as Hct, Hb, RBC, WBC values and biochemical indicators like glucose are widely utilized to assess the toxic stress of environmental contaminants [7]. Silver carp may be most abundant freshwater fish due to its fast growth [8]. In spite of the manufacture of silver carp is spreading out annually, the processing of this fish is limited. Silver carp rearing developed in many ponds that placed near agricultural fields.

The aim of this study was to investigate the effect of different diazinon concentrations on blood biochemical, hematological and hormonal indicators in silver carp (*hypophthalmichthys molitrix*).

Materials and methods
Preparation of fish in experimental condition
Fish with average weight 200 g of silver carp after 1-week adaptation to the new condition, were divided into 4 treatments (3 replicates for each treatment) and a control with

* Correspondence: Hedayati@gau.ac.ir
Department of Fisheries, Faculty of Fisheries and Environment, Gorgan
University of Agricultural Sciences and Natural Resources, Gorgan, Iran

21 fish in each 400 L tank. All fish were hand-fed with trading pellet twice a day. Fish were exposed to a concentrations of 50 % LC_{50} (3.93 ± 0.34) of diazinon for a period of 24, 48 and 96 h. Temperature, dissolved oxygen, conductivity and pH were measured during the experiment.

Blood sampling and hematological assay

At the beginning, fish were unconscious with 200 ppm clove powder. Blood samples quickly collected from tail blood vessel by heparinized syringes and immediately stored on ice. Then, computation of the blood parameters were carried out on fresh blood. Numbers of Blood erythrocytes and leukocytes were performed by diluting heparinized blood with Giemsa stain at 1:30 dilution and cells were counted using a hemacytometer Neubauer under the light microscope. The leukocyte differential calculation was performed in peripheral blood spots stained by Merck Giemsa, giving the Neutrophils quantity of differential neutrophils and the mononuclear quantity of differential lymphocytes, eosinophile and monocyte. Hematocrite percent (Ht%) was shortly computed after act of collecting representative samples by placing fresh blood in glass capillary tubes and centrifuged in a microhematocrit centrifuge for 5 min at 10,000 rpm (Hettich, Germany), afterward, taking of dimensions the packed cell capacity. Hematocrite explanation was carried out with the usage of a microhematocrit reader. Hemoglobin levels (Hb mg/l) were resulted colorimetrically of cyanomethemoglobin by measuring the formation [9]. Erythrocytes Indices (M.C.H. or Mean Corpuscular Hemoglobin, M.C.H.C. or Mean Cell Hemoglobin Concentration and M.C.V. or Mean Corpuscular Volume) were accounted from RBC, Ht, and Hb [9, 10].

Statistical analyses

To observation of significant differences to appraise, the effect of diazinon on blood parameters of silver carp was used an analysis of variance (ANOVA) with Duncan Post Hoc. Pearson coefficients of correlation (r) were calculated between half-lethal concentration of diazinon and blood parameters to examine coalition between bioaccumulation and its impacts. To determine the unity between diazinon concentration and blood parameters were utilized multiple regressions. Data were analyzed statistically at $p < 0.05$ by SPSS software version 16.

Ethical approval

This work was approved by ethical committee of GAU University numbered 6177515-5. To minimize suffering of the fish, all animals were exposed with clove essence, low dose for anesthesia; hypothermia prior to euthanasia and eventually spinal cord dislocation for euthanasia.

Results and discussion
Hematological parameters

In this study, we surveyed effects of 50 % LC50 diazinon pesticide on a series of hematological, biochemical and Immunological parameters of silver carp for 96 h.

Hematocrit content significantly elevated after subchronic exposure, exhibiting the importance of the route of pollution (Fig. 1). Also, Ht and Hb levels significantly increased that was similar with results of Ahmad [11] on Cyprinus carpio in exposed to diazinon but was conversed with the findings Shaluie et al. [12] on silver carp exposed to Nanocid. Also Banaee et al. [13] and Anees [14], respectively, to study on Cyprinus carpio and Bloch (*Channa punctatus*) concluded that Ht and Hb levels significantly increased in exposed to diazinon. In another studies, Oimoek Köprücü et al. [15] found out that Ht and Hb levels increased in fingerling European catfish (*Silurus glanis L.*) exposed to diazinon. Observed results were agreement with results of Chowdhury et al. [16], who stated an elevation of blood hemoglobin and hematocrit happen in exposed to environmental hypoxia and chronic doses of waterborne metals that elevate blood oxygen carrying capacity caused to disorder of gas exchange.

Our findings showed that MCV and MCH levels increased in fish exposed to 50 % LC50 diazinon at 48 h but MCHC level increased at 96 h (Fig. 2). Results of the study were similar with the findings of Banaee et al. [13] about MCV and MCH. Shaluie et al. [12] to survey on silver carp exposed to Nanocid expressed that MCH and MCHC levels raised. Also, Mohammad Nejad et al. [17] and Anees [14] to studying on *Rutilus frisii kutum* and Bloch (*Channa punctatus*) in exposed to diazinon found out that MCH, MCV and MCHC levels increased that were agreement with our study.

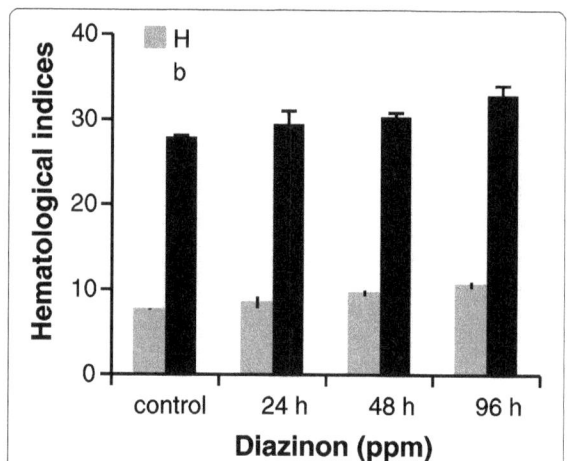

Fig. 1 Haematocrite and Hemoglobin changes of silver carp during exposure to LC_{50-96h} diazinon

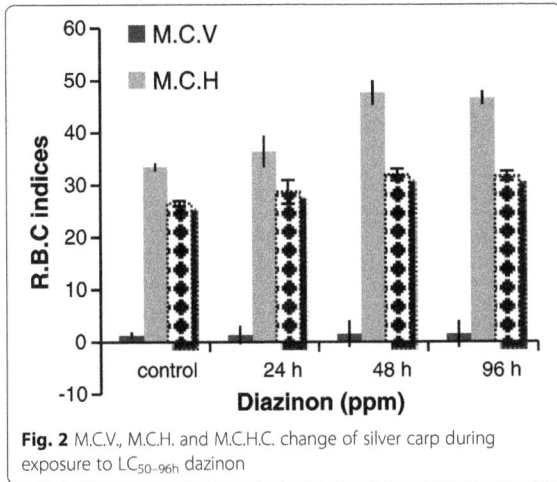

Fig. 2 M.C.V., M.C.H. and M.C.H.C. change of silver carp during exposure to LC_{50-96h} dazinon

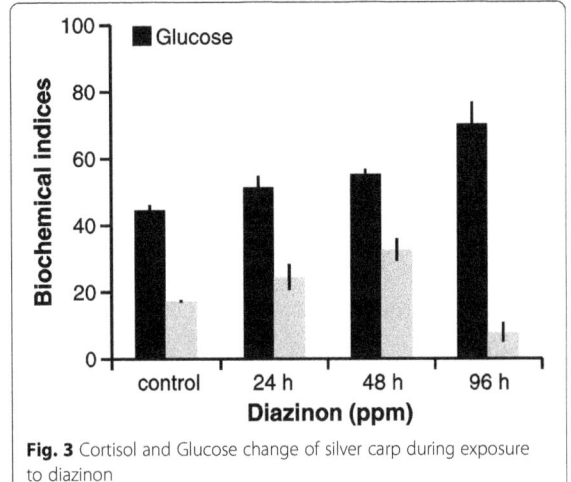

Fig. 3 Cortisol and Glucose change of silver carp during exposure to diazinon

In the present study, WBC level of silver carp blood significantly increased in exposed to diazinon. Whereas, Banaee et al. [13] and Mohammad Nejad et al. [17] obtained opposite results to this finding. The number of red blood cells (RBC) was less sensitive toward diazinon exposure so that the RBC count did not significantly changed in fish. This finding was not similar with Mohammad Nejad et al. [17], Banaee et al. [13], Khoshbavar Rostami and Soltani [18] and Anees's studies [14] on *Rutilus frisii kutum*, *Cyprinus carpio*, *Acipenser nudiventris* and Bloch (*Channa punctatus*) in exposed to diazinon. Plenty of WBC provides an exhibition of fish health and a high WBC number may exhibit a subclinical infection.

The correlation between diazinon with all parameters statistically examined by analyzing the data obtained during the diazinon exposed for 96 h. In this concentration (LC_{50-96h}) Ht, MCV, MCH and MCHC levels indicated significant positive correlation ($p < 0.01$) and WBC, RBC and Hb levels did not showed significant correlation with diazinon exposure.

Biochemical analysis

Glucose level significantly ($p < 0.05$) increased in 50 % LC_{50} concentration of diazinon in comparison of control group (Fig. 3) and showed significant positive correlation ($p < 0.01$) in exposure to diazinon.

It has been stated that the raised blood glucose is usually perceived in fish under unfavorable situations and it helps the animal by readying energy substrates to vital organs to afford with the increased energy demand [13]. Also, increased of blood glucose level was vastly used as a secondary marker of a stress response [19]. Increased in blood glucose levels have been reported in Silver carp [12] and *Cyprinus carpio* [13], after exposure to copper sulfate and diazinon, respectively, that was comparable to our findings of glucose on silver carp.

Blood cortisol is an important corticosteroid hormone in fish and may have a meaningful effect on its dynamics [20]. In the present study, cortisol level in concentration of 50 % LC50 for 48 h significantly increased toward control (Fig. 3) and did not show significant correlation in exposure to diazinon. The blood serum cortisol results were indicated in (Fig. 4), revealed a significant increasing at 48 h in the exposed fish. Bakhshwan et al. [21] and Shaluie et al. [12] obtained similar results to studying on Clarias gariepinus and silver carp in exposed to diazinon and Nanocid, respectively. The above data may be stated by the activation hypothalamo-pituitary-inter renal axis with their discharge of steroid cortisol in

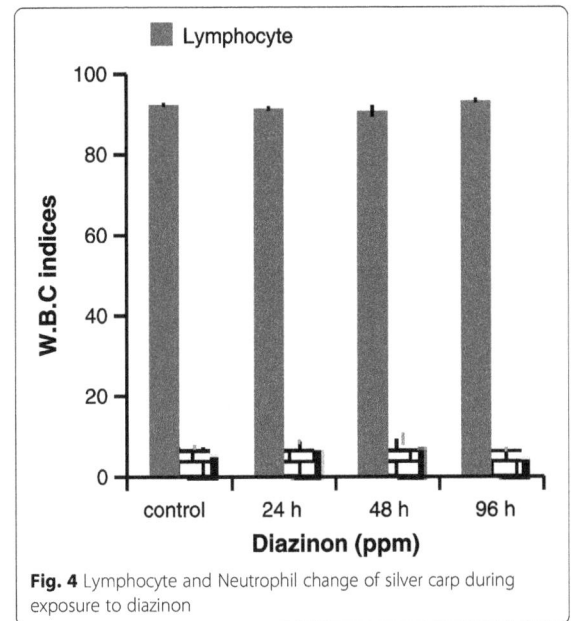

Fig. 4 Lymphocyte and Neutrophil change of silver carp during exposure to diazinon

blood circuit due to stress [22]. In addition, it could be ascribed to the raise in osmotic water-influx, which may reason a cortisol promotion, to mend the hydromineral balance [21].

Immunological indices

Lymphocyte level was significantly increased in exposed to diazinon for 96 h and neutrophil level was significantly increased for 48 h ($p < 0.05$) (Fig. 4). Eosinophiles and monocyte did not show significant correlation in 50 % LC50 concentration.

Results of current study showed that there were significant increases in lymphocyte and neutrophil counts. Hedayati and Jahanbakhshi [23] reported increasing in neutrophil and decreasing in lymphocyte to study on the great sturgeon Huso huso in exposed to diesel oil. In addition, Soltani and Khoshbavar Rostami [24] stated elevation neutrophil count and decreasing of lymphocyte count in *Acipenser guldenstadti*. In other studies, Svoboda et al. [25] and Banaee et al. [13] observed neutrophil count increased but lymphocyte count decreased. It is believed that neutrophils have phagocytic activity, which might describe their elevated percentage during infectious status.

To determine the relationship between diazinon concentration with hematological, biochemical and immunological activity were utilized curve estimation regressions data. Ht, MCV, MCH, MCHC, Lymphocyte and Glucose levels in fish exposed to diazinon indicated significant linear regression ($p < 0.01$) Y = a ± bX.

Conclusion

The present results indicated that diazinon can cause serious hematological and biochemical alterations in silver carp at the physiological level. Finally, the results got in this study exactly indicated under experimental conditions, blood parameters were sensitive to different prospect of diazinon exposure. Hematological and biochemical properties of this fish exposed to diazinon are poorly understood and there is not adequate knowledge concerning the metabolism of reference toxicants. More knowledge of these activities in fish is necessary before they can be employed as biochemical indicators of stress due to pollutions.

Competing interests

The authors declare that they have no competing interests.

Authors' contributions

The authors declare that they have no contributions in this paper. Both authors read and approved the final manuscript.

Acknowledgments

The authors thank the aquaculture research center and Fishery group for the supply of research material. This study was supported by the Gorgan University of Agricultural Sciences and Natural resources and was done at Veniro wet laboratory.

References

1. Tinoco R, Halperin D. Poverty, production and health inhibition of erythrocytes choline esterase via occupational exposure to organophosphate insecticides in Chiapas, Mexico. Arch Environ Health. 1998;53:29–35.

2. Burkepile DE, Moore MT, Holland MM. The susceptibility of Wvenontarget organisms to aqueous diazinon exposure. Bull Environ Contam Toxicol. 2000;64:114–21.

3. Kalender S, Ogutcu A, Uzunhisarcikli M, Acikgoz F, Durak D, Ulusoy Y, et al. Diazinon-induced hepatotoxicity and protective effect of Vitamin E on some biochemical indices and ultrastructural changes. Toxicology. 2005;211:197–206.

4. Shayeghi M, Darabi H, Abtahi H, Sadeghi M, Pakbaz F, Golestaneh SR. Assessment of persistence and residue of diazinon and malathion in three Rivers (Mond, Shahpour and Dalaky) of Bushehr province in 2004–2005 years. Iran South Med J. 2007;10(1):54–60.

5. Annual report of performance of the Ministry of Agriculture of Iran (ARMAI) (2005–2006). Ministry of Agriculture of Iran; 215–17 p.

6. Arjmandi R, Tavakol M, Shayeghi M. Determination of organophosphorus insecticide residues in the rice paddies. Int J Environ Sci Tech. 2010;7(1):175–82.

7. Kavitha C, Malarvizhi A, Senthil Kumaran S, Ramesh M. Toxicological effects of arsenate exposure on hematological, Biochemical and liver transaminases activity in an Indian major carp, *Catla catla*. Food Chem Toxicol. 2010;48:2848–54.

8. Aryanta RW, Fleet GH, Buckle KA. The occurrence and growth of microorganisms during the fermentation of fish sausage. Int J Food Microbio. 1991;13(2):143–55.

9. Lee RG, Foerster J, Jukens J, Paraskevas F, Greer JP, Rodgers GM. Wintrobe's clinical hematology. 10th ed. New York, USA: Lippincott Williams & Wilkins; 1988.

10. Plous S, Herzog H. Reliability of protocol reviews for animal research. Science. 2001;293:608–9.

11. Ahmad Z. Acute toxicity and hematological changes in common carp (*Cyprinus carpio*) caused by diazinon exposure. African J Biotech. 2011;10(63):13852–9.

12. Shaluie F, Hedayati A, Jahanbakhshi A, Miandareh HK, Fotovat M. Effect of subacute exposure to silver nanoparticle on hematological and plasma biochemical indices in silver carp (*Hypophthalmichthys molitrix*). Hum Exp Toxicol. 2013. doi:10.1177/0960327113485258.

13. Banaee M, Mirvagefei AR, Rafei GR, Majazi Amiri B. Effect of sub-lethal diazinon concentrations on blood plasma biochemistry. Int J Environ Res. 2008;2:189–98.

14. Anees MA. Haematological abnormalities in a freshwater teleost, *Channa punctatus*, (Bloch), exposed to sublethal and chronic levels of three organophosphorus insecticides. Int J Ecol Environ Sci. 1978;4:53–60.

15. Oimoek Köprücü S, Köprücü K, Sener Ural M, Ispir U, Pala M. Acute toxicity of organophosphorous pesticide diazinon and its effects on behavior and some hematological parameters of fingerling European catfish (*Silurus glanis*). Pest Biochem Physiol. 2006;86:99–105.

16. Chowdhury MJ, McDonald DG, Wood CC. Gastrointestinal uptake and fate of cadmium in rainbow trout acclimated to sublethal dietary cadmium. Aquat Toxicol. 2004;69:149–63.

17. Mohammad Nejad Shamoushaki M, Soltani M, Kamali A, Imanpoor MR, Sharifpour I, Khara H. Effects of organophosphate, diazinon on some haematological and biochemical changes in *Rutilus frisii kutum* male brood stocks. Iran J Fish Sci. 2012;11(1):105–17.

18. Khoshbavar Rostami HA, Soltani M. The effects of diazinon on haematological indices and LC50(96 h) of *Acipenser nudiventris*. Iran J Fish Sci. 2005;14(3):1–12.

19. Toal DG, Afonso LOB, Iwama GK. Stress response of juvenile rainbow trout (*Oncorhynchus mykiss*) to chemical cues released from stressed conspecifics. Fish Physiol Biochem. 2004;30:103–8.

20. Mommsen TP, Vijayan MM, Moon TW. Cortisol in teleosts: dynamics, mechanisms of action and metabolic regulation. Rev Fish Biol Fish. 1999;9:211–68.

21. Bakhshwan SA, Marzouk MS, Hanna MI, Heba Hamed S. Some investigations on the clinical and biochemical alterations associated with dizinon toxicity in *Clarias gariepinus*. Egypt J Aquat Biol Fish. 2009;13(2):173–97.

22. Reddy PK, Leatherland JF. Stress physiology. In: Leatherland JF, Woo PTK, editors. Fish disease and disorders, vol.2 non–infections disorders. UK: Cabi Pub; 1998. p. 279–301.

23. Hedayati A, Jahanbakhshi A. The effect of water-soluble fraction of diesel oil on some hematological indices in the great sturgeon Hus huso. Fish Physiol Biochem. 2012. doi:10.1007/s10695-012-9672-7.

24. Soltani M, Khoshbavar Rostami HA. Effect of Diazinon on some hematological and biochemical indices in Russian sturgeon (*Acipenser guldenstadti. Iranian*). J Mar Sci Tech Iran. 2001;4:1–11.

25. Svoboda M, Luskova V, Drastichova J, Zlabek V. The effect of Diazinon on haematological indices of common carp (*Cyprinus carpio*). Acta Vet Brno. 2001;70:457–65.

7

Hierarchical distance-based fuzzy approach to evaluate urban water supply systems in a semi-arid region

Tahereh Sadeghi Yekta[1], Mohammad Khazaei[1], Ramin Nabizadeh[2], Amir Hossein Mahvi[2,3], Simin Nasseri[2,4*] and Ahmad Reza Yari[1]

Abstract

Hierarchical distance-based fuzzy multi-criteria group decision making was served as a tool to evaluate the drinking water supply systems of Qom, a semi-arid city located in central part of Iran. A list of aspects consisting of 6 criteria and 35 sub-criteria were evaluated based on a linguistic term set by five decision-makers. Four water supply alternatives including "Public desalinated distribution system", "PET Bottled Drinking Water", "Private desalinated water suppliers" and "Household desalinated water units" were assessed based on criteria and sub-criteria.

Data were aggregated and normalized to apply Performance Ratings of Alternatives. Also, the Performance Ratings of Alternatives were aggregated again to achieve the Aggregate Performance Ratings. The weighted distances from ideal solution and anti-ideal solution were calculated after secondary normalization. The proximity of each alternative to the ideal solution was determined as the final step. The alternatives were ranked based on the magnitude of ideal solutions.

Results showed that "Public desalinated distribution system" was the most appropriate alternative to supply the drinking needs of Qom population. Also, "PET Bottled Drinking Water" was the second acceptable option. A novel classification of alternatives to satisfy the drinking water requirements was proposed which is applicable for the other cities located in semi-arid regions of Iran.

The health issues were considered as independent criterion, distinct from the environmental issues. The constraints of high-tech alternatives were also considered regarding to the level of dependency on overseas.

Keywords: Fuzzy logic, Drinking water, MCDM, Distribution system

Introduction

Evaluating the alternatives to satisfy the drinking water demands of societies is a complicated issue that usually should be relied on human judgments. Furthermore, Different criteria should be considered to evaluate the alternatives available for supplying the drinking water needs, especially in populations faced with fresh water scarcity which are relied on brackish water sources [1].

Various methods based on human decision-making have been used to evaluate the alternatives assigned for water supply systems such as Life cycle assessment [2, 3], MCDM approach [4], Five-parametric matrix [5], Multi-criteria decision aid (MCDA) approach [6], and consumer cooperatives [7].

The major concern related to the water supply systems in developing countries is the large scale projects such as trans-basin water transfer [8], and constructing the sophisticated water supply systems which may not be completed on time because of the financial deficiencies or changing in political considerations [9]. So, applying the available water supply systems as the viable alternatives can be helpful to deliver an obvious viewpoint for administrators as well as for the public sector [10]. Also, few studies, worked on evaluating the available alternatives, have drown the hierarchy of aspects directly from the other studies and did not consider the background

* Correspondence: naserise@tums.ac.ir
[2]Department of Environmental Health Engineering, School of Public Health, Tehran University of Medical Sciences, Poursina St, Keshavarz Blvd, PO BOX: 6446-14155, Tehran, Iran
[4]Center for Water Quality Research, Institute for Environmental Research, Tehran University of Medical Sciences, Tehran, Iran
Full list of author information is available at the end of the article

factors in their intrinsic society which may influence the arrangement of criteria and sub-criteria [4, 7, 11].

This paper outlines a methodology that evaluates the available alternatives to supply drinking water demands of Qom population, a city located in plains fed with brackish aquifers. The evaluation processes are according to a complete package of criteria and sub-criteria.

A simple-minded and well-known method of decision-making is adopted based on fuzzy logic to evaluate the alternatives. The presented method is known as hierarchical distance-based fuzzy multi-criteria group decision making (DBF –MCDM) approach. Applying DBF–MCDM enables the decision-making committee to improve the identification of discrepancies and similarities of their judgments [12]. Also, the DBF–MCDM process justifies both ideal and anti-ideal solutions simultaneously that help the decision-makers to have more obvious judgments [13]. A new arrangement of criteria and sub-criteria to evaluate the drinking water supply alternatives is also adopted using the MCDM method under fuzzy environment.

Methodology

Various aspects should be considered when a team or organization decides to make a decision among several available alternatives. The decision making process maybe comes more complicated if the number of alternatives and criteria be increased [14]. This section dedicates a short description about the principles of multi-criteria group decision making (MCDM) that is based on fuzzy set theory to resolve the decision making problems on the subject of drinking water supply alternatives.

Fuzzy sets theory
Definition 1 A fuzzy set can be defined as $\tilde{A} = (X, \mu_{\tilde{A}}(x))$, Where X is the space on which the fuzzy set is defined, and $\mu_{\tilde{A}}(x) \rightarrow [0, 1]$, $x \in X$, the membership function of the set [15].

Definition 2 As shown in Fig 1, a triangular fuzzy number $\tilde{A}$ can be depicted with a triplet (a_1, a_2, a_3) which its membership function are symbolized as follows [16]:

$$\mu_{\tilde{A}}(x) = \begin{cases} \dfrac{x-a_1}{a_2-a_1}, & a_1 \leq x \ a_2, \\ \dfrac{x-a_3}{a_2-a_3}, & a_2 \leq x \ a_3, \\ 0, & Otherwise. \end{cases} \quad (1)$$

Using the triangular fuzzy number is due to its simplicity compare with trapezoid or sigmoid fuzzy numbers and intuitively easy for decision-makers to utilize. Furthermore, modeling according to triangular fuzzy numbers is

a competent approach for organizing the decision-making problems [17, 16].

Definition 3 A linguistic variable is defined as a kind of variable whose values are expressed in linguistic terms. Because of the imprecise and vague nature of human judgments, it is preferred to express the expert judgments via linguistic terms. The linguistic terms are the study variables with the capability of describing the qualitative data. A linguistic variable comprises an ordinary word or phrase in natural language and so they are representatives of imprecise data whose values are not numbers. In situations that the study has been affected by ill defined or complex variables, a linguistic term can be a useful tool to prepare an approximate characterization [18].

Definition 4 The criteria $a_1, a_2 ..., a_3$ are defined as the evaluation tools of each alternative. This assumption must be taken into account that all criteria are relevant for various alternatives. The different alternatives are represented as $A_1, A_2 ... ., A_m$ For certain alternative A_i, the relative value of criteria a_i is allocated by a rating, identified as r_{ij}. Also, the relative importance of a given criterion a_j is allocated by a weighting coefficient, denoted as w_j. So, the alternative A_i obtains the weighted average rating as follows:

$$\bar{r}_i = \frac{\sum_{j=1}^{n} W_i r_{ij}}{\sum_{j=1}^{n} W_j} \quad (2)$$

Comparing and ranking the final ratings $\bar{r}_1, \bar{r}_2 ..., \bar{r}_m$ are performed to judge the relevant values of the different alternatives [14].

Definition 5 If $\tilde{n}$ be considered as a triangular fuzzy number and $n_{\ell}^{\propto} > 0, n_u^{\propto} \leq 1$ for $\propto \in [0, 1]$ then $\tilde{n}$ is called a normalized positive triangular fuzzy number [19].

Definition 6 The ideal solution $A^* = (r_1^*, r_2^*, ..., r_n^*)$ and also the anti-ideal solution $A^- = (r_1^-, r_2^- ..., r_n^-)$ are defined where $r_j^* = (1, 1, 1)$ and $r_j^- = (0, 0, 01)$ for $j = 1, 2 ..., n$ [20].

Definition 7 The distance measure $d_v(\tilde{A}\tilde{B})$ is applied to indicate the distance between the fuzzy numbers $\tilde{A} = (a_1, a_2, a_3)$ and $\tilde{B} = (b_1, , b_2, , b_3)$ as follows [21]:

$$d_v(\tilde{A}\tilde{B}) = \frac{1}{2}\{ \max \ (|a_1-b_1|, |a_3-b_3|) + |a_2-b_2|\} \quad (3)$$

The size of the trapezoidal area is obtained by the distance formula. The larger values of $|a_1 - b_1| \ or |a_3 - b_3|$ are the lower trapezoid base. The values of $|a_2 - a_2|$ determine the upper trapezoid base, and the trapezoid height is

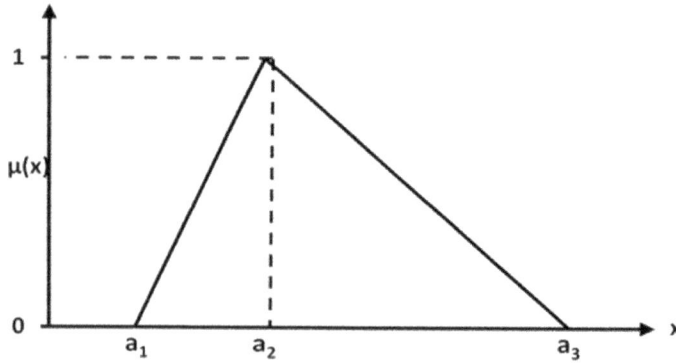

Fig. 1 A triangular fuzzy number $\tilde{A}$

equal to one. The Closer triangular numbers $\tilde{A}$ *and* $\tilde{B}$ the smaller trapezoidal area.

Hierarchical distance-based fuzzy Multi-criteria group decision making (DBF –MCDM) approach

The fuzzy multi-criteria group decision making approach has the ability of addressing the decision problems including a multi-level hierarchical structure which has been equipped with attributes of qualitative performance [22]. The distance-based fuzzy MCDM approach has been introduced by Karsak (2002) for selecting the technology alternative [23]. The DBF-MCDM is constructed according to the closeness to the ideal alternative concept. Also, DBF-MCDM has the potential of including both crisp and fuzzy data.

Usually, the performance attributes can be organized in multi-level hierarchy when they are in large numbers. The multi-level hierarchy enables the analysis to be done more efficiently.

Here, a subversion known as "multi-expert" from the algorithm of hierarchical DBF-MCDM which originally introduced by Karsak and Ahiska (2005) and later represented by Dursun (2011.a) is applied. Figure 2 illustrates a brief representation of hierarchical DBF-MCDM approach.

The following successive steps present the hierarchical DBF-MCDM approach implementation:

Step 1. Establish a decision- makers team of z experts ($l = 1,2..., z$). Introduce the alternatives, necessary criteria, and attributed sub-criteria.

Step 2. Assemble the decision matrices that comprise the importance weights of criteria and attributed sub-criteria. The decision matrices also, should be included the fuzzy assessments in relation with sub-criteria for each decision-maker.

Step 3. Introduce the mathematical signs used for representation the criteria, sub-criteria, decision makers and alternatives and their relationships as depicted in Table 1.

Step 4. Calculate the aggregated fuzzy assessments of alternatives $\left(\tilde{X}_{ijkl}\right)$, the aggregated importance weight of sub-criteria $\left(\tilde{W}_{jkl}\right)$ and the aggregated importance weight of criteria $\left(\tilde{W}_{jl}\right)$ based on follows:

$$\tilde{W}_j = \sum_{l-1}^{z} \nu_l \tilde{W}_{jl} \tag{4}$$

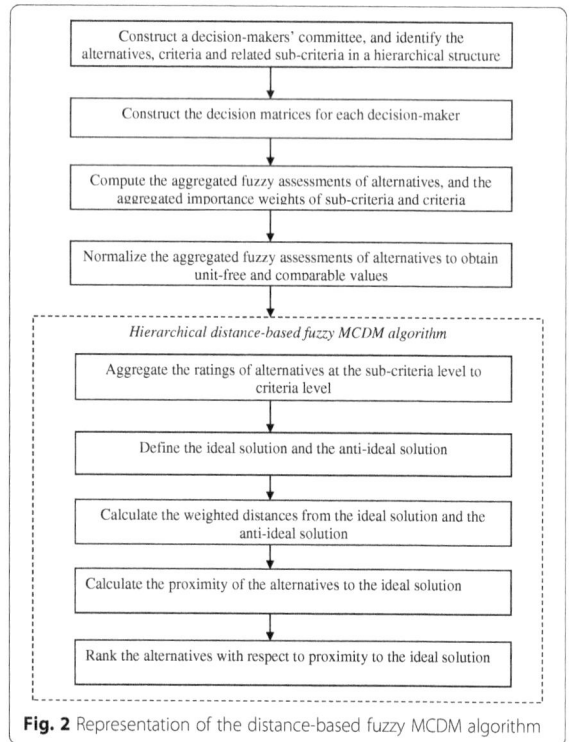

Fig. 2 Representation of the distance-based fuzzy MCDM algorithm

Table 1 Mathematical signs used for representing the equations

Definition	Description
$i = (1, 2 ..., m)$	Set of alternatives
$j = (1, 2 ..., n)$	Set of criteria
$k = (1, 2 ..., p)$	Set of sub-criteria
$l = (1, 2 ..., z)$	Set of decision makers
$\tilde{X}_{ijkl} = \left(X_{ijkl}^1, X_{ijkl}^2, X_{ijkl}^3 \right)$	Alternative i attributed to sub-criterion k of criterion j.
$\tilde{W}_{jkl} = \left(W_{jkl}^1, W_{jkl}^2, W_{jkl}^3 \right)$	Importance weight of sub-criterion k of criterion j.
$\tilde{W}_{jl} = \left(W_{jl}^1, W_{jl}^2, W_{jl}^3 \right)$	Importance weight of criterion j for the lth decision-maker

$$\tilde{W}_{jk} = \sum_{l-1}^{z} v_l \tilde{W}_{jkl} \qquad (5)$$

$$\tilde{X}_{ijk} = \sum_{l-1}^{z} v_l \tilde{X}_{ijkl} \qquad (6)$$

Where $v_l \in [0, 1]$ represents weight assigned to the lth decision-maker.

Also, $\sum_{l=1}^{z} v_l = 1$.

So, by using above equations, aggregated ratings of alternatives with respect to each sub-criterion ($\tilde{X}_{ijk}$), aggregated importance weights of sub-criteria $\tilde{W}_{jk}$ and aggregated importance weights of criteria ($\tilde{W}_j$) can be computed as $(X_{ijk}^1, X_{ijk}^2, X_{ijk}^3)$, $(W_{jkl}^1, W_{jkl}^2, W_{jk.}^3)$ and (W_j^1, W_j^2, W_j^3) respectively.

Step 5. To obtain the unit-free and comparable sub-criteria values, the aggregated decision matrix resulted from step 4 should be normalized. Among various methods used for data normalization [24, 17] a linear scale transformation is selected. Based on this approach, first the sub-criteria are categorized in two groups known as benefit-related (BR) and cost related (CR) ones as identified in Fig. 3. Then, the linear scale transformation is used for data normalization as follows:

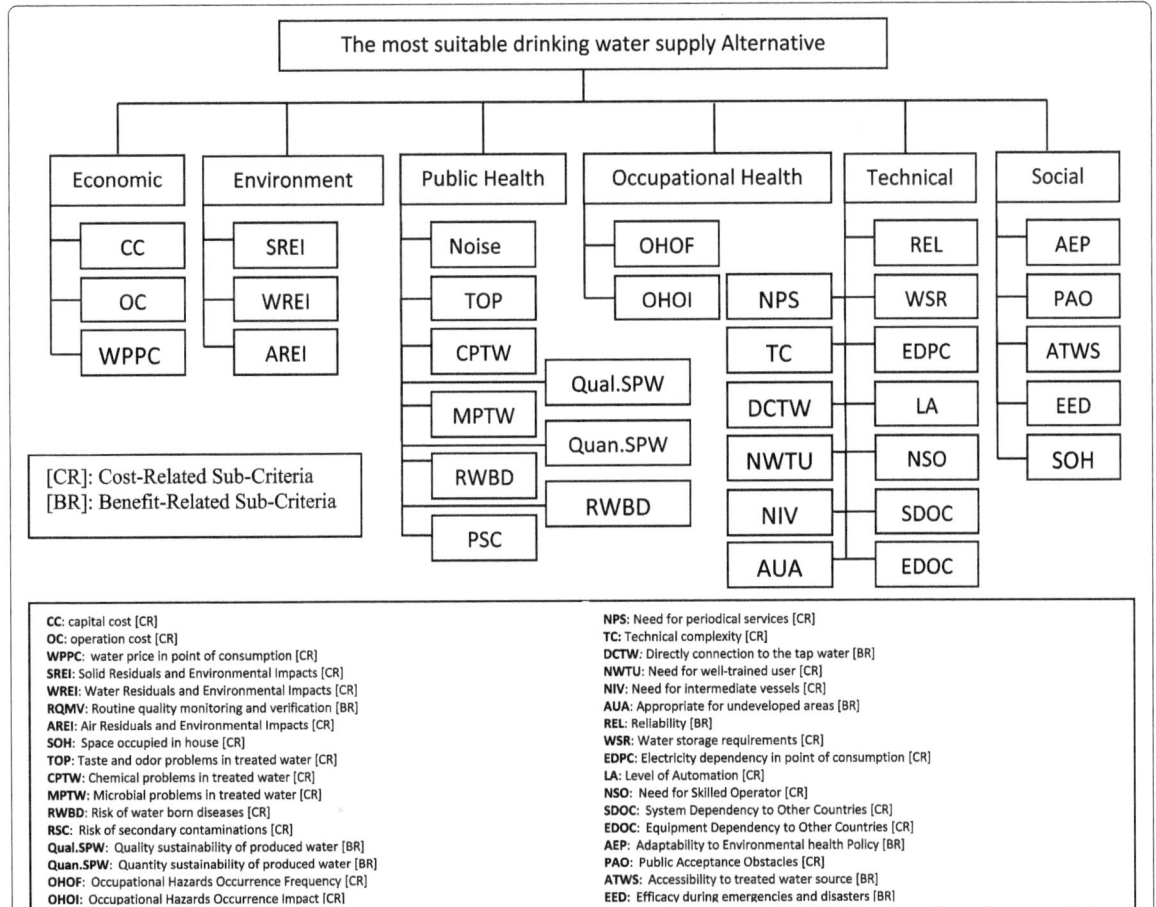

CC: capital cost [CR]
OC: operation cost [CR]
WPPC: water price in point of consumption [CR]
SREI: Solid Residuals and Environmental Impacts [CR]
WREI: Water Residuals and Environmental Impacts [CR]
RQMV: Routine quality monitoring and verification [BR]
AREI: Air Residuals and Environmental Impacts [CR]
SOH: Space occupied in house [CR]
TOP: Taste and odor problems in treated water [CR]
CPTW: Chemical problems in treated water [CR]
MPTW: Microbial problems in treated water [CR]
RWBD: Risk of water born diseases [CR]
RSC: Risk of secondary contaminations [CR]
Qual.SPW: Quality sustainability of produced water [BR]
Quan.SPW: Quantity sustainability of produced water [BR]
OHOF: Occupational Hazards Occurrence Frequency [CR]
OHOI: Occupational Hazards Occurrence Impact [CR]

NPS: Need for periodical services [CR]
TC: Technical complexity [CR]
DCTW: Directly connection to the tap water [BR]
NWTU: Need for well-trained user [CR]
NIV: Need for intermediate vessels [CR]
AUA: Appropriate for undeveloped areas [BR]
REL: Reliability [BR]
WSR: Water storage requirements [CR]
EDPC: Electricity dependency in point of consumption [CR]
LA: Level of Automation [CR]
NSO: Need for Skilled Operator [CR]
SDOC: System Dependency to Other Countries [CR]
EDOC: Equipment Dependency to Other Countries [CR]
AEP: Adaptability to Environmental health Policy [BR]
PAO: Public Acceptance Obstacles [CR]
ATWS: Accessibility to treated water source [BR]
EED: Efficacy during emergencies and disasters [BR]

Fig. 3 Hierarchical structure of the problem and identifying the CR and CB nature of criteria and sub-criteria

$$\tilde{r}_{ijk} = \left(r^1_{ijk}, r^2_{ijk}, r^3_{ijk} \right)$$

$$= \begin{cases} \left(\dfrac{x^1_{ijk}-x^-_{jk}}{x^*_{jk}-x^-_{jk}}, \dfrac{x^2_{ijk}-x^-_{jk}}{x^*_{jk}-x^-_{jk}}, \dfrac{x^3_{ijk}-x^-_{jk}}{x^*_{jk}-x^-_{jk}} \right), & k \in BR_j; \ i=1,2...,m; \\ & j=1,2...,n \\[2mm] \left(\dfrac{x^*_{jk}-x^3_{ijk}}{x^*_{jk}-x^-_{jk}}, \dfrac{x^*_{jk}-x^2_{ijk}}{x^*_{jk}-x^-_{jk}}, \dfrac{x^*_{jk}-x^1_{ijk}}{x^*_{jk}-x^-_{jk}} \right), & k \in CR_j; \ i=1,2...,m; \\ & J=1,2...,n \end{cases}$$

$$(7)$$

Where, $\tilde{r}_{ijk}$ is the normalized value of $\tilde{x}_{ijk}$, x^*_{jk} denotes $\max_i x^3_{ijk}$ and x^-_{jk} is $\min_i x^1_{ijk} BR_j$ is the set of benefit-related sub-criteria of criterion j for which the higher the efficiency value the more performance of it and CR_j is the sets of cost-related sub-criteria of criterion j for which the higher the efficiency value the less preference of it. Also, m identifies the number of alternatives and n denotes the number of criteria.

Step 6. The performance ratings of alternatives at the sub-criteria stage to criteria stage should be aggregated to compute the aggregate performance ratings (APRs) as follows:

$$\tilde{y}_{ij} = \left(y^1_{ij}, y^2_{ij}, y^{31}_{ij} \right) = \frac{\sum_{k=1}^{p} \tilde{w}_{jk} \otimes \tilde{r}_{ijk}}{\sum_{k=1}^{p} \tilde{w}_{jk}}, i = 1,2...,m;$$

$$j = 1,2...,n$$

$$(8)$$

Where, $\tilde{y}_{ij}$ is served as the APR of alternative i in relation with criterion j. It should be added that $\otimes$ is the multiplication operator in fuzzy logic.

Step 7. The APRs are normalized at criteria stage with linear normalization method again. Based on this approach and as can be recognized from the following equation, the best results acquire the value equal 1 and the worst ones obtain the value equal 0.

$$\tilde{\tilde{y}}_{ij} = \left(\tilde{\tilde{y}}^1_{ij}, \tilde{\tilde{y}}^2_{ij}, \tilde{\tilde{y}}^3_{ij} \right) = \left(\frac{y^1_{ij}-y^-_j}{y^*_j-y^-_j}, \frac{y^2_{ij}-y^-_j}{y^*_j-y^-_j}, \frac{y^3_{ij}-y^-_j}{y^*_j-y^-_j} \right),$$

$$i = 1,2 \ ..., \ m; \ j = 1,2 \ ..., \ n \qquad (9)$$

Where, $\tilde{\tilde{y}}_{ij}$ is the normalized APR of alternative i with respect to criterion j. $y^*_j = \max_i y^3_{ij}$ and $y^-_j = \min_i y^1_{ij}$.

Step 8. The weighted distances (WDs) from ideal solution and anti-ideal solution may be represented as D^*_i and D^-_i respectively. The value of WD for each alternative can be computed as follows:

$$oD^*_i = \sum_{j=1}^{n} \frac{1}{2} \left\{ \max \left(\tilde{w}^1_j | \tilde{\tilde{y}}^1_{ij} - 1 |, \tilde{w}^3_j | \tilde{\tilde{y}}^3_{ij} - 1 | \right) + \tilde{w}^2_j | \tilde{\tilde{y}}^2_{ij} - 1 | \right\},$$

$$i = 1,2 \ ..., \ m$$

$$(10)$$

$$D^-_i = \sum_{j=1}^{n} \frac{1}{2} \left\{ \max \ \tilde{w}^1_j | \tilde{\tilde{y}}^1_{ij} - 0 |, \tilde{w}^3_j | \tilde{\tilde{y}}^3_{ij} - 0 | \right) + \tilde{w}^2_j | \tilde{\tilde{y}}^2_{ij} - 0 | \right\},$$

$$i = 1,2 \ ..., \ m$$

$$(11)$$

Step 9. The proximity of the alternatives to the ideal solution is represented with Ω^*_i and can be calculated as follows:

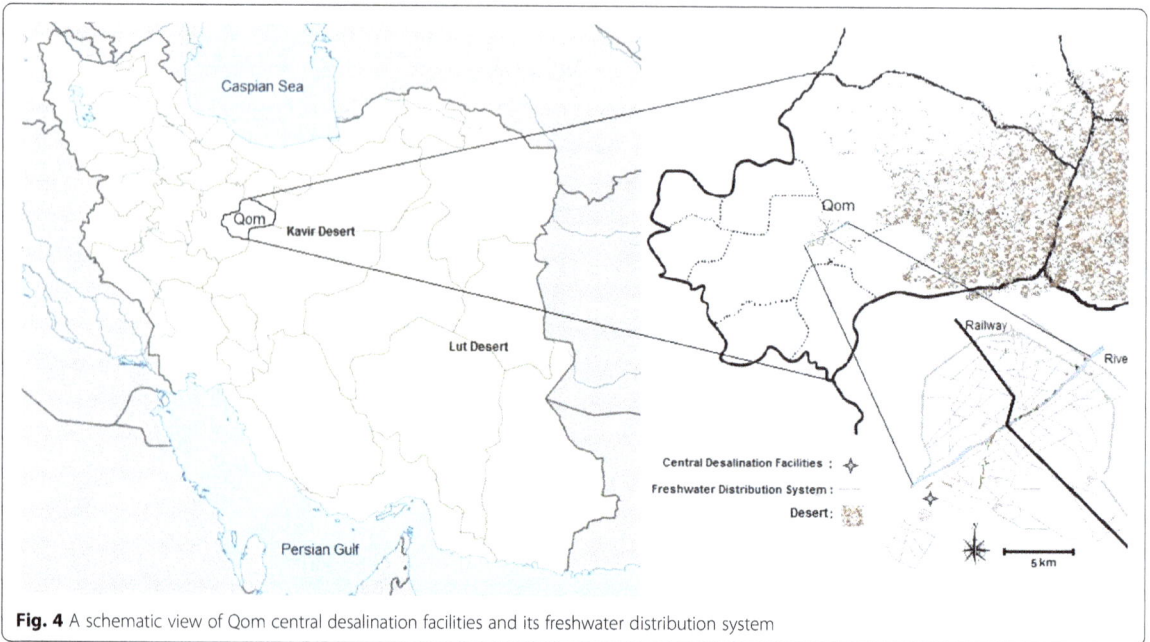

Fig. 4 A schematic view of Qom central desalination facilities and its freshwater distribution system

Table 2 Linguistic term set for criteria and sub-criteria

Linguistic term	Fuzzy value		
Very low(VL)	0	0	0.25
Low(L)	0	0.25	0.5
Moderate(M)	0.25	0.5	0.75
High(H)	0.5	0.75	1
Very High(VH)	0.75	1	1

$$\Omega_i^* = \frac{D_i^-}{D_i^* + D_i^-}, \qquad i = 1, 2..., m. \qquad (12)$$

By using the Ω_i^* concept, the distances from ideal and anti-ideal solutions are computed.

Step10. If the results of Ω_i^* are sorted from largest to the smallest values, the best alternative is one which has obtained the highest Ω_i^* value and therefore is located in the top of the descending ranking of alternatives.

Study area

As shown in Fig. 4, Qom province has been located in central part of Iran. Qom is the only city of province and has the population more than 1 million permanent inhabitants. Qom is the second city in Iran after Mashhad as a pilgrimage center [25], so its population has noticeable annually fluctuations because of religious tourists reception [26].

Qom province has low annual precipitation and also salty marls are prevalence geological structures [27] in its plains. Consequently, like the other cities located in central part of Iran, Qom population has engaged with both water quality and quantity crisis [28, 29]. Local water sources of Qom which are flowed in public salty distribution system (PSDS) contain relatively high levels of salt and are considered only for non-drinking purposes. Dissolved solids concentration (TDS) of surface water and groundwater sources of the province is around 1800 and 4500 mg/L, respectively. To improve the quality of these brackish water sources for drinking demands, some programs have been carried out since past decades, such as Public desalinated distribution system (PDDS), Private desalinated water suppliers (PDWS), and Household desalinated water units (HDWU) [29].

Evaluating drinking water supply alternatives using DBF –MCDM approach

The following methods were considered as capable alternatives to supply the drinking water demands of Qom:

A$_1$: Public desalinated distribution system (PDDS)
A$_2$: PET Bottled Drinking Water (PBDW)
A$_3$: Private desalinated water suppliers (PDWS)
A$_4$: Household desalinated water units (HDWU)

Six and 35 evaluation criteria and sub-criteria were defined, respectively which illustrated in Fig. 3. Also, sub-criteria were classified to Cost-Related and Beneficial-Related groups. The benefit-related sub-criteria are those for which the higher the performance value the more its preference, and the cost-related sub-criteria are considered as sub-criteria for which the higher the performance value the less its preference (Fig 3).

The evaluation was performed by a team of five decision-makers which are identified as DM_1, DM_2, DM_3, DM_4 and DM_5. DM_1 is a professor of environmental health engineering. DM_2 is a technical advisor specialized in water desalination facilities, DM_3 is a professor in epidemiology, DM_4 is a water treatment expert from Qom Water and Sewage Company (QWSC), and DM_5 is a socio-economic advisor specialized in urban water management. Decision-makers used the linguistic term set shown in Table 2 which also has illustrated as a fuzzy triangular depiction in Fig. 5.

The linguistic terms assigned by decision-makers to each criterion and sub-criterion for determining their importance are represents in Table 3. Tables 3 and 4 depict the importance allocated by decision-makers with respect to criteria and sub-criteria, respectively. Table 5 represents the ratings of alternatives assigned by decision-makers with respect to sub-criteria.

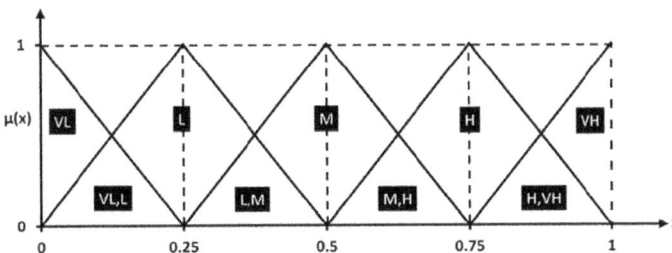

Fig. 5 Linguistic term set in fuzzy depiction

Table 3 Importance of criteria

Criteria	DM_1	DM_2	DM_3	DM_4	DM_5
Economic	M	H	M	H	H
Environmental	VH	H	VH	H	H
Public Health	VH	VH	VH	VH	VH
Occupational Health	VH	VH	H	H	H
Technical	H	H	H	H	VH
Social	VH	H	VH	H	H

Table 4 Importance of sub-criteria

Decision maker	DM_1	DM_2	DM_3	DM_4	DM_5
Sub-criteria					
CC*	H	M	M	H	H
OC	VH	H	H	H	H
WPPC	H	VH	H	H	H
SREI	VH	H	VH	H	H
WREI	VH	H	H	H	H
AREI	M	L	VL	M	L
Noise	M	M	L	L	VL
TOP	VH	H	VH	H	H
CP	VH	H	H	VH	H
MP	VH	VH	VH	VH	VH
RQMV	H	VH	H	H	M
RSC	H	VH	VH	H	H
Qual.SPW	H	H	VH	H	H
Quan.SPW	H	H	H	H	H
RWBD	VH	VH	VH	VH	VH
OHOF	VH	VH	VH	H	H
OHOI	H	VH	VH	H	H
REL	VH	VH	VH	H	H
WSR	H	VH	VH	H	H
EDPC	H	M	H	M	H
LA	H	M	M	M	H
NSO	M	M	L	VL	VL
NWTU	M	L	VL	L	VL
SDOC	M	M	L	VL	VL
EDOC	H	M	H	M	L
TC	M	M	L	VL	VL
DCTW	H	VH	H	H	H
NIV	H	VH	H	M	H
AUA	H	H	M	M	M
NPS	H	H	H	H	H
AEP	VH	H	VH	H	H
PAO	H	VH	VH	H	H
SOH	M	M	L	L	L
ATWS	VH	H	VH	H	H
EED	VH	VH	VH	H	H

Results and discussion

Equations 4 and 5 were employed to aggregate the importance of criteria (see Table 3) and sub-criteria (see Table 4) and results were represented in Tables 6 and 7 for criteria and sub-criteria, respectively. The ratings of alternatives (see Table 5) were aggregated using Eq. 6 and results were shown in Table 8. It should be noted that in this study, the decision-makers were considered with equal weights v_l. Thus $v_1 = v_2 = v_3 = v_4 = v_5 = \frac{1}{5}$, as previously denoted by Dursun (2011a).

Normalized ratings of alternatives with respect to sub-criteria were computed using Eq. 8 which is based on the linear scale transformation approach (results were not shown). Then, aggregate performance ratings (APRs) of alternatives with respect to sub-criteria are calculated by Eq. 9 (results were not shown). Eq. 9 was applied to aggregate the sub-criteria values to criteria level according to the findings of Karsak (2002). Normalized APRs were calculated by using Eq. 10 and results are illustrated in Table 9, in which, 0 implies the worst value and 1 represents the best value.

The weighted distances from ideal solutions (D_i^*) and anti-ideal solutions (D_i^-) were computed using Eq. 11 and 12, respectively. Then, the proximity of the alternatives to the ideal solution (Ω_i^*) was calculated by using Eq. 12. The results of the D_i^*, D_i^- and Ω_i^* values are presented in Table 10.

After sorting the alternatives according to the magnitude of Ω_i^* values, the following ranking order was achieved:

$$A_1 > A_2 > A_4 > A_3$$

As can be inferred from Table 10 the Public Desalinated Distribution System (A_1) is the best alternative as drinking water source for Qom population.

Abrishamchi and co-workers (2004) denoted a small potable water network (less than 30 km) with public valves (water standpipes) at several points across the city of Zahidan. They considered the "Extension of the small drinking water distribution network with public standpipes" as an alternative to supply the drinking water needs of population.

Public Desalinated Distribution System (PDDS) has several benefits such as simple operation of treatment facilities and ease of health inspection process. Now, more than 180 km of potable water network has been constructed in the city of Qom which have connected to 260 public valve (water standpipes) and supply more than 4500 cubic meter of desalinated water per day [29]. The only noticeable problem dealing with the PDDS is the low extension of distribution system which tends to handle the water containers from public valves to houses by people.

Table 5 Ratings of the alternatives with respect to the sub-criteria (The full form of abbreviations was represented in Fig. 3)

Decision Maker	DM$_1$				DM$_2$				DM$_3$			
Alternative	A$_1$(PDDS)	A$_2$(PBDW)	A$_3$(PDWS)	A$_4$(HDWU)	A$_1$(PDDS)	A$_2$(PBDW)	A$_3$(PDWS)	A$_4$(HDWU)	A$_1$(PDDS)	A$_2$(PBDW)	A$_3$(PDWS)	A$_4$(HDWU)
Sub-criteria												
CC*	VL	VL	VL	VH	L	VL	L	VH	L	L	L	H
OC	VL	L	VL	VH	VL	M	VL	H	L	M	L	VH
WPPC	VL	H	M	L	VL	VH	L	VL	VL	VH	M	VL
SREI	L	VH	L	H	VL	VH	VL	VH	VL	VH	VL	M
WREI	L	L	L	H	L	VL	L	VH	VL	VL	VL	H
AREI	VL	VL	VL	VL	VL	VL	VL	L	VL	VL	VL	VL
Noise	VL	VL	VL	M	VL	VL	VL	M	VL	L	VL	H
TOP	L	VL	H	H	L	VL	VH	H	VL	VL	H	H
CP	L	VL	VH	H	L	VL	H	M	L	VL	H	M
MP	VH	VL	VH	H	VH	VL	VH	M	H	VL	VH	H
RQMV	VH	VH	H	L	VH	H	M	VL	H	VH	H	VL
RSC	VH	L	VH	VH	VH	L	VH	H	VH	M	VH	H
Qual.SPW	VH	H	L	L	VH	VH	L	VL	VH	VH	VL	L
Quan.SPW	L	M	M	VH	M	M	M	VH	M	M	L	H
RWBD	H	VL	H	M	M	VL	VH	M	H	L	H	L
OHOF	VH	L	VH	L	VH	L	VH	M	VH	M	VH	L
OHOI	H	VL	H	M	H	VL	H	H	H	VL	H	H
REL	H	M	L	H	VH	H	VL	H	H	H	L	M
WSR	H	VH	H	M	VH	H	VH	L	VH	VH	VH	M
EDPC	M	VL	M	VH	H	VL	H	VH	H	VL	H	VH
LA	H	VL	VL	H	H	VL	VL	H	H	VL	VL	H
NSO	L	VL	L	VH	L	L	L	VH	VL	VL	VL	H
NWTU	H	VL	L	VH	M	VL	VL	H	M	VL	VL	VH
SDOC	M	VL	L	H	M	L	L	VH	L	L	VL	H
EDOC	L	VL	L	VH	L	VL	VL	H	M	L	VL	VH
TC	L	VL	VL	VH	M	L	VL	H	M	L	VL	VH
DCTW	VL	VL	VL	VH	VL	VL	VL	VH	VL	VL	VL	VH
NIV	VH	L	VH	VL	VH	L	VH	VL	H	M	H	L
AUA	H	H	H	L	H	H	VH	VL	VH	VH	VH	VL
NPS	L	L	L	VH	L	VL	L	VH	L	VL	L	VH
AEP	M	VH	L	H	H	VH	L	M	M	VH	M	M
PAO	H	H	VH	M	M	H	VH	L	L	H	M	L
SOH	M	VL	M	H	L	L	L	H	L	VL	L	VH
ATWS	M	L	M	VH	L	L	L	VH	M	M	M	VH
EED	M	VH	VH	VL	L	VH	H	VL	M	VH	M	VL

Table 5 Ratings of the alternatives with respect to the sub-criteria (The full form of abbreviations was represented in Fig. 3)

Decision Maker	DM$_4$				DM$_5$			
Alternative	A$_1$(PDDS)	A$_2$(PBDW)	A$_3$(PDWS)	A$_4$(HDWU)	A$_1$(PDDS)	A$_2$(PBDW)	A$_3$(PDWS)	A$_4$(HDWU)
Sub-criteria								
CC*	VL	VL	VL	VH	VL	VL	VL	VH
OC	L	H	VL	VH	VL	M	VL	H
WPPC	L	VH	H	L	VL	VH	M	VL
SREI	L	H	L	H	VL	VH	VL	H
WREI	L	VL	L	H	L	VL	L	VH
AREI	VL	VL	VL	VL	VL	VL	VL	VL
Noise	VL	VL	VL	M	L	VL	L	M
TOP	L	VL	H	H	L	L	H	M
CP	M	VL	VH	L	L	L	H	M
MP	VH	VL	H	H	VH	L	H	H
RQMV	H	H	M	VL	VH	H	M	VL
RSC	H	L	VH	H	VH	VL	VH	H
Qual.SPW	H	VH	L	L	VH	VH	L	L
Quan.SPW	L	H	L	H	L	M	M	M
RWBD	H	VL	H	M	M	L	VH	M
OHOF	H	L	VH	L	H	L	H	M
OHOI	H	VL	VH	H	M	L	H	M
REL	VH	VH	VL	M	M	VH	VL	H
WSR	VH	H	VH	L	H	M	H	L
EDPC	H	VL	H	VH	M	VL	M	H
LA	H	VL	VL	H	M	VL	VL	M
NSO	VL	VL	VL	VH	VL	VL	VL	H
NWTU	H	VL	VL	H	H	VL	L	H
SDOC	L	VL	L	VH	L	VL	L	H
EDOC	VL	VL	VL	VH	L	VL	VL	H
TC	L	VL	L	VH	L	L	L	VH
DCTW	VL	VL	VL	VH	VL	VL	VL	VH
NIV	VH	L	VH	VL	VH	L	VH	VL
AUA	VH	VH	VH	L	VH	VH	VH	L
NPS	VL	VL	VL	H	L	VL	L	VH
AEP	L	H	L	M	M	H	M	L
PAO	M	H	H	M	H	H	H	L
SOH	M	VL	M	H	L	L	L	H
ATWS	M	L	M	VH	M	L	M	H
EED	L	H	H	L	M	H	H	L

Jafaripour estimated that over 36000 houses in Qom use the Household desalinated water units (HDWU) which cover more than 15 % of all population. Based on the findings of Jafaripour, more than 1000 m^3 of brine water and up to 550 discarded filter are produced by using of Household desalinated water units (HDWU) [30].

Yari reported that 24 Private desalinated water suppliers (PDWS) are operated in the city of Qom. Their results showed that the chemical characteristics of potable water produced by PDWS could not meet the national standard criteria. Also, transferring the water containers by vendees is the other constraint of PDWS. Purchased

Table 6 Aggregated Importance weights of criteria

Criteria/Sub-criteria	Aggregated weights
Economic	(0.40, 0.50, 0.90)
Environmental	(0.60, 0.70, 1)
Public Health	(0.75, 0.80, 1)
Occupational Health	(0.60, 0.70, 1)
Technical	(0.55, 0.60, 1)
Social	(0.60, 0.70, 1)

Table 7 Aggregated Importance weights of sub-criteria

Sub-criteria	Aggregated weights
CC	(0.40, 0.50, 0.9)
OC	(0.55, 0.65, 1)
WPPC	(0.55, 0.65, 1)
SREI	(0.60, 0.70, 1)
WREI	(0.60, 0.70, 1)
AREI	(0.10, 0.25, 0.55)
Noise	(0.10, 0.30, 0.55)
TOP	(0.60, 0.70, 1)
CP	(0.60, 0.70, 1)
MP	(0.75, 0.80, 1)
RQMV	(0.50, 0.65, 0.95)
RSC	(0.60, 0.70, 1)
Qual.SPW	(0.55, 0.65, 1)
Quan.SPW	(0.50, 0.60, 1)
RWBD	(0.75, 0.80, 1)
OHOF	(0.65, 0.75, 1)
OHOI	(0.60, 0.70, 1)
REL	(0.65, 0.75, 1)
WSR	(0.60, 0.70, 1)
EDPC	(0.60, 0.70, 1)
LA	(0.60, 0.70, 1)
NSO	(0.60, 0.70, 1)
NWTU	(0.60, 0.70, 1)
SDOC	(0.60, 0.70, 1)
EDOC	(0.30, 0.50, 0.80)
TC	(0.10, 0.25, 0.50)
DCTW	(0.55, 0.65, 1)
NIV	(0.50, 0.60, 0.95)
AUA	(0.35, 0.50, 0.85)
NPS	(0.50, 0.60, 1)
AEP	(0.60, 0.70, 1)
PAO	(0.60, 0.70, 1)
SOH	(0.10, 0.30, 0.6)
ATWS	(0.60, 0.70, 1)
EED	(0.65, 0.75, 1)

water containers may stored in homes for a long time in uncontrolled health condition [31].

More than 18 various brands of PET Bottled Drinking Water (PBDW) are sold in the retails of Qom city [32]. Noticeable merits of PBDW are Chemical and biological acceptable quality which serve as an alternative beside the other water supply system. High price and lack of coverage for all population, in the other hand, are the essential drawbacks of PBDW.

A significant factor that should be considered in the judgment process of purchasing high-tech equipment is the level of dependency to the foreign suppliers. A more appropriate strategy is to encourage the use of the alternative technologies available within the country. Hence, except for the household desalinated water units (HDWU), the other alternatives could not obtain higher levels of linguistic terms by decision-makers for SDOC and EDOC sub-criteria.

Considering the occupational and public health criteria independent of the environmental and technical criteria significantly improved the precision of the results.

Conclusions

An efficient analysis was performed by applying the evaluation criteria and their associated sub-criteria on a hierarchical structure. Thirty five sub-criteria associated with six criteria were structured in a multi-level hierarchy and the decision processes allowed the decision-makers to employ linguistic concepts, and thus, decreased the cognition problems during the evaluation process.

In this study, hierarchical distance-based fuzzy multi-criteria group decision making (DBF –MCDM) approach was presented to avoid the problems that may occurred when the classical decision-making approaches are employed for evaluating the water supply alternatives.

New arrangement of criteria and sub-criteria was proposed in this study. Traditionally, four criteria including financial, environmental, technical, and social aspects have been proposed in similar works. Using a new hierarchy containing the public health and occupational health aspects as the independent criteria enabled the decision-making process to assign more effective evaluations.

System and equipment dependency to other countries (SDOC and EDOC) were added to the technical aspects as sub-criteria for obtaining a state of compatibility with the socioeconomic condition which restrict the level of dependency on the foreign companies.

The DBF–MCDM method proposed in this research is a simple approach that can be used for similar environmental management issues only with some modifications.

Table 8 Aggregated ratings of alternatives with respect to sub-criteria

Sub-criteria	A₁	A₂	A₃	A₄
CC	(0.00, 0.10, 0.35)	(0.00, 0.05, 0.30)	(0.00, 0.10, 0.35)	(0.70, 0.95, 1)
OC	(0.00, 0.10, 0.35)	(0.25, 0.50, 0.75)	(0.00, 0.50, 0.30)	(0.65, 0.90, 1)
WPPC	(0.00, 0.05, 0.30)	(0.70, 0.95, 1)	(0.25, 0.50, 0.75)	(0.00, 0.10, 0.35)
SREI	(0.00, 0.01, 0.35)	(0.70, 0.95, 1)	(0.00, 0.10, 0.35)	(0.50, 0.75, 0.95)
WREI	(0.00, 0.02, 0.45)	(0.00, 0.05, 0.30)	(0.00, 0.20, 0.45)	(0.60, 0.85, 1)
AREI	(0.00, 0.00, 0.25)	(0.00, 0.00, 0.25)	(0.00, 0.00, 0.25)	(0.00, 0.05, 0.30)
Noise	(0.00, 0.05, 0.30)	(0.00, 0.25, 0.30)	(0.00, 0.05, 0.30)	(0.30, 0.55, 0.80)
TOP	(0.00, 0.20, 0.45)	(0.00, 0.05, 0.30)	(0.55, 0.80, 1)	(0.45, 0.70, 0.95)
CP	(0.05, 0.30, 0.55)	(0.00, 0.05, 0.30)	(0.60, 0.85, 1)	(0.25, 0.50, 0.75)
MP	(0.70, 0.95, 1)	(0.00, 0.05, 0.30)	(0.65, 0.90, 1)	(0.45, 0.70, 0.95)
RQMV	(0.65, 0.90, 1)	(0.60, 0.85, 1)	(0.35, 0.60, 0.85)	(0.00, 0.05, 0.30)
RSC	(0.70, 0.95, 1)	(0.05, 0.25, 0.5)	(0.75, 1, 1)	(0.55, 0.80, 1)
Qual.SPW	(0.70, 0.95, 1)	(0.70, 0.95, 1)	(0.00, 0.20, 0.45)	(0.00, 0.20, 0.45)
Quan.SPW	(0.01, 0.35, 0.60)	(0.30, 0.55, 0.80)	(0.15, 0.40, 0.65)	(0.55, 0.80, 0.95)
RWBD	(0.40, 0.65, 0.90)	(0.00, 0.10, 0.35)	(0.60, 0.85, 1)	(0.20, 0.45, 0.70)
OHOF	(0.65, 0.90, 1)	(0.05, 0.30, 0.55)	(0.70, 0.95, 1)	(0.10, 0.35, 0.60)
OHOI	(0.45, 0.70, 0.95)	(0.00, 0.05, 0.30)	(0.55, 0.80, 1)	(0.40, 0.65, 0.90)
REL	(0.55, 0.80, 0.95)	(0.55, 0.80, 0.95)	(0.00, 0.10, 0.35)	(0.40, 0.65, 0.90)
WSR	(0.65, 0.90, 1)	(0.45, 0.65, 0.75)	(0.65, 0.90, 1)	(0.10, 0.35, 0.60)
EDPC	(0.40, 0.65, 0.90)	(0.00, 0.00, 0.25)	(0.40, 0.65, 0.90)	(0.70, 0.95, 1)
LA	(0.45, 0.70, 0.95)	(0.00, 0.00, 0.25)	(0.00, 0.00, 0.25)	(0.45, 0.70, 0.95)
NSO	(0.00, 0.10, 0.35)	(0.00, 0.05, 0.30)	(0.00, 0.10, 0.35)	(0.65, 0.90, 1)
NWTU	(0.40, 0.65, 0.90)	(0.00, 0.00, 0.25)	(0.00, 0.10, 0.35)	(0.60, 0.85, 1)
SDOC	(0.05, 0.20, 0.35)	(0.00, 0.10, 0.35)	(0.00, 0.20, 0.45)	(0.60, 0.85, 1)
EDOC	(0.05, 0.25, 0.50)	(0.00, 0.05, 0.30)	(0.00, 0.05, 0.30)	(0.65, 0.90, 1)
TC	(0.10, 0.35, 0.60)	(0.00, 0.15, 0.40)	(0.00, 0.10, 0.35)	(0.70, 0.95, 1)
DCTW	(0.00, 0.00, 0.25)	(0.00, 0.00, 0.25)	(0.00, 0.00, 0.25)	(0.75, 1, 1)
NIV	(0.70, 0.95, 1)	(0.05, 0.30, 0.55)	(0.70, 0.95, 1)	(0.00, 0.05, 0.30)
AUA	(0.65, 0.90, 1)	(0.65, 0.90, 1)	(0.70, 0.95, 1)	(0.00, 0.15, 0.40)
NPS	(0.00, 0.20, 0.45)	(0.00, 0.05, 0.30)	(0.00, 0.20, 0.45)	(0.70, 0.95, 1)
AEP	(0.25, 0.50, 0.75)	(0.65, 0.90, 1)	(0.10, 0.35, 0.60)	(0.25, 0.50, 0.75)
PAO	(0.30, 0.55, 0.8)	(0.50, 0.75, 1)	(0.55, 0.80, 0.95)	(0.10, 0.35, 0.60)
SOH	(0.10, 0.35, 0.60)	(0.50, 0.10, 0.35)	(0.10, 0.35, 0.60)	(0.55, 0.80, 1)
ATWS	(0.20, 0.45, 0.70)	(0.05, 0.30, 0.55)	(0.20, 0.45, 0.70)	(0.70, 0.95, 1)
EED	(0.15, 0.40, 0.65)	(0.65, 0.90, 1)	(0.50, 0.75, 0.95)	(0.00, 0.10, 0.35)

Table 9 Normalized the aggregated performance ratings

Criteria/Sub-criteria	Aggregated weights
Economic	(0.40, 0.50, 0.90)
Environmental	(0.60, 0.70, 1)
Public Health	(0.75, 0.80, 1)
Occupational Health	(0.60, 0.70, 1)
Technical	(0.55, 0.60, 1)
Social	(0.60, 0.70, 1)

Table 10 Ranking of the drinking water alternatives

Alternative	D_i^*	D_i^-	Ω_i^*	Rank
A₁: Public Desalinated Distribution System (PDDS)	2.131	3.346	0.611	1
A₂: PET Bottled Drinking Water (PBDW)	2.212	3.405	0.606	2
A₄: Household Desalinated Water Units (HDWU)	2.279	3.482	0.604	3
A₃: Private Desalinated Water Suppliers (PDWS)	2.384	3.01	0.558	4

Competing interests

The authors declare that they have no competing interests.

Authors' contributions

TSY has participated in conducting the experiments, analyzing the data and preparation of the manuscript. MK participated in data collection and carried out fuzzy analysis and manuscript preparation. RN carried out technical analysis of data and participated in healthcare waste study.AHM participated in the intellectual helping in different stages of the study. SN participated in design of the study, final deeply revision of the manuscript and intellectual helping thorough the study. ARY participated in data collection and carried out technical analysis and manuscript preparation. All authors read and approved the final manuscript.

Acknowledgments

We would like to thank the professors and experts of Qom University of Medical Sciences (QUMS) and Qom Water and Wastewater Organization (QWWO) who support the study as decision-makers.

Author details

[1]Research Center for Environmental Pollutants, Qom University of Medical Sciences, Qom, Iran. [2]Department of Environmental Health Engineering, School of Public Health, Tehran University of Medical Sciences, Poursina St, Keshavarz Blvd, PO BOX: 6446-14155, Tehran, Iran. [3]Center for Solid Waste Research, Institute for Environmental Research, Tehran University of Medical Sciences, Tehran, Iran. [4]Center for Water Quality Research, Institute for Environmental Research, Tehran University of Medical Sciences, Tehran, Iran.

References

1. Blair DA, Spronz WD, Ryan KW. Brakish groundwater desalination: a community's solution to water supply and aquifer protection 1. 1999. Wiley Online Library.
2. Stokes J, Horvath A. Life cycle energy assessment of alternative water supply systems (9 pp). Int J Life Cycle Assess. 2006;11(5):335–43.
3. Mithrarratne N, Vale R. Life-cycle resource efficiency of conventional and alternative water supply systems. 12th Annual International Sustainable Development Research Conference. 2006.
4. Abrishamchi A, Ebrahimian A, Tajrishi M, Mariño MA. Case study: application of multicriteria decision making to urban water supply. J Water Res Plan Manag. 2005;131(4):326–35.
5. Rak J, Tchórzewska-Cieślak B. Review of matrix methods for risk assessment in water supply system. J Konbin. 2006;1(1):67–76.
6. Lai E, Lundie S, Ashbolt N. Review of multi-criteria decision aid for integrated sustainability assessment of urban water systems. Urban Water J. 2008;5(4):315–27.
7. Ruiz-Mier F, van Ginneken M. Consumer cooperatives: An alternative institutional model for delivery of urban water supply and sanitation services. Water Supply and Sanitation Board. 2006.
8. Gohari A, Eslamian S, Mirchi A, Abedi-Koupaei J, MassahBavani A, Madani K. Water transfer as a solution to water shortage: a fix that can Backfire. J Hydrology. 2013;491:23–39.
9. Foltz RC. Iran's water crisis: cultural, political, and ethical dimensions. J Agr Environ Ethics. 2002;15(4):357–80.
10. Faramarzi M, Abbaspour KC, Schulin R, Yang H. Modelling blue and green water resources availability in Iran. Hydrol Processes. 2009;23(3):486–501.
11. Rak J, Tchorzewska-Cieslak B. Five-parametric matrix to estimate the risk connected with water supply system operation. Environ Protection Eng. 2006;32(2):37.
12. Muralidharan C, Anantharaman N, Deshmukh S. A multi 2010 criteria group decisionmaking model for supplier rating. J Supply Chain Manag. 2002;38(4):22–33.
13. Zeleny M, Cochrane JL. Multiple criteria decision making. New York: McGraw-Hill; 1982.
14. Baas SM, Kwakernaak H. Rating and ranking of multiple-aspect alternatives using fuzzy sets. Automatica. 1977;13(1):47–58.
15. Dubois D, Prade H. Operations on fuzzy numbers. Int J Systems Sci. 1978;9(6):613–26.
16. Zimmermann HJ. Fuzzy set theory-and its applications. Norwell. Massachusett: Kluwer Academic Publishers; 2001.
17. Kahraman C. Fuzzy multi-criteria decision making: theory and applications with recent developments. New York: Springer; 2008.
18. Zadeh LA. The concept of a linguistic variable and its application to approximate reasoning—I. Inform Sci. 1975;8(3):199–249.
19. Chen C-T. Extensions of the TOPSIS for group decision-making under fuzzy environment. Fuzzy Set Syst. 2000;114(1):1–9.
20. Karsak EE, Ahiska SS. Fuzzy multi-criteria decision making approach for transport projects evaluation in Istanbul. Computational Science and Its Applications–ICCSA 2005.Springer. 2005. p. 301–11.
21. Bojadziev G, Bojadziev M. Fuzzy sets, fuzzy logic, applications. London: World Scientific; 1995.
22. Dursun M, Karsak EE, Karadayi MA. Assessment of health-care waste treatment alternatives using fuzzy multi-criteria decision making approaches. Res Conserv Recycling. 2011;57:98–107.
23. Karsak E. Distance-based fuzzy MCDM approach for evaluating flexible manufacturing system alternatives. Int J Prod Res. 2002;40(13):3167–81.
24. Murofushi T, Sugeno M. Fuzzy Measures and Integrals: Theory and Applications. New York: Springer; 2000.
25. Ebrahimzadeh I, Kazamizd S, Eskandari SM. Strategic planning for tourism development, emphasizing on religious tourism (case study: Qom City). Hum Geogr Res Quart. 2011;76(2):19–21.
26. Heidarabadi SM. Strategies for planning domestic and international tourism development of Qom Province with emphasis on religious Tourism. 2008.
27. Talbot C, Aftabi P. Geology and models of salt extrusion at Qum Kuh, central Iran. J Geological Soc. 2004;161(2):321–34.
28. Ardakanian R. Overview of water management in Iran. Water conservation, reuse, and recycling, Proceeding of an Iranian American workshop. 2005.
29. Khazaei M, Mahvi AH, Fard RF, Izanloo H, Yavari Z, Tashayoei HR. Dental caries prevalence among Schoolchildren in Urban and Rural areas of Qom Province, Central part of Iran. Middle-East J Sci Res. 2013;18(5):584–91.
30. JafaripourMohammadreza SAM, Abbas Z, Davoudi R. Health, sanitary and economic evaluation of home-like systems of water treatment (RO) in Qom city water and wastewater. Water Wastewater. 2011;22(2):15–21.
31. Yari A. The physical, chemical and microbial quality of treated water in Qom s desalination plants. Qom Univ Med Sci J. 2012;1(1):45–54.
32. Bidgoli MS, Ahmadi E, Yari AR, Hashemi S, Majidi G, Nazari S, et al. Concentration of nitrate in bottled drinking water in Qom, Iran. Arch Hygiene Scie. 2013;2(4):122–6.

8

Soil to plant transfer of alpha activity in potato plants: impact of phosphate fertilizers

Rishi Pal Chauhan[*] and Amit Kumar

Abstract

Background: Radionuclides in the phosphate fertilizers belonging to ^{232}Th and ^{238}U and ^{40}K are the major contributors to the outdoor terrestrial natural radiation. These radionuclides are transferred from fertilizer to food through soil.

Materials and methods: Present work deals with the alpha activity in the different parts of the potato (*Solanum Tuberosum*) plants grown under controlled pots experiment using different amounts of phosphate fertilizers and urea. Alpha activities have been measured by track etch technique using the solid-state nuclear track detectors (LR-115).

Results: Translocation factor for the fruit (edible Part) varied from 0.13 (for DAP) to 0.73 (for PF) with an average of 0.40 ± 0.26 for the plant grown with 20 g of fertilizers. Translocation factors increased with the increase in amount of fertilizers having value 0.51 ± 0.31 for the plant grown with 50 g of fertilizers. The translocation factor for the lower and the upper part of leaves varied from 0.44 to 0.67 and 0.22 to 0.83 with an average value 0.55 ± 0.15 and 0.45 ± 0.23 respectively. The transfer factor (*TF's*) for the potato plants varied from 1.5×10^{-2} to 1.03×10^{-1} for root, from 1.3×10^{-2} to 1.23×10^{-1} for stem, from 2.1×10^{-3} to 4.5×10^{-2} for fruit and from 5.4×10^{-3} to 5.8×10^{-3} for lower part of the leaves after 105 days of the plantation.

Conclusions: The results revealed that the alpha activity in the potato plants was higher in case of the plants grown with the use of phosphate fertilizers than with other fertilizers.

Keywords: Potato, Alpha radioactivity, LR-115, Phosphate fertilizers

Background

The accumulation of the radioactive elements in the environmental air and soil is of concern in relation to health of general public. The causes of radioactive contamination include accidental spills and emissions from nuclear-fuel operations, fallout from nuclear testing, and accidents at nuclear reactors. Radionuclides scattered into the environment are transported by air and water, ultimately reaching the soil and sediment, where they become bound [1]. The radioactive elements present in biosphere will involve with human through his food chain. Transportation of these radionuclides through soil to plant and then to food leads to uptake of radioactive elements to human. In biosphere these radioactive elements and their decay products get into the plants both through the above ground parts like leaves, and stem by external contamination and through the root spring up into the soil containing radioactive elements. The external contamination on aboveground parts is of less concerned as these are used in food after washing with water. The radioactive elements transfer from soil to plants root and then trans-located in different parts of plants depending upon the metabolism and condition of growth of plants [2]. ^{238}U, ^{226}Ra and ^{232}Th concentrations in the soil and phosphate fertilizers are of vital importance, because through several pathways these radionuclides and their decays products will reach to human and increase the ingestion radiation doses. The pathways through which these radionuclides enter in biosphere are the use of fertilizers having high radioactivity contents [3]. The radionuclide contents of fertilizers varied directly with phosphate (P_2O_5) content [4].

* Correspondence: chauhanrpc@gmail.com
Department of Physics, National Institute of Technology, Kurukshetra 136119, India

Fertilizers Impact in agriculture

In order to increase crop production and improving the properties of the nutrient-deficient lands, various chemical and phosphate fertilizers are now used. Miranzadeh et al. [5] reported that interactive application nitrogen fertilizer could have beneficial effects on wheat grain yield under similar agro-climatic conditions. The negative effect of the phosphate fertilizers on the agriculture use is the contamination of cultivated lands by trace metals (Cd, Cu and Zn) and increase in radioactivity in the vegetations and food [6–8]. Phosphate rock may be sedimentary, volcanic or biological origin is the starting material for all phosphate products including phosphate fertilizers [9]. The uranium, thorium and potassium contents of phosphate fertilizer depend upon the origin of phosphate rock varied from 10 to 5022 Bq/kg, 10 to 394 Bq/kg and 8 to 397 Bq/Kg respectively [10]. Phosphate rocks contain high concentration of ^{238}U, ^{226}Ra and ^{232} and their decays products due to accumulation of dissolved uranium during its formation [3]. The phosphate ore are used to produce phosphoric acid in Wet process by attack of the sulfuric acid (H_2SO_4). The ^{238}U remains concentrated into phosphoric acid while ^{226}Ra, ^{210}Po, ^{232}Th and ^{210}Pb precipitated out as sulfate salt concentrated in phosphogypsum as the byproduct [11]. The uranium either in form of $[U(SO_4)_2]$ or $[UO_2(SO_4)]$ in phosphoric acid are water soluble, remain in phosphoric acid which is used for the fertilizers production, thus the uranium content of fertilizers expected to be high Khater [9]. Da Conceicao and Bonotto [12] reported the high concentration of ^{232}Th and ^{226}Ra in phosphoric acid than phosphogypsum.

There exist a linear relation between the P_2O_5 content and uranium in phosphate fertilizers [13]. The phosphorus is one of the 17 nutrients essential for plants growth. It involves in many chemical reactions occurring in plants for their growth like plant energy transfer, photosynthesis, genetic transfer and nutrient transports. It enters the plants through roots hair, root tips and outermost layers of root cells in the form of primary orthophosphate ion ($H_2PO_4^-$) or secondary Orthophosphate ($H_2PO_4^{2-}$). The stored phosphorus in plants root transports to other parts of plants, causing the mobility of uranium called translocation. The consumption of various parts of plants in food by man leads to ingestion of radiation dose. According to UNSCEAR [14] the uranium contents in the food varies from 1.2 Bq/kg to 400 Bq/kg for grain products and 9.8 Bq/kg to 400 Bq/kg for leafy vegetables. Plants metabolism plays important role in regulating the trace radioactive elements in the soil-plant-animal food chain system. Once non-nutrient radioactive trace elements are solubilized in water, plants root actively accumulate them depending upon chemical activity of the element in soil solution, the presence of competing ions, the redox potential and absorption capacity of the root. After absorption in the plant, trace elements are translocated, metabolized and stored in different part of plants depending upon properties of the element as well as plant [15]. The transfer of elements from soil to plants assumed to be constant at low content in soil and vary for high content [16].

Need of the study

The presence of uranium, radium, thorium and potassium in the plants enhance the gamma and alpha dose to the users: animal and man. The gamma spectroscopy or alpha spectroscopy applied for the measurements of these radio-activity in soil and different parts of plant enhanced by the use of phosphate fertilizers. The gamma spectroscopy uses the High Purity Germanium detector for determination of contents of different radionuclides like radium, polonium and lead ([17]. The alpha emitting nature of radium, polonium and bismuth make it possible to use alpha spectrometry for the determination of radioactivity contents used by Rodrıguez et al. [18]. The collective dose from the plants may be calculated by the total dose received by all the radionuclides present in the various parts of plants. The radionuclides present in the plants cause an alpha activity, which is measured by the track etch technique using solid state nuclear tracks detector (LR-115) used in present study.

Assessment parameters

The assessment of the impact of these radionuclides on human can be derived from the estimation of plant uptake from the soil and applying food-chain model. The important factor that quantifies the soil to plant uptake is termed as transfer factor (TF_{sp}) defined for the present study as the ratio of alpha activity caused by radionuclides from plants to that of soil. In present work we assumed the following;

1. The alpha activity from the plants is directly proportional to radioactivity contents
2. The range of alpha particle in samples is larger than the thickness of sample so that more than 90 % of alpha produced emit out from the samples.
3. The radioactivity contents in the plants assumed to be small, thus a linear law for transfer factor is applicable.

The transfer factor for soil to plant (TF_{sp}) is given by Freundlich equation, which is defined by the non-linear relationship [16, 19]

$$C_i = TLF_{ri} \cdot C_r \tag{1}$$

C_p and C_s is the alpha track density produced by soil and plants, TF_{sp} the transfer factor and g is factor depends upon the radioactivity contents and hence the alpha activity of soil. For low activity concentration the value of g should be taken equal to one. For the mobility of uranium and their decays products in the plants, another factor called translocation factor from root to other parts of plant (TLF_{ri}) factor defined by

$$C_P = TF_{SP} \cdot C_s^g \qquad (2)$$

Where i = L,S,F stands for leaf, stem and fruit part of plants. In present work the transfer factor was determined for the alpha track density enhanced by the use of phosphates fertilizers in different age of potato plants grown under control laboratory condition. The potato plant is selected because its pabulum roots are used as major vegetable products all around the world.

Materials and methods

The potato (Solanum Tuberosum) plants were grown in earthen pots in control laboratory condition. The soil of the local region was dried at 100^0 C for 24 h in microwave oven so that most of the moisture of soil was removed and any organic matter present was destroyed. Then the soil left undisturbed to cool for more than a week. A 15 kg soil filtered through 100 micron sieve was filled in 14 different pots of same dimensions having 20 and 50 g of different fertilizer (Diammounium phosphate, Nitrogen phosphate potassium, Phosphate Fertilizer, Single super phosphate, Zinc sulphate and urea) thoroughly mixed in soil. Small seed potatoes were planted a whole, while larger tubers were cut into pieces with 2 to 4 "eyes" on each piece. The seed potatoes gardened at the bottom of a 6 to 10 cm depth of soil from surface and 10 to 15 cm above from lower side of pots. All pots were kept at same open environment condition and similar watering was provided at regular interval of times. Fig. 1 shows the potato plants grown at the age of 30 days. After the 30 and 60 days of plantation, healthy leaves of same size and same portion of the plants were plugged out form each plant, washed thrice with fresh water, so that any contamination on leaves were removed and then dried in normal sunlight for two days. Then leaves were oven dried for 48 hrs at 60 °C, so that most of the moisture was removed. Then each leaf was sandwiched between two SSNTD's (LR-115, Type-2, Strippable, size 3×3 cm^2), wrapped into aluminum foil sheet and kept into sealed plastic canisters so that it remained isolated from the environment [20]. After the exposure of 60 days the LR-115 detectors were removed and chemically etched in 2.5 N alkaline NaOH solutions at 60^0 C for 90 mins. The tracks produced by the alpha particles, were counted by spark counter techniques. The etched detectors were first pre sparked at 900 V twice and then tracks were counted at 500 V thrice. The background counts were subtracted from each measurement for alpha track density. The same procedure was repeated for the leaves, stem, roots and fruit part of potato plants at the age of 60 and 105 days. The alpha track density from the fertilized soil and fertilizers was measured by placing SSNTD's (LR-115) of dimension 2×2 cm^2 over the fertilized soil in pots and fertilizes separately. The measurement of alpha track density measurement from soil using LR-115 was performed three times at age of 30, 60 and 120 days during the growth of plant.

Result and discussion

Alpha track density from leaves of potato plant at different ages.

Fig. 1 The potato plants at the age of 30 days

Two samples were analyzed for given amount of each fertilizer added to the soil. The alpha track density from the leaves of the potato plants ($T\,cm^{-2}d^{-1}$) measured at the age of 30 days and 60 days are listed in Tables 1 and 2. The alpha track density at 30 days age varied from 0.12 ± 0.01 (0.18 ± 0.04) to 0.43 ± 0.01 (0.81 ± 0.17) for the upper (Lower) part of leaves grown using 20 g of different fertilizers while that for 50 g it varied from 0.13 ± 0.0 (0.24 ± 0.07) to 0.30 ± 0.03 (0.55 ± 0.0) for upper (Lower) part of leaves. The alpha track densities for the leaf part increased with increase in the amount of fertilizers used. This may be because there exist a linear relation between the content of P_2O_5 and uranium as both form of uranium $[U(SO_4)_2]$ or $[UO_2(SO_4)]$ in phosphoric acid (raw materials of fertilizers) is soluble in water [9]. With increase in the amount of fertilizers in pots, increased the phosphorus content in soil [21, 22]. hence, uranium into the plants. Also the alpha track density from the lower parts of the leaves was found to be higher than that of the upper parts, this may be due to high trichrome density in the lower parts of leaves as compared to upper parts [23, 24].

The alpha track density ($T\,cm^{-2}d^{-1}$) at 60 days age varied from 0.19 ± 0.03 (0.29 ± 0.02) to 0.52 ± 0.03 (0.62 ± 0.03) for the upper (lower) part of leaves grown using 20 g of different fertilizers while that for 50 g it varied from 0.23 ± 0.06 (0.37 ± 0.03) to 0.46 ± 0.15 (0.88 ± 0.01) for upper (Lower) part of leaves. The alpha track density in leaves was found to be higher in case of 50 g fertilizers. Also, the alpha track density of the potato plants leaves increased with increase in life of plants. This may be due to longer half life of ^{235}U, ^{226}Ra and ^{210}Pb and available for the transfer from soil to roots and then leaves. The alpha track density from leaves were found to be minimum for the plants grown with urea because the phosphorous content of urea fertilizers was found to be minimum as compare to other fertilizers and maximum for Zinc sulphate (ZnS), diammonium phosphate (DAP) and Nitrogen Phosphorous Potassium (NPK) fertilizers due to their high phosphorous contents measured by Chauhan et al. [24].

The alpha track density from various parts of the potato plants at the age of 105 days of plantation are listed in Table 3. The alpha track density for the root parts (The part of plant underground other than edible part) varied from 0.62 ± 0.16 (0.66 ± 0.19) to 1.52 ± 0.049 (1.64 ± 0.06) for the plants grown using 20 g (50 g) of different fertilizers and found to be minimum for urea fertilized plants and maximum for diammonium phosphate and potash fertilized plants. For the stem parts the maximum value of track density was 1.37 ± 0.19 (1.84 ± 0.17), for single phosphate fertilizer using 20 g (50 g) amount.

The alpha track density ($T\,cm^{-2}d^{-1}$) in the fruit part (edible parts) of potato plants varied from 0.04 ± 0.01 to 0.46 ± 0.04 for 20 g and 0.15 ± 0.07 to 1.43 ± 0.04 for 50 g amount of fertilizers and again found to be minimum for urea fertilizer. The track density in the leaves of plants showed the similar trends as listed in Tables 1 and 2 and confirmed the accumulations of uranium and their decays products with time. The alpha track density for the plants grown without the use of fertilizers was less than that with fertilizers, which reflects the increase in alpha activity by use of fertilizers.

Translocation factor

The behaviour of different parts of plants with respect to photo-chemical reaction is not same, therefore the translocation of different nutrient (N, P, K, Fe, etc.) is also not same. The phosphorus is one of the 17 essential nutrients for the growth of leaves, stem as well fruit. The nutrient transfer from soil to roots of the plants is further translocated in the different parts of the plant, which may act as a carrier for uranium and decays products. The translocation factor is the ratio of the activity concentration in the different parts of the plants to that of roots. The calculation of the translocation factor was made to evaluate the ability of potato plants to translocate the radionuclide from root to different part of plants. The leaves of the potato plants are used to feed animals and edible parts (fruit) is used as vegetable all

Table 1 Alpha Track density per day at the age of 30 days in leaves of potato plants

| Fertilizer used | Alpha track density per day in leaves ($T\,cm^{-2}d^{-1}$) | | | |
| | Fertilizer used (20 g) | | Fertilizers used (50 g) | |
	Upper face	Lower face	Upper face	Lower face
Urea	0.12 ± 0.01	0.18 ± 0.04	0.14 ± 0.07	0.24 ± 0.07
Single Super Phosphate	0.23 ± 0.01	0.31 ± 0.03	0.3 ± 0.02	0.45 ± 0.06
NPK	0.2 ± 0.02	0.52 ± 0.07	0.13 ± 0.0	0.55 ± 0.0
Zinc Sulphate	0.43 ± 0.01	0.81 ± 0.17	0.30 ± 0.04	0.34 ± 0.01
Potash fertilizer	0.27 ± 0.05	0.38 ± 0.06	0.28 ± 0.03	0.48 ± 0.04
Diammonium Phosphate	0.41 ± 0.26	0.45 ± 0.23	0.23 ± 0.03	0.47 ± 0.12

Table 2 Alpha Track density per day at the age of 60 days in leaves parts of potato plants

| Fertilizer used | Alpha track density per day in leaves (T cm^{-2}d^{-1}) | | | |
| | Fertilizer used (20 g) | | Fertilizers used (50 g) | |
	Upper face	Lower face	Upper face	Lower face
Urea	0.19 ± 0.03	0.29 ± 0.02	0.23 ± 0.06	0.37 ± 0.03
Single Super Phosphate	0.39 ± 0.02	0.43 ± 0.11	0.34 ± 0.02	0.47 ± 0.09
NPK	0.34 ± 0.09	0.35 ± 0.12	0.46 ± 0.15	0.48 ± 0.04
Zinc Sulphate	0.37 ± 0.08	0.62 ± 0.03	0.39 ± 0.06	0.53 ± 0.14
Potash fertilizer	0.38 ± 0.03	0.41 ± 0.11	0.38 ± 0.08	0.88 ± 0.01
Diammonium Phosphate	0.52 ± 0.03	0.29 ± 0.07	0.31 ± 0.03	0.44 ± 0.05

over the world, thus the translocation factor for fruit and leaves part were calculated by using Eq. 1 listed in Table 4.

The translocation factor for the fruit (edible Part) varied from 0.13 (for DAP) to 0.73 (for PF) with an average of 0.40 ± 0.26 for the plant grown with 20 g of fertilizers. The translocation factor increased with increase in amount of fertilizers having value 0.51 ± 0.31 for the plant grown with 50 g of fertilizers. The translocation factor for the lower part of leaves grown with 20 g and 50 g fertilizers varied from 0.33 to 0.67 and 0.59 to 0.88 with an average value 0.55 ± 0.15 and 0.69 ± 0.14 respectively. The fruit part of the plants have translocation factor less than that of leaves. This may be because the accumulation of radioactivity in the plant is time dependent. The leaves in the potato plants get developed after 10 days of plantation and accumulation of radionuclide was continuing upto the sampling. On the other hand, the fruit parts of plant get developed after at the age of 40 days, thus, have low alpha activity. The large

variation (30-60 %) in the translocation factor for these fertilizers may be due to different chemical behaviour of plants, which may arise due to different pH value of soil. The translocation factors listed in Table 4 represented the combined effect of uranium, radium and their decays products. Rodriguez et al. [19] reported the value of translocation factor 0.00068 ± 0.00009 for uranium and 0.86 ± 0.03 for radium for the sun flower plants. The translocations factor values from present study lies within these two.

Soil to plant root transfer factor (TF's)

The measurement of alpha track density was carried out in duplicate to ascertain the statistical error. The alpha track density per day from the fertilizers and fertilized soil was measured using LR-115 detector and listed in Table 5. The alpha track density from the fertilizers varied from 8.8 ± 0.4 for urea to 97 ± 4 for Diammonium phosphate fertilizers. While track density per day from the fertilized soil samples varied from 49 ± 3 for Zinc

Table 3 Alpha Track density per day at the age of 105 days in root, stem, fruit and leaves of potato plants

Fertilizers	Alpha track density per day in root, stem, fruit and leaves (T cm^{-2}d^{-1})									
	Root		Stem		Fruit		leaves			
							Upper face	Lower face	Upper face	Lower face
	Fertilizer used (20 g)	Fertilizer used (50 g)	Fertilizer used (20 g)	Fertilizers used (50 g)	Fertilizer used (20 g)	Fertilizer used (50 g)	Fertilizer used (20 g)		Fertilizers used (50 g)	
Urea	SNR	0.66 ± 0.19	SNR	1.01 ± 0.25	0.04 ± 0.01	0.15 ± 0.07	0.74 ± 0.36	0.55 ± 0.02	1.17 ± 0.03	0.51 ± 0.13
Single Super Phosphate	0.77 ± 0.20	1.5 ± 0.12	1.37 ± 0.19	1.84 ± 0.17	0.44 ± 0.19	1.43 ± 0.04	0.48 ± 0.13	0.34 ± 0.03	0.77 ± 0.26	0.96 ± 0.69
NPK	0.98 ± 0.24	1.64 ± 0.06	1.27 ± 0.09	0.51 ± 0.12	0.11 ± 0.03	0.35 ± 0.07	0.44 ± 0.01	0.48 ± 0.03	1.01 ± 0.01	0.96 ± 0.04
Zinc Sulphate	0.91 ± 0.46	1.22 ± 0.05	1.09 ± 0.07	1.47 ± 0.07	0.39 ± 0.04	0.55 ± 0.07	0.63 ± 0.04	0.51 ± 0.13	1.18 ± 0.31	1.04 ± 0.29
Potash fertilizer	0.62 ± 0.16	1.03 ± 0.65	0.89 ± 0.3	1.05 ± 0.09	0.46 ± 0.04	0.72 ± 0.04	0.41 ± 0.01	0.49 ± 0.01	1.34 ± 0.01	0.69 ± 0.04
Di-ammonium Phosphate	1.52 ± 0.04	SNR	1.29 ± 0.37	1.49 ± 0.15	0.2 ± 0.05	0.5 ± 0.1	0.51 ± 0.04	0.40 ± 0.01	0.84 ± 0.13	0.52 ± 0.04
Without fertilizers	SNR	SNR	57 ± 1		0.06 ± 0 .00		0.31 ± 0.14		0.23 ± 0.09	

SNR Sample not retrieved

Table 4 Translocation factors* for fruit and leaves of potato plants at age of 105 days

Fertilizers used for plantation	Translocation factor(TLF's) in			
	Fruit part		Leaf (lower part)	
	20 g fertilizer used	50 g fertilizers used	20 g fertilizer used	50 g fertilizer used
Urea	SNR	0.22	SNR	0.72
Single Super Phosphate	0.57	0.95	0.63	0.64
NPK-1	0.12	0.21	0.48	0.59
Zinc Sulphate	0.42	0.45	0.67	0.85
Potash fertilizer	0.73	0.69	0.66	0.67
Diammonium Phosphate	0.13	-	0.33	-
Average ± SD	0.40 ± 0.26	0.51 ± 0.31	0.55 ± 0.15	0.69 ± 0.14

*Translocation factors in Table 4 were determined by the average value of track density from various parts thus, standard deviation is not included in Table 2

sulphate to 98 ± 3 for Diammonium phosphate fertilizers and showed a good correlation ($R^2 = 0.77$) with the fertilizer alpha track density (Fig. 2). The transfer factors defined by Eq. 2 were calculated in order to measure the ratio of radionuclides transfer from soil to various parts of plants (Table 6). The measurement of the transfer factor was carried out from the average value of track density in Tables 1, 2, 3 and 5, therefore, the standard deviation and errors are not reported in Table 6.

The TF's value for potato plants varied from 1.5×10^{-2} to 1.0×10^{-1} for root, from 1.3×10^{-2} to 1.2×10^{-1} for stem, from 2.1×10^{-3} to 4.5×10^{-2} for fruit and from 5.4×10^{-3} to 5.8×10^{-2} for lower part of leaves at the age of the 105 days of the potato plants using different fertilizers. A comparison of the TF's value from various parts of plant grown using different fertilizers is shown in Fig. 3. The TF's value from the lower part of the leaves of potato plant at different age of the plants were also compared and shown in Fig. 4.

The alpha activity from the fruit (edible) parts of the potato plants was smaller than the roots in spite of their underground existence. The root's hair under the ground provide the large surface to volume ratio for the uptake of element and water and the edible parts life under the ground was smaller to accumulate the radioactivity

content than that of the root. Due to same reason the alpha activity from the stem and leaves parts are also higher than fruit parts. The increase in the alpha activity in the leaves of potato plants evidence the time accumulation of radioactivity. Different plant reaction to the uptake of elements is different depending upon the condition of growth, chemistry and climate regime. The uranium which is non essential element usually accumulates in the roots of the plants due to *casparian strip* [25], thus evidence the occurrence of high alpha activity in the roots in present study. While due to the different chemical behaviour of radium as that of uranium and long time necessary for successful *phytoextraction*, it migrates in the different parts of plant and causes high alpha activity in stem and leaves part. Soudek et al. [26] and Masri et al. [27] reported the TF's value from 0.9 to 7.8 for ^{40}K, 0.3×10^{-2} to 1.2×10^{-1} for ^{238}U and 2.8×10^{-2} to 9.1×10^{-2} for ^{210}Po and ^{210}Pb for stem, leaves and fruit of different plants species. The TF's value reported for leaves parts was higher than that of fruit parts for olives, apple and grapes. Soudek et al. [26] reported the TF's value from 1.3×10^{-2} to 1.6×10^{-1} for 28 different species grown in green house and confirmed the nearly linear behaviours of element from soil to plant uptake. The alpha activity in the different parts of plant

Table 5 Alpha track density from the fertilizers and fertilize soil

Fertilizers used	Alpha track density from the		
	Fertilizers	Fertilized soil (20 g)	Fertilized soil (50 g)
Diammonium Phosphate	97 ± 4	57.5 ± 3	98 ± 3
NPK	44 ± 10	32 ± 10	84 ± 1
Potash fertilizer	40 ± 2	41 ± 1	65 ± 7.5
Single super phosphate	21 ± 3	33.5 ± 5.5	34.5 ± 6.5
Urea	8.8 ± 0.4	60.5 ± 13.5	59.5 ± 6.5
Zinc Sulphate	8.8 ± 1.1	35.5 ± 4	49 ± 3

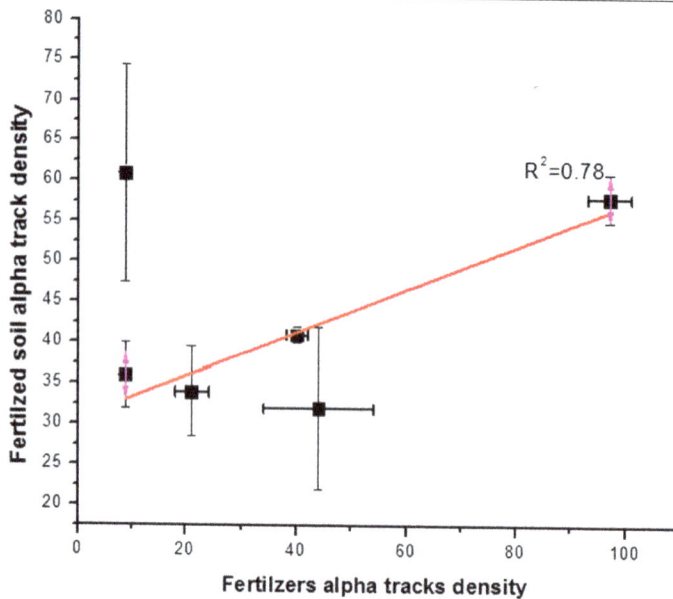

Fig. 2 Correlation between the alpha track densities from fertilizers and fertilized soil

considered in the present study was assumed to be caused by the radionuclide uranium and radium along with their decay products thus the TF's value lie in the range reported in previous studies. The measurement of TF's value for five boreal forest species was carried out by Tuovinen et al. [16] and reported that the assumption for the linear increase of element (Constant TF's value) did not hold. The element concentration ratio remains symmetrical and decreased with increase in the element concentration in soil. The applicability of linear and non linear assumptions is not unique for specific plants and elements. Sheppard [28] considered linearity assumption valid for range of elements concentrations in the soil is 5 orders of magnitude. Sheppard and Sheppard [29] studied the transfer factor for natural uranium, and concluded that the linearity assumption is valid only for concentrations higher than 1.8×10^6 Bq/kg. Blanco Rodrıguez et al.

[18] reported that the linearity assumption for ^{238}U, ^{230}Th and ^{226}Ra is valid for two orders of magnitude of range concentrations of elements. In present study the alpha activity from the potato plants varied two orders of magnitude, thus linearity assumption for transfer factor was applied. The result of TF's value for potato plants agreed with the range published in literature for different plant species.

A linear and non linear assumption for the soil to root uptake of stable and radioactive elements is not only the function of elements concentration in soil but the other factors like the physiology of plants with reference to particular elements, condition of growth, and metabolism mechanism altered by the usage of fertilizers, soil property and uptake of water also affect the TF's value thus, includes a lot of uncertainty [2]. This means a single model for transfer factor cannot be

Table 6 Transfer factor from different parts of potato plants at age of 105 days

Fertilizers	Root	Stem	Fruit	Lower side of leaves
Diammonium Phosphate	1.6×10^{-2}	1.3×10^{-2}	2.1×10^{-3}	5.4×10^{-3}
NPK	2.2×10^{-2}	2.9×10^{-2}	2.5×10^{-3}	2.2×10^{-2}
Potash fertilizer	1.5×10^{-2}	2.6×10^{-2}	1.2×10^{-2}	1.7×10^{-2}
Single super phosphate	3.7×10^{-2}	6.5×10^{-2}	2.1×10^{-2}	4.6×10^{-2}
Urea	SNR	SNR	4.5×10^{-2}	5.8×10^{-2}
Zinc Sulphate	1.0×10^{-1}	1.2×10^{-1}	4.4×10^{-2}	1.9×10^{-2}

SNR Sample not retrieved

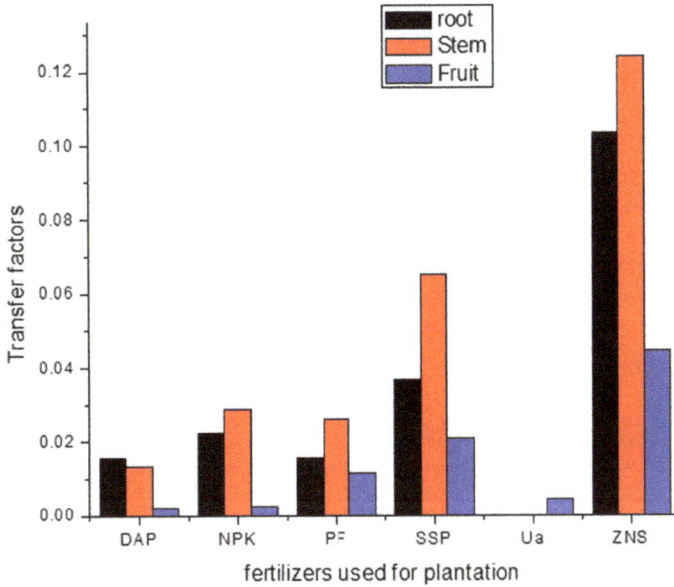

Fig. 3 Transfer factor from different parts of potato plant at the age of 105 days

applied for group of plants species, even belonging to same family. In the absence of broadly applicable and low-variance models for plant uptake, site-specific models should be derived for definitive risk assessments intended to support remediation decisions which would be applicable to the soil type(s), variety of plant taxa, and variety of chemical forms at a contaminated site [30].

Conclusions

Radioactivity is always present in the soil and inspite of the low concentration of these elements; they significantly increase the radioactivity in the ecosystem. Therefore the present study has been regarded as of vital importance. The alpha activity from various parts of the potato plants was found to be higher in case of

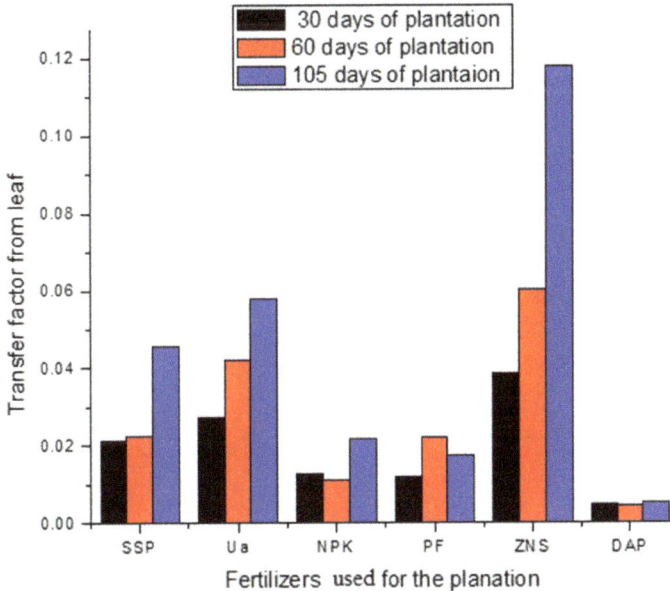

Fig. 4 TF's value from lower part of leaf at different age of potato plant

phosphate fertilizers used compared with others. This may have been caused by the presence of uranium and radium in the phosphate rocks; raw materials used for the fertilizers production. The translocation factors from root to leaves and fruit (edible parts) were also calculated. The translocation factors for the fruit (edible parts) of the potato plants grown with different fertilizers were found to be lower than root hair. This may be explained on the basis of small residency time of the fruit as compared to root hair, remaining underground along with the plants.

Competing interests
The authors declare that they have no competing interests.

Authors' contribution
Both authors have made contribution to the review/finalization of this manuscript and approved the final manuscript.

Acknowledgments
The authors are thankful to University grant commission, New Delhi, Govt. of India for providing financial assistance in the terms of major research project.

References
1. Ozaki T, Ambe S, Abe T, Francis AJ. Adsorption of Radionuclides on Silica and Their Uptake by Rice Plants from Silica-Multitracer Solutions: The Effects of pH. Biol Trace Elem Res. 2002;90:273–81.
2. Ehlken S, Kirchner G. Environmental processes affecting plant root uptake of radioactive trace elements and variability of transfer factor data: a review. J Environ Radioact. 2002;58:97–112.
3. Ahmed NK, El-Arabi AGM. Natural radioactivity in farm soil and phosphate fertilizer and its environmental implications in Qena governorate, Upper Egypt. J Environ Radioact. 2005;84:51–64.
4. Spalding RF, Sackett WM. Uranium in runoff from the Gulf of Mexico distribution province Anomalous concentration. Science. 1972;175:629–31.
5. Miranzadeh H, Emam Y, Pilesjö P, Seyyedi H. Water use efficiency of four dryland wheat cultivars under different levels of nitrogen fertilization. J Agriculture Sci Technol. 2011;13:843–54.
6. Lambert R, Grant C, Sauve C. Cadmium and zinc in soil solution extracts following the application of phosphate fertilizers. Sci Total Environ. 2007;378:293–305.
7. Cheraghi ML, Merrikhpour H. Investigation of the Effects of Phosphate Fertilizer Application on the Heavy Metal Content in Agricultural Soils with Different Cultivation Patterns. Biol Trace Elem Res. 2012;145:87–92.
8. Grant CA, Sheppard SC. Fertilizer Impacts on Cadmium Availability in Agricultural Soils and Crops. Hum Ecol Risk Assess. 2008;14:210–28.
9. Khater AEM, AL-Sewaidan HA. Radiation exposure due to agricultural uses of phosphate fertilizers. Radiat Meas. 2008;43:1402–7.
10. Taher AE, Makhluf S. Natural activity level in phosphate fertilizers and its environment implication in assuit governorate, Upper Egypt. Indian J Pure Appl Physics. 2010;48:697–702.
11. Mazzilli B, Palmiro V, Saueia C, Nisti MB. Radiochemical characterization of Brazilian phosphogypsum. J Environ Radioact. 2000;49:113–22.
12. Da Conceicao TF, Bonotto DM. Radionuclides, heavy metals and fluorine incidence at Tapira phosphate rocks, Brazil, and their industrial by- products. Environ Pollution. 2006;139:232–43.
13. Bouwer FJ, MeKlveen JW, McDowell WJ. Uranium assay of phosphate fertilizers and other phosphatic materials. Health Phys. 1978;34:345–52.
14. UNSCEAR. Report of the United Nations Scientific Committee on the effects of atomic radiation to the general assembly. New York: Exposures from natural radiation sources, United Nations; 2000.
15. Cataldo DA, Wildung RE. The role of soil and plant metabolic processes in controlling trace element behavior and bioavailability to animals. Sci Total Environ. 1983;28:159–68.
16. Tuovinen TS, Roivainen P, Makkonen S, Kolehmainen M, Holopainen T, Juutilainen J. Soil-to-plant transfer of elements is not linear: Results for five elements relevant to radioactive waste in five boreal forest species. Sci Total Environ. 2011;410–411:191–7.
17. Chen SB, Zhua YG, Hu QH. Soil to plant transfer of ^{238}U, ^{226}Ra and ^{232}Th on a uranium mining-impacted soil from southeastern China. J Environ Radioact. 2005;82:223–36.
18. Blanco Rodriguez P, Vera Tome F, Lozano JC. About the assumption of linearity in soil-to-plant transfer factors for uranium and thorium isotopes and 226Ra. Sci Total Environ. 2002;284:167–75.
19. Blanco Rodriguez P, Vera Tome F, Perez Fernandez F, Lozano JC. Linearity assumption in soil-to-plant transfer factors of natural uranium and radium in Helianthus annuus L. Sci Total Environ. 2006;361:1–7.
20. Chauhan P, Chauhan RP. Elemental analysis of fertilizers using X-ray fluorescence and their impact on alpha radioactivity of plants. J Radioanal Nucl Chem. 2013;295:1097–105. doi:10.1007/s10967-012-2244-6.
21. Qian P, Schoenau JJ, Mooleki P. Phosphorus Amount and Distribution in a Saskatchewan Soil after Five Years of Swine and Cattle Manure Application. Candian J Soil Sci. 2004;84:275–81.
22. Al-Mohammadi F, Al-Zubi Y. Soil chemical properties and yield of tomato as influenced by different levels of irrigation water and fertilizer. J Agriculture Sci Rechnol. 2011;13:289–99.
23. Jin S, Verma AD, Lange T, Daniell H. Release of Hormones from Conjugates: Chloroplast Expression of β-Glucosidase Results in Elevated Phytohormone Levels Associated with Significant Increase in Biomass and Protection from Aphids or Whiteflies Conferred by Sucrose Esters. Plant Physiol. 2011;155:222–35.
24. Chauhan P, Chauhan RP, Gupta M. Estimation of naturally occurring radionuclides in fertilizers using gamma spectrometry and elemental analysis by XRF and XRD techniques. Microchem J. 2013;106:73–8.
25. Denny H. Plant mineral nutrition. In: Ridge I, editor. Plants. New York: Oxford University Press; 2002. p. 167–220.
26. Soudek P, Petrova S, Benesova D, Kotyza J, Vagner M, Vankova R, et al. Study of soil–plant transfer of 226Ra under greenhouse conditions. J Environ Radioact. 2010;101:446–50.
27. Al-Masri MS, Al-Akel B, Nashawani A, Amin Y, Khalifa KH, Al-Ain F. Transfer of ^{40}K, ^{238}U, ^{210}Pb, and ^{210}Po from soil to plant in various locations in south of Syria. J Environ Radioact. 2008;99:322–31.
28. Sheppard MI. The environmental behaviour of uranium and thorium. Canada: AECL, At Energy Canada Ltd; 1980. p. 6795.
29. Sheppard MI, Sheppard SC. The plant concentration concept as applied to natural uranium. Health Phys. 1985;48:494–500.
30. Efroymson RA, Sample BE, Suter GW. Bioaccumulation of Inorganic Chemicals from Soil by Plants: Spiked Soils vs. Field Contamination or Background. Hum Ecol Risk Assess. 2004;10:1117–27.

9

Determination of toxic (Pb, Cd) and essential (Zn, Mn) metals in canned tuna fish produced in Iran

Seyed Vali Hosseini[1], Soheil Sobhanardakani[2*], Hamed Kolangi Miandare[3], Mohammad Harsij[4] and Joe Mac Regenstein[5]

Abstract

Background: Metal pollution of waterways directly affects human health and can impact the food chain. Seafood living in polluted water can accumulate trace metals. The purpose of this study was to analyze the toxic metals Pb and Cd and the dietary essential metals Zn and Mn in 120 cans of tuna species from four different brands processed in Iran and purchased in 2012.

Results: The mean level of metals for each brands of canned fish obtained in mg/kg were as follows: yellowfin tuna (Pb: 0.19 ± 0.015, Zn: 5.77 ± 4.17, Mn: 0.08 ± 0.07, Cd: 0.15 ± 0.12), Kilka (Pb: 0.95 ± 0.88, Zn: 30.47 ± 29.82, Mn: 1.01 ± 0.73, Cd: 0.07 ± 0.05), Kawakawa (Pb: 0.28 ± 0.23, Zn: 6.77 ± 5.21, Mn: 0.17 ± 0.12, Cd: 0.12 ± 0.09), longtail tuna (Pb: 1.59 ± 1.56, Zn: 7.44 ± 6.11, Mn: 0.04 ± 0.03, Cd: 0.06 ± 0.04). Pb, Zn and Cd levels were generally higher than the FAO/WHO permissible limits (Pb: 0.50 mg/kg, Zn: 50.0 mg/kg and Cd: 0.50 mg/kg) and the European Union acceptable dietary limits.

Conclusions: Based on the United States Environmental Protection Agency health criteria, there is no health risk associated with Mn concentration in the samples analyzed. The limits of detection of the method for Pb, Zn, Mn and Cd in mg/kg were 0.01, 0.5, 0.01 and 0.01, respectively. The result of the one-way analysis of variance suggested significant variations ($p < 0.05$) in the concentration of the metals in the different types of canned fish with the following being outside of compliance levels.

Keywords: Canned fish, Food safety, Lead, Cadmium, Zinc, Manganese

Introduction

During the last few decades there has been a growing interest in determining the level of toxic metals in marine and fresh water environments with additional emphasis on the measurement of contamination levels in the food supply, particularly fish [1–12] including canned fish. For example, Tuzen and Soylak determined the Cu, Zn, Mn, Fe, Se, Al, Cr, Ni, Pb and Cd concentrations in canned fish marketed in Turkey [1]. Mol analyzed the levels of Fe, Zn, Cu, Cd, Sn, Hg and Pb in canned bonito, sardines, and mackerel [5], canned tuna fish [6] and canned anchovies and canned rainbow trouts [7] produced in Turkey. Hosseini et al.

determined Hg, Se and Sn concentrations in canned fish marketed in Iran [10] while Mahalakshmi et al. determined Al, Cd, Pb and Hg in canned tuna fish available in Canada and India [11]. Trace metals are important for both their necessity and toxicity. Some elements like Mn and Zn are essential functional and structural elements in biological systems [13–18] often catalyzing reactions by binding to substrates, there by favoring various reactions such as the mediation of oxidation–reduction reactions, or redox reactions, through reversible changes in the oxidation state of the metal ions [15, 16, 19–21]. For Mn and Zn, often called micronutrients, there are fixed allowed levels that provide for an adequate dietary intake according to the World Health Organization (WHO). In adults from 5.0 to

* Correspondence: s_sobhan@iauh.ac.ir
[2]Department of the Environment, College of Basic Sciences, Hamedan Branch, Islamic Azad University, Hamedan, Iran
Full list of author information is available at the end of the article

22.0 mg are recommended for Zn and from 2.0 to 20.0 mg are recommended for Mn [22, 23]. At high concentrations, Zn causes nephritis, anuria and extensive lesions in the kidneys [24, 25].

Pb and Cd are very toxic to humans. They are only tolerated at extremely low concentrations and excesses are associated with many adverse health effects [12, 26]. They may injure the kidney and cause symptoms of chronic toxicity, including impaired organ function, poor reproductive capacity, hypertension, tumors and hepatic dysfunction [24]. Moreover, Pb can also affect brain function by interfering with neurotransmitter release and synapse formation. Exposure to Pb has been associated with reduced IQ, learning disabilities, slow growth, hyperactivity, antisocial behaviors and impaired hearing [27]. Generally Pb-poisoning is ranked as the most common environmental health hazard [28].

Use of Cd in agriculture and industry has been identified as a major source of its wide dispersion in the environment and food. The major route of exposure to Cd for non-smokers is via food; the contribution from other pathways to total uptake is small [28]. Certain marine vertebrates contain markedly elevated Cd concentrations [29].

Trace and toxic metals may contaminate fish, mainly during the growth phase but also, sometimes, due to contamination during transportation and storage. Therefore, most countries monitor the levels of toxic metals in seafood [5, 30, 31]. Levels of toxic metals in fish depend on many factors like the duration of exposure of fish to contaminants in the water, the feeding habit of the fish, the concentrations of contaminants in the water column, and sometimes to the water chemistry, contamination of fish during handling and processing, age, sex, weight, season, fish species, catching area, transportation, and storage, etc. [21, 32, 33]. According to Taha'n et al., the pH of the canned product, the quality of the lacquer coatings of canned products, oxygen concentration in the headspace, quality of the coating and storage place may also affect metal levels in canned fishes [34].

Many species of commercially caught marine fish, especially in Iran, are canned, thus making them more available for human consumption to those living far from the sea [35].

Tuna fish are long living organisms prone to accumulate pollutants. Canned tuna fish is eaten regularly in many countries including Iran (globally over the 10×10^{10} tonne per year) [12, 36–39]. In this study the levels of the toxic and essential metals (Pb, Zn, Mn and Cd) in four different commercial types of canned tuna fish (longtail tuna, Kawakawa, Kilka and yellowfin tuna) commonly consumed in Iran were determined using GFAAS . This study will help to generate the data needed for surveillance programs aimed at ensuring the safety of the food supply and minimizing human exposure to toxic metals.

Materials and methods
Sample collection
During the year 2012, 120 samples (185 g each) of four different commercial types of canned fish commonly consumed in Iran (30 samples for each type: "Famila Co." (Tehran): yellowfin tuna (YT) (*Thunnus albacares*); "Shilaneh Co." (Qazvin): common Kilka (CK) (*Clupeonella cultriventris* caspia); "Pars Tuna Co." (Bushehr): Kawakawa (Ka) (*Euthynnus affinis*), and "Hiltune Co." (Tehran): longtail tuna (LT) (*Thunnus tonggol*)) were purchased at markets within Tehran.

Chemical analyses
All glassware was cleaned by soaking overnight in 10 % nitric acid, followed by rinsing with distilled water. The acids used for wet digestion were of analytical reagent (Merck, Darmstadt, Germany) grade, while the distilled water was further deionized. The blank values were below the detection limits of the instrument. Working standards were made from the stock by dilution of the measured aliquots with 1.0 M nitric acid. Each sample was analyzed in triplicate and the results, which mostly agreed within ±1.0 %, were averaged. A reagent blank determination was carried out with every batch of 10 samples.

After opening each can oil/broth was drained off and the meat was homogenized thoroughly in a food blender (Hongdun HWT, Zhangqiu, China). Samples were then digested without delay. About 5 g of each sample was digested in a quartz Erlenmeyer flask with 15 ml of suprapure nitric:perchloric:sulphuric acid (25 + 25 + 1 v:v:v) mixture, using a hot plate at 150 °C. Further aliquots of nitric acid were added until a complete colorless solution was obtained. After evaporation using the Perkin Elmer Multiwave 3000, the residue was dissolved in 10 ml of water with 1 ml of concentration of suprapure HCl at 100 °C. Finally, the volume was made up to 25 ml with deionized-distilled water. Determination of Pb and Cd were done on a stabilized temperature graphite furnace atomic absorption spectrometer (GFAAS) (4110 ZL, Perkin Elmer). Zn and Mn were done by direct aspiration of the sample solution into the air-acetylene flame [38]. All analyses were performed with three replications. The recovery of various metals from canned fish samples is presented in Table 1.

Statistical analysis
One-way analyses of variance (ANOVA) and Tukey's test were used to determine whether Pb, Zn, Mn and Cd concentrations varied significantly between specimens, with probability values less than 0.05 ($p < 0.05$) considered statistically significant. The statistical calculations were done using SPSS 15.0 version (SPSS Inc., Chicago, IL, USA) statistical package.

Table 1 Recovery of various metals from canned fish samples

Sample	Pb			Zn			Mn			Cd		
	Added (mg/kg)	Found (mg/kg)	Recovery (%)	Added (mg/kg)	Found (mg/kg)	Recovery (%)	Added (mg/kg)	Found (mg/kg)	Recovery (%)	Added (mg/kg)	Found (mg/kg)	Recovery (%)
YT	30	30	100	30	29.5	98.3	30	29.9	99.6	30	28.5	95.0
CK	30	29.5	98.3	30	29.8	99.3	30	29.5	98.3	30	29.3	97.6
Ka	30	29.3	97.6	30	28.0	93.3	30	30	100	30	29.3	97.6
LT	30	28.8	96.0	30	29.9	99.6	30	29.5	98.3	30	28.9	96.3

Results

The concentrations of Pb, Zn, Mn and Cd in canned fish are presented in Table 2 along with relevant statistical parameters. Thirty samples for each canned fish were analyzed. Cadmium mean concentration in all samples analyzed exceed the EU limit of 0.05 mg kg^{-1}, whereas Pb levels in 90 of 120 samples (75.0 %) exceeded the EU acceptable limit of 0.2 mg/kg [40].

The comparative levels based on the average of Pb, Zn, Mn and Cd in the various types of canned fish shows that the average concentrations of Pb in LT are higher than the other species by 1.7, 5.7 and 8.3 times as compared to CK, Ka and YT, respectively. The average Zn content in CK was much higher than other the canned fish, as compared to LT, Ka and YT, respectively. Similar behavior is shown by Mn. The average Pb content in LT was much higher than other the canned fish, as compared to YT, Ka and CK, respectively. The average Cd content in LT was about 3 times lower than YT.

Discussion

Eboh et al. reported that Pb in the muscle, gills and liver tissue of five common commercially available fish species in Nigeria (catfish, tilapia, ilisha, bonga and mudskipper) were found in the range of 0.001-0.002 mg/kg but did not find any toxic metal residues in salmon and mackerel species [41]. On the other hand, the mean concentrations of Cd (0.37-0.79 mg/kg) and Pb (4.27-6.12 mg/kg) reported by Canli and Atli in muscle tissues of six different fish species (*Sparus auratus, Atherina hepsetus, Mugil cephalus, Trigla cuculus, Sardina pilchardus*

and *Scomberesox saurus*) are higher than the current results [42]. Turkmen et al. determined the metal levels in the muscles and livers of 12 fish species from the Aegean and Mediterranean Seas and reported that the levels of Cd, Mn, Pb and Zn in muscles of fish were <0.01-0.39, 0.18-2.78, 0.21-1.28 and 3.51-53.5 mg/kg, respectively [43]. Boadi et al. determined Pb, Zn, Fe, Cd, Mn and Hg in 46 canned fish samples of nine different brands marketed in Ghana. The trace metals were found in the range of 0.058- 0.168 mg kg^{-1} for Pb, 0.010-0.370 mg/kg for Zn, 0.001-0.057 mg/kg for Mn and Cd concentration, were below their detection limit for all the samples. Zn levels were generally below the FAO recommended limit of 40 mg kg^{-1}. The concentration of Pb in the canned fish was also below the MAFF (1995) guidelines of 2.0 mg kg^{-1} [33, 44]. Also, based on the U.S. EPA health criteria for carcinogens, there are no health risks associated with Pb concentrations in the canned fishes they analyzed [33]. The Cd concentrations in tuna fish samples from the Persian Gulf area of Iran were reported as 0.0046-0.0720 mg/kg with mean values of 0.0223 mg/kg while the Pb concentration ranged as 0.0162-0.0726 mg/kg with mean values of 0.0366 mg/kg [12]. These were lower than the current results. Comparison of the results of this study with other studies is shown in Table 3.

Several organizations, such as the FAO and WHO, provide guidelines on the intake of trace elements by humans. The provisional tolerable weekly intake (PTWI) recommended by the Joint FAO/WHO Expert Committee (1972) for Cd and Pb are 7 μg Cd/kg body weight per week and 25 μg Pb/kg body weight per week respectively

Table 2 Metals content (mg/kg) for various species of canned fish*

Metal	Yellowfin tuna		Kilka		Kawakawa		Longtail tuna	
	Range	Mean ± SD	Range	Mean ± SD	Range	Mean ± SD	Range	Mean ± SD
Pb	0.02-0.93	0.19 ± 0.015[a]	0.02-3.77	0.95 ± 0.88[c]	0.02-1.17	0.28 ± 0.23[b]	0.07-5.50	1.59 ± 1.56[d]
Zn	4.05-8.22	5.77 ± 4.17[a]	24.72-36.22	30.47 ± 29.82[d]	4.92-10.02	6.77 ± 5.21[b]	4.60-13.37	7.44 ± 6.11[c]
Mn	0.02-0.40	0.08 ± 0.07[b]	0.35-1.40	1.01 ± 0.73[d]	0.02-0.37	0.17 ± 0.12[c]	0.02-0.30	0.04 ± 0.03[a]
Cd	0.02-0.27	0.15 ± 0.12[d]	0.02-0.20	0.07 ± 0.05[b]	0.000-0.27	0.12 ± 0.09[c]	0.02-0.37	0.06 ± 0.04[a]

*Number of each canned fish sample = 30

Vertically, letters a, b, c and d show statistically significant differences of metals content between canned fish specimens according to one way ANOVA analysis (p < 0.05)

Table 3 Comparison of present mean values in specimens with other studies result

Specimen/Fish Species	Area	Tissue	Pb (mg/kg)	Zn (mg/kg)	Mn (mg/kg)	Cd (mg/kg)
Yellowfin tuna, Kilka, Kawakawa, longtail tuna	Present Study	Muscle	0.02-5.50	4.05-36.22	0.02-1.40	0.00-0.27
Anguilla Anguilla, Mugil cephalus, Oreochromis niloticus	Turkey[a]	Muscle	0.21-1.45	34.76-134.76	2.34-18.97	0.04-0.23
		Liver	0.56-2.10	1.28-587.23	4.12-23.81	0.11-4.19
		Gills	0.23-2.36	1.28-169.87	4.39-753.74	0.01-0.52
Otolithes ruber, Pampus argenteus, Parastromateus niger, Scomberomorus commerson, Onchorynchus mykiss	Iran[b]	Muscle	0.007-0.075	0.005-0.40		0.001-0.45
		Liver	0.005-0.10	0.012-0.45		0.002-0.58
		Gills	0.031-0.11	0.03-0.062		0.004-0.09
Canned salmon, sardine and tuna fish	KSA [c]	Muscle	0.03-1.97	3.80-23.90		0.01-0.69
Canned fish	USA[d]		0.0-0.03	0.14-97.8	0.01-2.55	0.0-0.05
Canned sardines	Nigeria[e]			0.09-4.63	0.64-1.37	0.11-0.26
Canned tuna	Turkey[f, g, h]		0.09-0.45	8.20-12.4		0.01-0.02
Canned tuna			0.09-0.45	8.20-12.40		0.01-0.02
Canned anchovies, canned rainbow trout			0.178	17.06		0.145
Canned tuna	Canada[i]		0.011-0.089			0.020-0.025

[a]Yilmaz [49]
[b]Sobhanardakani et al. [50, 51]
[c]Ashraf et al. [37]
[d]Ikem and Egeibo [38]
[e]Iwegbue et al. [25]
[f]Mol [5]
[g]Mol [6]
[h]Mol [7]
[i]Mahalakshmi et al. [11]

[45, 46]. Therefore, the provisional tolerable weekly intake of Cd and Pb for a 60 kg adult (halfway between the 70 kg male and 50 kg female normally taken as the standard) was estimated to be 420 and 1500 µg/person/week, respectively. The maximum Cd level found in this study was 0.37 mg/kg and therefore a 60 kg adult could safely consume 1135 g portions of fish per week with that level. The maximum Pb concentration observed was 5.50 mg/kg and thus consumption of more than 272 g of fish per week exceeds the tolerable weekly intake of Pb. Toxic metal concentrations in this study were considerably higher than those found previously found in fish by Ikem and Egeibo; Iwegbue et al.; Mol; Boadi et al. and Mahalakshmi et al. [5, 11, 25, 33, 38].

Conclusion

The results of this study suggested that significant differences existed in the element concentrations across four different species of canned fish. Also, analytical data obtained from this study shows that the metal concentrations for the varieties of canned fishes especially Cd and Pb were generally higher than the FAO/WHO, FDA and U.S. EPA recommended limits for fish [47, 48]. Therefore both low-risk groups (adolescents and adults) and high-risk groups (pregnant mothers and children) must, based on the results obtained, reduce their consumption of canned fish as frequent consumption may result in bioaccumulation of the metals and increased health risks. Globally, further reduction in the levels of environmental contaminants emanating from power plants and other industrial emissions and effluent discharges are needed to reduce contaminant inputs into the aquatic environment. More research and assessments of seafood quality is needed in many countries to provide more data and help safeguard the health of humans. Therefore, it was concluded that toxic metals in canned fish must be monitored comprehensively and periodically with respect to the consumer health.

Competing interests
The authors declare that they have no competing interests.

Authors' contributions
All authors had equal role in design, work and manuscript writing. All authors read and approved the final manuscript.

Acknowledgements
The authors are grateful to the University of Tehran for providing facilities to conduct and complete this study.

Author details
[1]Department of Fisheries, College of Agriculture & Natural Resources, University of Tehran, Karaj, Iran. [2]Department of the Environment, College of Basic Sciences, Hamedan Branch, Islamic Azad University, Hamedan, Iran. [3]Department of Fisheries, Gorgan University of Agricultural Sciences and Natural Resources, Gorgan, Iran. [4]Department of Natural Resources, Gonbad Kavous University, Gonbad Kavous, Iran. [5]Department of Food Science, Cornell University, Ithaca, New York, USA.

References

1. Tuzen M, Soylak M. Determination of trace metals in canned fish marketed in Turkey. Food Chem. 2007;101:1378–82.

2. Ruelas-Inzunza J, Meza-Lopez G, Paez-Osuna F. Mercury in fish that are of dietary importance from the coasts of Sinaloa (SE Gulf of California). J Food Comp Anal. 2008;21(3):211–8.

3. Bhattacharyya S, Chaudhuri P, Dutta S, Chandra SS. Assessment of total mercury level in fish collected from East Calcutta Wetlands and Titagarh sewage fed aquaculture in West Bengal, India. Bull Environ Contam Toxicol. 2010;84:618–22.

4. Sae S, Dehkordi S, Fallah AA, Nematollahi A. Arsenic and mercury in commercially valuable fish specied from the Persian Gulf: influence of season and habitat. Food Chem Toxicol. 2010;48(10):2945–50.

5. Mol S. Levels of heavy metals in canned bonito, sardines, and mackerel Produced in Turkey. Biol Trace Elem Res. 2011;143:974–82.

6. Mol S. Levels of selected trace metals in canned tuna fish produced in Turkey. J Food Comp Anal. 2011;24:66–9.

7. Mol S. Determination of trace metals in canned anchovies and canned rainbow trouts. Food Chem Toxicol. 2011;49(2):348–51.

8. Ordiano-Flores A, Galván-Magaña F, Rosiles-Martínez R. Bioaccumulation of Mercury in Muscle Tissue of yellowfin tuna, Thunnus albacares, of the Eastern Pacific Ocean. Biol Trace Elem Res. 2011;144(1-3):606–20.

9. Rezayi M, Esmaeli AS, Valinasab T. Mercury and selenium content in Otolithes ruber, and Psettodes erumei from Khuzestan Shore, Iran. Bull Environ Contam Toxicol. 2011;86:511–4.

10. Hosseini SV, Aflaki F, Sobhanardakani S, Tayebi L, Babakhani Lashkane A, Regenstein JM. Analysis of mercury, selenium and tin concentrations in canned fish marketed in Iran. Environ Monit Assess. 2013;185(8):6407–12.

11. Mahalakshmi M, Balakrishnan S, Indira K, Srinivasan M. Characteristic levels of heavy metals in canned tuna fish. J Toxicol Environ Health Sci. 2012;4(2):43–5.

12. Emami Khansari F, Ghazi-Khansari M, Abdollahi M. Heavy metals content of canned tuna fish. Food Chem. 2005;93:293–6.

13. Narin I, Soylak M, Elci L, Dog̃an M. Determination of trace metal ions by AAS in natural water samples after preconcentration of pyrocatechol violet complexes on an activated carbon column. Talanta. 2000;52:1041–6.

14. Soylak M, Saracog̃lu S, Tuzen M, Mendil D. Determination of trace metals in mushroom samples from Kayseri, Turkey. Food Chem. 2005;92:649–52.

15. Ibrahim HS, Ibrahim MA, Samhan FA. Distribution and bacterial bioavailability of selected metals in sediments of Ismailia Canal, Egypt. J Hazard Mater. 2009;168:1012–6.

16. Ibrahim M, Shaltout AA, Atta DE, Jalbout AF, Soylak M. Removal of COOH, Cd and Pb using water hyacinth: FTIR and flame atomic absorption study. J Iran Chem Soc. 2009;6:364–72.

17. Sounderajan S, Kumar GK, Udas AC. Cloud point extraction and electrothermal atomic absorption spectrometry of Se (IV) –3,3 diaminobenzidine for the estimation of trace amounts of Se (IV) and Se (VI) in environmental water samples and total selenium in animal blood and fish tissue samples. J Hazard Mater. 2010;175(1-3):666–72.

18. Ghaedi M, Shokrollahi A, Kianfar AH, Mirsadeghi AS, Pourfarokhi A, Soylak M. The determination of some heavy metals in food samples by flame atomic absorption spectrometry after their separation-preconcentration on bis salicyl aldehyde, 1,3 propan diimine (BSPDI) loaded on activated carbon. J Hazard Mater. 2008;154:128–34.

19. Hlihor RM, Gavrilescu M. Removal of some environmentally relevant heavy metals using low-cost natural sorbents. Environ Eng Manag J. 2009;8:353–72.

20. Kadi MW. Soil pollution hazardous to environment: a case study on the chemical composition and correlation to automobile traffic of the roadside soil of Jeddah city, Saudi Arabia. J Hazard Mater. 2009;168:1280–3.

21. Mendil D, Demirci Z, Tuzen M, Soylak M. Seasonal investigation of trace element contents in commercially valuable fish species from the Black Sea, Turkey. Food Chem Toxicol. 2010;48(3):865–70.

22. World Health Organization (WHO). Health criteria and other supporting information. In: Guidelines for drinking water quality. 2nd ed. Geneva: World Health Organization; 1995. p. 31–388.

23. Tarley CRT, Coltro WKT, Matsushita M, De Souza NE. Characteristic levels of some heavy metals from Brazilian canned sardines (Sardinella brasiliensis). J Food Comp Anal. 2001;14:611–7.

24. Abou-Arab AAK, Ayesh AM, Amra HA, Naguib K. Characteristic levels of some pesticides and heavy metals in imported fish. Food Chem. 1996;57(4):487–92.

25. Iwegbue CMA, Nwajei GE, Arimoro FO, Eguavoen O. Characteristic levels of heavy metals in canned sardines consumed in Nigeria. Environmentalist. 2009;29:431–5.

26. Castro-González MI, Méndez-Armenta M. Heavy metals: implications associated to fish consumption. Environ Toxicol Pharmacol. 2008;26(3):263–71.

27. Dahiya S, Karpe R, Hegde AG, Sharma RM. Lead, cadmium and nickel in chocolate and candies from suburban areas of Mumbai, India. J Food Comp Anal. 2005;18:517–22.

28. Goyer RA. Toxic effects of metals. In: Amdur MO, Douli J, Klansmen CD, editors. Caserrett and Doull's Toxicology: The Basic Science of Poisons. Fourthth ed. New York: Pergamon Press; 1991. p. 623–80.

29. Jarup L, Berglund M, Elinder CG, Nordberg G, Vahter M. Health effects of cadmium exposure- a review of the literature and a risk estimate. Scand J Work Environ Health. 1998;24:1–52.

30. Ashraf W. Levels of selected heavy metals in tuna fish. Arab J Sci Eng. 2006;31(1A):89–92.

31. Ganjavi M, Ezzatpanah H, Givianrad MH, Shams A. Effect of canned tuna fish processing steps on lead and cadmium contents of Iranian tuna fish. Food Chem. 2010;118:525–8.

32. Kagi JH, Schaffer A. Biochemistry of metallothionein. Biochem. 1998;27:8509–15.

33. Boadi NO, Twumasi SK, Badu M, Osei I. Heavy metal contamination in canned fish marketed in Ghana. Amer J Sci Indus Res. 2011;2(6):877–82.

34. Taha'n JE, Sanchez JM, Granadillo VA, Cubillan HS, Romero RA. Concentration of total Al, Cr, Cu, Fe, Hg, Na, Pb, and Zn in commercial canned seafood determined by atomic spectrometric means after mineralization by microwave heating. J Agric Food Chem. 1995;43:910–5.

35. Dabeka RW, McKenzi AD, Albort RH. Atomic absorption spectrophotometric determination of tin in canned foods, using nitric acid-hydrochloric acid digestion and nitrous oxide-acetylene flame: collaborative study. J Assoc Off Anal Chem. 1985;68:209–13.

36. Voegborlo RB, El-Methnani AM, Abedin MZ. Mercury, cadmium and lead content of canned tuna fish. Food Chem. 1999;67:341–5.

37. Ashraf W, Seddigi A, Abulkibash A, Khalid M. Levels of selected metals in canned fish consumed in Kingdom of Saudi Arabia. Environ Monit Assess. 2006;117:271–9.

38. Ikem A, Egeibor NO. Assessment of trace elements in canned fishes (mackerel, tuna, salmon, sardines and herrings) marketed in Georgia and Alabama (United States of America). J Food Comp Anal. 2005;18:771–87.

39. Lourenço HM, Afonso C, Martin MF, Lino AR, Nunes ML. Levels of toxic metals in canned seafood. J Aqua Food Prod Technol. 2004;13(3):117–25.

40. European Union (EU). Commission regulation as regards heavy metals. Amending Regulation 466/2001; 2005., no 78/2005.

41. Eboh L, Mepba HD, Ekpo MB. Heavy metal contaminants and processing effects on the composition, storage stability and fatty acid profiles of five common commercially available fish species in Oron Local Government, Nigeria. Food Chem. 2006;97(3):490–7.

42. Canli M, Atli G. The relationships between heavy metal (Cd, Cr, Cu, Fe, Pb, Zn) levels and the size of six Mediterranean fish species. Environ Pollut. 2005;121:129–36.

43. Turkmen M, Turkmen A, Tepe Y, Tore Y, Ates A. Determination of metals in fish species from Aegean and Mediterranean Seas. Food Chem. 2009;113:233–7.

44. Ministry of Agriculture, Forestry and Fisheries (MAFF). Monitoring and surveillance of non-radioactive contaminants in the aquatic environment and activities regulating the disposal of wastes at sea, Aquatic Environment Monitoring; 1995. Report No. 44.

45. Food and Agriculture/World Health Organization (FAO/WHO). Evaluation of certain food additives and the contaminants mercury, cadmium and lead. Geneva: WHO Technical Report Series; 1972. p. 505.

46. Anonymous: IARC (International Agency for Research on Cancer). Cadmium and cadmium compounds (Group 1), IARC monographs.58. Lyon, France: International Agency for Research on Cancer; 1993.

47. Anonymous. Joint FAO/WHO expert committee on food additives, Sixty-first Meeting Rome, 10-19 June 2003.

48. Aucoin J, Blanchard R, Billiot C. Trace metals in fish and sediments from Lake Boeuf, South Eastern Louisiana. Microchem J. 1999;62:299–307.

49. Yilmaz F. The comparison of heavy metal concentrations (Cd, Cu, Mn, Pb, and Zn) in tissues of three economically important fish (*Anguilla anguilla*, *Mugil cephalus* and *Oreochromis niloticus*) inhabiting Köycegiz Lake-Mugla (Turkey). Turk J Sci Tech. 2009;4(1):7–15.

50. Sobhanardakani S, Tayebi L, Farmany A. Toxic metal (Pb, Hg and As) contamination of muscle, gill and liver tissues of *Otolithes rubber*, *Pampus argenteus*, *Parastromateus niger*, *Scomberomorus commerson* and *Onchorynchus mykiss*. World Appl Sci J. 2011;14(10):1453–6.

51. Sobhanardakani S, Tayebi L, Farmany A, Cheraghi M. Analysis of trace elements (Cu, Cd and Zn) in muscle, gill and liver tissues of some fish species using anodic stripping voltammetry. Environ Monit Assess. 2012;184(11):6607–11.

10

Fluoride adsorption on $\gamma - Fe_2O_3$ nanoparticles

Lakmal Jayarathna[1,2*], Athula Bandara[3†], W.J. Ng[4†] and Rohan Weerasooriya[5†]

Abstract

Background: Fluoride contamination of groundwater, both anthropogenic and natural, is a major problem worldwide and hence its removal attracted much attention to have clean aquatic systems. In the present work, removal of fluoride ions from drinking water tested using synthesized γ-Fe$_2$O$_3$ nanoparticles.

Methods: Nanoparticles were synthesized in co-precipitation method. The prepared particles were first characterized by X-ray diffraction (XRD) and Transmission Electron Microscope (TEM). Density functional theory (DFT) calculations on molecular cluster were used to model infrared (IR) vibrational frequencies and inter atomic distances.

Results: The average size of the particles was around 5 nm initially and showed a aggregation upon exposure to the atmosphere for several hours giving average particle size of around 5–20 nm. Batch adsorption studies were performed for the adsorption of fluoride and the results revealed that γ-Fe$_2$O$_3$ nanoparticles posses high efficiency towards adsorption. A rapid adsorption occurred during the initial 15 min by removing about 95 ± 3 % and reached equilibrium thereafter. Fluoride adsorption was found to be dependent on the aqueous phase pH and the uptake was observed to be greater at lower pH. Fourier transform infrared spectroscopy (FT-IR) was used for the identification of functional groups responsible for the adsorption and revealed that the direct interaction between fluoride and the γ-Fe$_2$O$_3$ particles.

Conclusions: The mechanism for fluoride removal was explained using the dehydoxylation pathway of the hydroxyl groups by the incoming fluoride ion. FT-IR data and other results from the ionic strength dependence strongly indicated that formation of inner-spherically bonded complexes. Molecular clusters were found to be good agreement with experimental observations. These results show direct chemical interaction with fluoride ions.

Keywords: γ-Fe$_2$O$_3$ nanoparticles, Fluoride, FTIR, Adsorption and removal, High efficiency, DFT, Molecular modeling, Gaussian 09

Background

Fluorine is a naturally occurring element in minerals, geochemical deposits and natural water systems and that enters into food chains through either drinking water or eating plants and cereals [1]. Elevated concentrations of fluoride in soil and ground water arising from both natural and anthropogenic activities harm living beings around the world including Sri Lanka. Chemical weathering of some fluoride containing minerals leads to fluoride enrichment in soils and ground water. Discharge of fluoride from some industries, for example semiconductors, steel, etc. are among the main anthropogenic sources of fluoride pollution [2].

Removal of fluoride from water is one of the most important issues due to the effect on human health and environment. But as a necessary dilute element in human body fluoride in drinking water may be beneficial or detrimental depending on its concentration. Namely, dietary intake of fluoride with the concentration of 1 mg/L can prevent particularly skeletal and dental problems [3]. When the fluoride concentration is above this level, it leads to many bone diseases, mottling of teeth and lesions of the endocrine glands thyroid, liver and other organs. Owing to these adverse effects of fluoride, World Health Organization (WHO) accepted the drinking water with fluoride concentration of 1.5 mg/L [4]. In the literature, it was reported that many countries have regions where the water containing more than 1.5 mg/L of fluoride including north central province in Sri Lanka [5].

* Correspondence: lakmalipj@yahoo.co.uk

†Equal contributors

[1]Material Technology Section, Industrial Technology Institute, No 363, Bauddhaloka Mawatha, Colombo 07, Sri Lanka

[2]Chemical and Environmental System Modeling group, Institute of Fundamental Studies, Hanthana Road, Kandy, Sri Lanka

Full list of author information is available at the end of the article

Recently, removal of fluoride from ground water and wastewater has been paid high attention in literature and different materials and methods have been tested. The mostly tested methods are adsorption [6–9], ion exchange [8], precipitation [10, 11], Donna dialysis [12], electrolysis [12] and nanofiltration [10, 12].

Among these methods, adsorption is the most widely used method for the removal of fluoride from water. Though these techniques have been extensively used in worldwide, but due to high cost, that methods are not suitable for field application [1].

Therefore, in recent years considerable attention has been devoted to the study of different types of low-cost and effective materials such as different clays, spent bleaching earths, alum sludge, red mud etc. in this approach, a large number of low-cost materials have been examined for the fluoride removal [5, 13–15]. However, to date, adsorbent of magnetic nanoparticles were reported very little, if any, to removal of fluoride from water solution where as magnetic nanoparticles adsorbent with excellent controllable properties can be developed for separation and removal ions from even very dilute aqueous solutions as the nanoparticles usually undergo modification of its geometric and electronic properties compared to bulk systems leading different pathways for the adsorption of molecules or atoms. Further, if the particles or the adsorbent possess magnetic properties then the main advantage is that the adsorbent can be easily separated using the external magnetic field and will be reused [16–20].

The most important solid surfaces for fluoride adsorption in water are the surfaces of Iron and Aluminum hydroxides, for example magnetite and gibbsite. In turn, adsorption of ions on hydroxide surfaces can affect the pH by influencing adsorption of protons. In the case of fluoride, adsorption of the negative ions enhances proton adsorption and tends to increase the pH. Although the amount of background electrolyte ions involved in this adsorption is generally minimal relative to the amount present in the solution. These effectively uncouple the adsorption of protons and fluoride and make the adsorption of fluoride at variable pH a multi-component process [21, 22].

The electronic and optical properties and the chemical reactivity of small clusters are completely different from the known properties of bulk or at extended surfaces. To overcome such difficulties, complex quantum mechanical models are required to predict the properties with particle size, and typically well defined conditions are needed to compare experiment results with theoretical predictions. The most important techniques in computational modeling are *ab-initio*, semi-empirical and molecular mechanics [23, 24].

Density functional theory (DFT) is a one of the newest approaches in computational modeling. In this method, the energy of the molecule and all of its derivative values depend on the determination of the wavefunction. Even though the wavefunction does not exist as a physically, observable property of an atom or a molecule, the mathematical determination of the wavefunction (within the atomic and molecular orbitals) is a good predictor of energy and other actual properties of the molecule [25].

This in turn was adapted by Kohn and Sham into a practical version of the density functional theory as follows,

$$E[\rho] = T_s[\rho] + \int \rho(r)V(r)d^3r$$

$$+ \frac{1}{2} \int \int \frac{\rho(r_1)\rho(r_2)}{r_{12}} d^3r_1 d^3r_2 + E_{xc}[\rho]$$

Where, T_s is the kinetic energy of the non-interacting system; the second term is the nuclear attraction energy and the third is the classical coulomb self-energy; the last term is the E_{xc} energy. Each of these terms is a function of the function ρ, the electron density, which is itself a function of the three positional coordinates (x, y, and z) [26].

In this work, simple chemical method was used to synthesize magnetic iron oxide nanoparticles and employed to remove fluoride from solutions. Effects of pH and the background electrolyte were studied in the batch process. The FTIR spectroscopy was mainly used to characterize the systems in order to understand the adsorption mechanism of fluoride ions on the nanoparticles. Molecular modeling of the adsorbate-adsorbent interaction is very important to understand the surface complexation. Density functional theory, a type of *ab-initio* methods, applied to examine the atomistic and molecular level understanding of fluoride-γ-Fe_2O_3 interactions.

Methods

All the chemicals used were in analytical grade.

Ferromagnetic iron oxide nanoparticles were synthesized by using modified co-precipitation of ferrous and ferric ions in alkaline medium [20]. Briefly an aqueous solution of Fe ions with molar ratio Fe(II)/Fe(III) = 0.5 was prepared by dissolving 3.25 g $FeCl_3$ and $FeCl_2.4H_2O$ powder in 60 mL of aqueous HCl acid (50 mL deionized water + 10 mL of 1 M HCl) solution. The resulting solution was added drop wise in to 100 mL of 1 M of NaOH solution under vigorous stirring. After all the Fe ions solution was added, the reaction mixture was stirred further to prevent coagulation of particles. Then, obtained colloidal solution was centrifuged at 2500 rpm, and precipitate was washed with deionized water with several times. Finally, precipitate was dried under normal atmospheric conditions.

Characterization of iron oxide particles

Iron oxide particles were characterized by an X-ray diffraction (XRD) with an X-ray diffractometer equipped with a copper anode generating Cu K_α radiation (λ-1.5406 Å). The surface structure, size and morphology were investigated by Transmission Electron Microscope (TEM). Fourier transform-infrared spectroscopy (Nicolet 6700 FT-IR) was also conducted on the particles. The surface area of iron oxide nanoparticles was estimated as 16.5 ± 2.5 m^2/g according to Sears' method, comparable with literature (20.40 m^2/g) [27].

Adsorption characteristics

Batch adsorption studies were conducted by contacting 10 g/L suspension of iron oxide particles with 20 mL of fluoride solution at varying concentrations (10–100 ppm) in polystyrene high-density tubes shaking for a 12 h, which had been shown in preliminary study to ensure equilibrium to be reached. Temperature of adsorption test was ~25 ^{0}C while the pH of the reaction mixture was adjusted in range of 2–12 using 0.1 M NaOH or 0.1 M HNO_3. After shaking, the suspension was subjected for centrifugation and final fluoride ion concentration of the suspension was measured with a specific fluoride ion selective electrode (Orion 9409BN) by using an Orion EA960 auto-titrator. FTIR measurements in DRIFT mode were done on the residue solids obtained from each experiment in order to get insights into the mechanism of the fluoride adsorption on the iron oxide particles.

Molecular modeling

Molecular modeling calculations were performed with "Gaussian 03" computer codes [28]. Models were built with GaussView tools. Molecular structures were determined by searching the potential energy surface for minima with respect to each atomic coordinates using density functional theory (DFT) calculations. Two ferrous atoms hydroxide octahedral cluster used as a basic of γ-Fe$_2$O$_3$ surface because this fragment is large enough to describe fluoride adsorption. These two octahedra connected by two OH bridge as they were in the crystal structure. Cluster configurations was performed using the DFT hybride B3LYP (Becke 3-term correlation functional; Lee, Yang, and Parr exchange function) function with 6-31G (d, p) basic set. Minimum energy structures were verified by calculating IR spectra for any imaginary frequencies (i.e., unstable vibrational modes). Calculated frequencies vs. experimental frequencies was plotted to examine the best-fit scale-factor "m". It can be calculated as,

$$\nu(scaled) = m.\nu(DFT/basissets)$$

Where, m is the scale factor obtained from the slope of the plot, and ν is the calculated frequencies for selected theory/basis set.

Results and discussion

Characterization of nanoparticles: XRD and TEM

First, the crystal structure of synthesized nano particles was investigated by XRD using Cu K_α radiation. Figure 1(a) illustrates the XRD pattern, which matches well with that of γ-Fe$_2$O$_3$ [29]. Six characteristic peaks for γ-Fe$_2$O$_3$ (2θ =31.7^0, 36.7^0, 41.1^0, 53.4^0, 57.0^0 and 62.6^0) marked by their Miller indices (220), (311), (400), (422), (511) and (440), respectively, were observed [16]. As such the prepared particles showed high degree of crystalinity. Figure 1(b) shows the TEM image of the synthesized γ – Fe$_2$O$_3$ nanoparticles. As shown in Fig. 1(b), the powder consists of uniformly distributed spherical nanoparticles with particle size of 5–20 nm range, which is close to the calculated value (14.3 nm) from the XRD pattern. In bulk form, γ-Fe$_2$O$_3$ nanoparticles are spaniel cubic type and TEM figures illustrate high crystalline of the nanoparticles. However, particles are dry, it prefer to agglomerate with neighboring particles to reduce their surface charges and hence increasing the average size. Energy dispersive spectroscopy (EDS) also conformed that the ratio of Fe:O is in 2:3 ratio [30–33].

Measurements on fluoride adsorption

The synthesized γ-Fe$_2$O$_3$ nanoparticles were first characterized by the surface titration in order to get an idea on the point of zero charge (ZPC) and the resulted titration curve is shown in Fig. 2. The observed value of zero point charge (pH$_{zpc}$ = 8.13) suggests the presence of some weakly acidic groups on the surface of the adsorbent γ-Fe$_2$O$_3$ nanoparticles. According to literature data, the calculated pH$_{ZPC}$ $[= (^1/_2)(pK_1 + pK_2)]$ is comparable with experimentally measured pH$_{ZPC}$ (8.13) [34].

Effect of pH of the solution on fluoride removal in different ionic strength

After characterizing the particles with ZPC the effect of pH on the adsorption of Fluoride was investigated. Fluoride adsorption by iron oxide was found to be strongly pH dependent. Adsorption amount decreased with increasing pH up to 4.5 and then remain more or less constant in the pH range of 6.0–10.0 and also that the adsorption remains almost constant regardless of ionic strength, but decreased slightly after pH >10.0. This may indicate the formation of inner-spherically bonded complexes [31, 35].

These results indicate that the adsorbent exhibits a commendable removal capacity in wide range of pH. At lower pH, below pH$_{zpc}$, most of the surface sites are positively

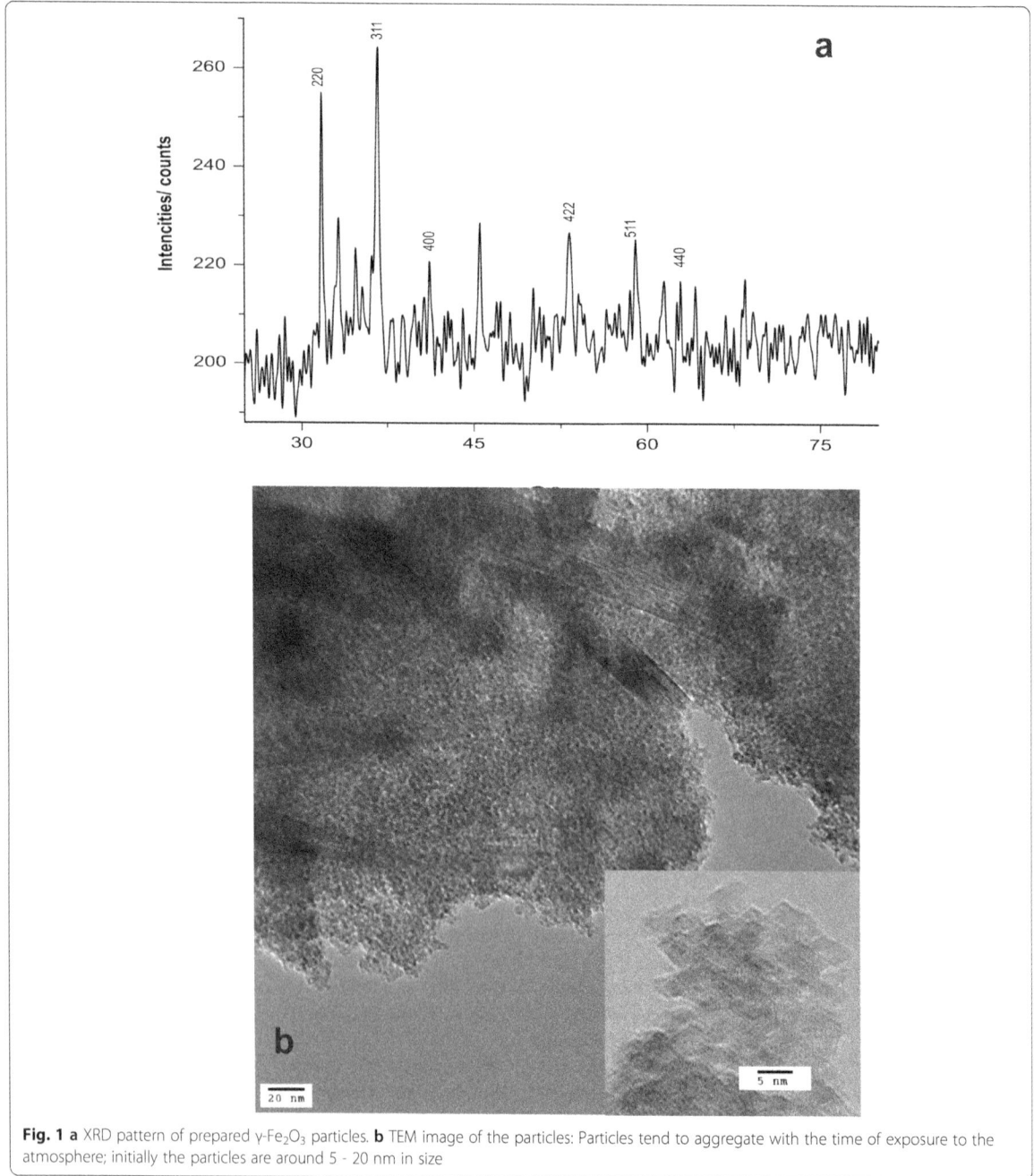

Fig. 1 a XRD pattern of prepared γ-Fe$_2$O$_3$ particles. **b** TEM image of the particles: Particles tend to aggregate with the time of exposure to the atmosphere; initially the particles are around 5 - 20 nm in size

charged and attract negatively charged fluoride easily by electrostatic interaction. However, at very high pH, the removal capacity decreases due to the competition between hydroxide and fluoride ions in this medium [36].

It has also been observed that the removal of fluoride is very rapid in the first 15 min and then reaches a maximum. The percent fluoride removal after 15 min was found to be 95% at the pH 3.6 to 6. The change in the rate of removal might be due to the fact that initially all adsorbent sites were vacant and the solute concentration gradient is high. After 15 min, the fluoride uptake rate by adsorbent had been decreased due to the decrease in number of adsorbent sites. This removal percentage is remarkably higher than the systems reported earlier by using different types of clays [5]. This nature might be due to the modification of particles geometrically and electronically due to its nano size

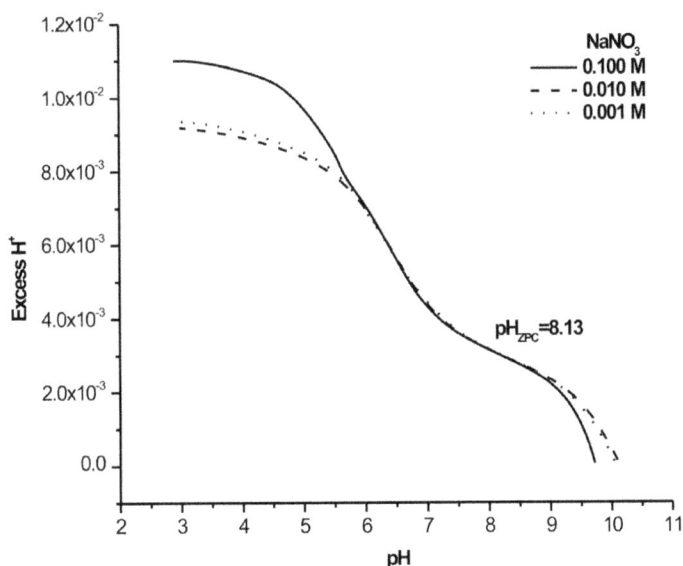

Fig. 2 The graph showing zero-point-charge of the prepared γ-Fe$_2$O$_3$ particles obtained by surface titrations at different NaNO$_3$ ionic strengths

and also the high surface area of the small particles leads high affinity towards adsorption.

FTIR measurements on the bare and fluoride adsorbed particles

FTIR measurements were carried out on the synthesized and fluoride adsorbed particles in order to characterize the systems with their nature of bonding. At first, the FTIR measurements were done on the bare particles to observe the existing functionalities and that the spectrum is shown by a dashed line in Fig. 3. IR absorption bands observed in the range 450–750 cm^{-1} are due to Fe-O bond vibrations and two sharp peaks at ~800 and ~900 cm^{-1} are due to the bending vibrations of O$^{...}$Fe$^{...}$O groups. The broad peak at around 3400 cm^{-1} is due to the hydrogen bonded OH as the surface adsorbed

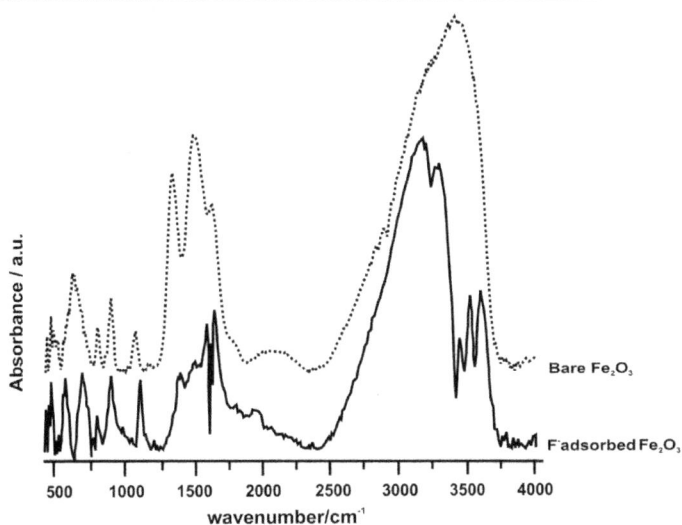

Fig. 3 FTIR spectra of as prepared γ-Fe$_2$O$_3$ particles (dotted line) and the particles treated with saturation of fluoride solution at pH 6 (solid line). These two spectra compare the changes in the system upon adsorption of fluoride in the whole range of 500–4000 cm^{-1}

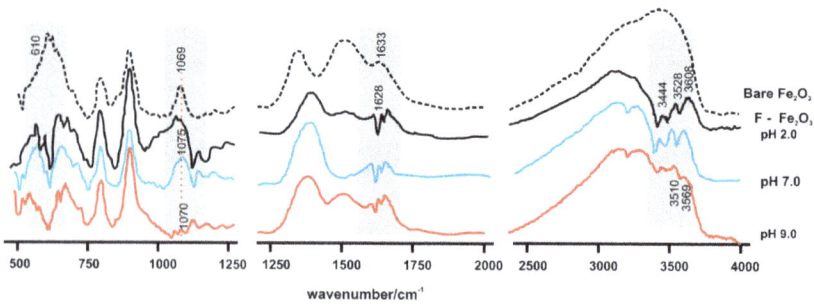

Fig. 4 Series of FTIR spectra collected for the bare γ-Fe₂O₃ particles and those treated with fluoride under 0.001 M NaNO₃. The wavenumber axis is broken into three different ranges of 500–1250, 1250–2000 and 2500–4000 cm⁻¹ for the clarity to see the changes upon fluoride adsorption. The important areas are shaded to emphasis the major changes observed with marking the peak positions where the changes observed

water is present on the particles and this further supported by the bands appearing around 1600 cm^{-1} [29, 32, 37].

Solid line in Fig. 3 shows the FTIR spectrum of fluoride adsorbed iron oxide nanoparticles. Adsorption bands at above 3000 cm^{-1} are well resolved now in comparison with the spectrum of bare-iron oxide. The IR absorption bands between 3300 and 3600 cm^{-1} are due to surface bound OH groups having their characteristic isolated nature. Indeed, a negative band at around 1620 cm^{-1} appeared and another band appeared at 1660 cm^{-1}, revealing the presence of characteristic absorption bands for unbound surface water layer. The spectra shown in Fig. 3 are expanded to three different spectral regions in order to see the changes clearly and the following section deals with the major changes observed in the IR spectra upon the fluoride adsorption.

According to the FT-IR spectral data, several significant changes can be observed in 500–1200 cm^{-1}, 1200–2500 cm^{-1} and 2500–4000 cm^{-1} regions and these are shown in Fig. 4 measured at different pH with the background electrolyte of NaNO$_3$ of 0.001 M. It has to be noted here that the spectra measured without background electrolyte for the fluoride adsorption showed similar pH dependent features except the remaining of some weak OH bands at around 3600 cm^{-1} at pH 9. Some negative absorption peaks appeared in 1200–2000 cm^{-1} region, while distinct features

due to OH stretching modes appeared in the 2500–4000 cm^{-1} region upon the adsorption of fluoride on γ – Fe$_2$O$_3$ nanoparticles. Around 3000 to 3600 cm^{-1}, five new peaks (at around 3182, 3295, 3442, 3528 and 3603 cm^{-1}) observed after adsorbing fluoride. In the bare γ – Fe$_2$O$_3$ spectrum, a broad feature appeared in 2500 to 3700 cm^{-1} is due to hydrogen bonding behavior of surface adsorbed (physisorbed) water and surface > FeOH groups. However, after adsorbing fluoride, hydrogen-bonding strength was diminished and OH groups become much free from hydrogen bonding character and therefore, isolated surface OH vibrations are visible around 3600 cm^{-1}. This nature is further evident by the splitting of the broad band around 600 cm^{-1} reflecting the clear Fe-O-H stretching vibrations free from hydrogen bonding and in addition, the appearance of negative absorption characteristics observed around 1630 cm^{-1} further supports the removal of hydrogen bonded OH sites. When considering the region 500 to 1200 cm^{-1}, absorption band at 1069 cm^{-1} in bare γ – Fe$_2$O$_3$ shifted to higher wavenumbers upon the fluoride adsorption at neutral and acidic pH and the effect was minimum at pH 9. This effect decreased with increasing initial pH and the IR spectrum of the sample of pH 9 shows similar observation made with the bare γ – Fe$_2$O$_3$. Peaks at 1344, 1506 and 1634 cm^{-1} are

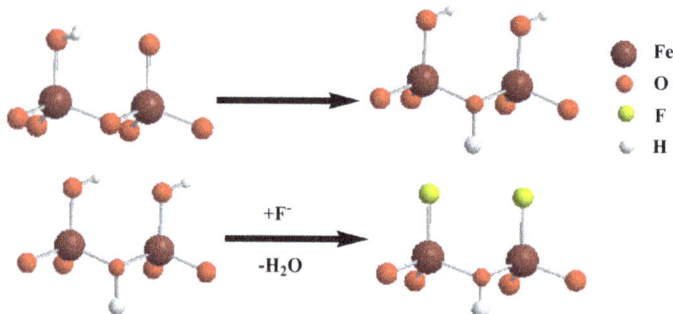

Fig. 5 Proposed mechanism for the adsorption of fluoride on γ-Fe₂O₃ particles

Fig. 6 Fully optimized configurations of $Fe_2(OH)_6(H_2O)_4$, $Fe_2(OH)_5(H_2O)_4F$ and $Fe_2(OH)_4(H_2O)_4F_2$ clusters. Blue color in Fe, Red color in O, Off white color in H and Off green color in F

assign to bending modes of surface adsorbed water in bare iron oxide particles. The remaining characteristic absorption band at around 1660 cm^{-1} for fluoride adsorbed γ − Fe_2O_3 reveals the presence of some isolated OH groups even after the adsorption of fluoride ions. With the collection of all the features observed in the IR spectral ranges described above it is clearly evident that the fluoride adsorption occurs in inner-sphere mechanism as the observations clearly suggest that the direct bonding between surface > FeOH sites and incoming fluoride ion occur by replacing outer-spherically bonded water. This can also be proved by considering adsorption of fluoride in the presence of background electrolyte of $NaNO_3$, where the amount of adsorption of fluoride does not depend on the ionic strength of the system. Further the nature of the particles as they have already modified electronically due to their nano dimensions may also influence the affinity to incoming ions with high efficiency. The changes observed in the IR spectra are negligible over the pH 9 and it is obvious that the particles start to dissolve when the point of zero charge is passed. By combining the pH dependence of IR data and other observations the following scheme is predicted for the adsorption of fluoride on the presently synthesized iron oxide particles [10, 18, 38].

In this scheme (Fig. 5), first, in the hydration step, surface adsorbed H_2O molecules are formed on > FeOH sites of γ-Fe_2O_3 surface. Then, fluoride ions directly attached to the Fe atoms by replacing the OH groups bound to Fe by leaving lattice OH groups (>Fe_2OH). As the size of F^- and OH^- are comparable and having the same charge, removal of OH^- by F^- may not be difficult in this event. Also the higher affinity of F^- than that of the OH^- towards Fe makes the replacement easier. As the Fe⋯F vibrations give rise IR bands in the region below 450 cm^{-1}, that was not observed in the measured region, however, the appearance of negative IR bands around 1600 cm^{-1} and 3640 cm^{-1} clearly supports the above mechanism [34].

Molecular modeling

Optimized clusters of $Fe_2(OH)_6(H_2O)_4$, $Fe_2(OH)_5(H_2O)_4F$ and $Fe_2(OH)_4(H_2O)_4F_2$ were showed in Fig. 6. Fully optimized cluster for $Fe_2(OH)_6(H_2O)_4$ predicts average Fe-Fe and Fe-O bond length of 2.842 and 1.914 respectively.

These bond distances are very much closer to literature data [39].

Table 1 shows optimized average bond distances for $Fe_2(OH)_6(H_2O)_4$, $Fe_2(OH)_5(H_2O)_4F$ and $Fe_2(OH)_4(H_2O)_4F_2$ clusters. Fluorine is highly electronegative and wishes to obtain additional electron density. It attempts to draw it from the other atoms which moves closer together in order to share the remaining electrons more easily as a result.

Cluster models were applied assuming the adsorption process is local phenomena. Such a model describing the surface adsorption sites can give important insights about the structure of the surface complexation. However, two ferrous atoms-hydroxide octahedral cluster ($Fe_2(OH)_6(H_2O)_4$) implemented as a basis of γ-Fe_2O_3 because this fragment is large enough to describe the fluoride adsorption and to avoid any complex time consuming calculation steps.

Comparison between calculated and experimental vibration spectra of $Fe_2(OH)_6(H_2O)_4$, $Fe_2(OH)_5(H_2O)_4F$ and $Fe_2(OH)_4(H_2O)_4F_2$ clusters were shown in Fig. 7. Computed frequencies of the DFT calculation are closely related to the experimental FT-IR vibrations. Calculated OH bending frequencies of iron oxide cluster ($Fe_2(OH)_6(H_2O)_4$) in good agreement with the experimental observation bellow 1000 cm^{-1}. However, Oh stretching frequencies is different due to H-bonding effects. The argument is noteworthy because a significant error is expected due to the fact that harmonic frequencies are calculated whereas anharmonic frequencies are observed. Generally, for more accuracy the scale factor (m) should be closer to one. In some cases, scale factor are actually found to be slightly greater than or

Table 1 Calculated bond parameters for $Fe_2(OH)_6(H_2O)_4$, $Fe_2(OH)_5(H_2O)_4F$ and $Fe_2(OH)_4(H_2O)_4F_2$ clusters

Computed bond distances Å			
Bond type	$Fe_2(OH)_6(H_2O)_4$	$Fe_2(OH)_5(H_2O)_4F$	$Fe_2(OH)_4(H_2O)_4F_2$
Fe-Fe	2.842	2.843	2.489
Fe-O (Bridging)	1.914	1.914	1.869
Fe-O(H)	1.863	1.842	1.880
Fe-O(H_2)	2.066	2.053	2.047
Fe-F		1.895	1.874

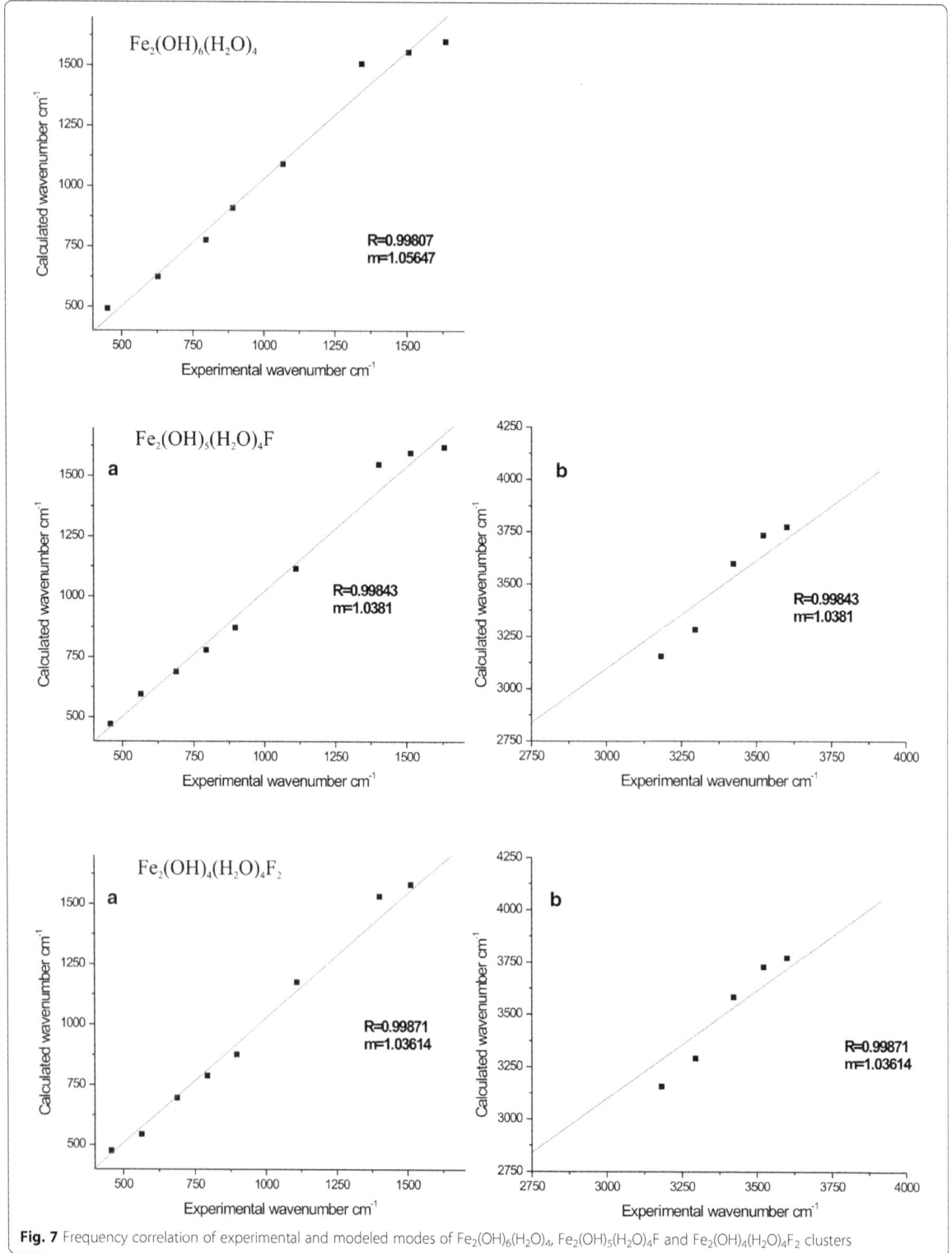

Fig. 7 Frequency correlation of experimental and modeled modes of $Fe_2(OH)_6(H_2O)_4$, $Fe_2(OH)_5(H_2O)_4F$ and $Fe_2(OH)_4(H_2O)_4F_2$ clusters

lower than 1.0. However, in this calculations scale factor very much closer to one [25, 40, 41].

According to the calculated IR spectra of the $Fe_2(OH)_6(H_2O)_4$, $Fe_2(OH)_5(H_2O)_4F$ and $Fe_2(OH)_4(H_2O)_4F_2$ clusters, spectra show higher wavenumber shifting for $Fe_2(OH)_5(H_2O)_4F$ and $Fe_2(OH)_4(H_2O)_4F_2$ clusters. It is good indication for decreases of hydrogen bond strength for fluoride adsorbed clusters.

Conclusion

Magnetic γ – Fe_2O_3 nanoparticles were synthesized by the co-precipitation method and this work confirmed that magnetic γ – Fe_2O_3 nanoparticles possess remarkably high efficiency for the removal of fluoride from drinking water and wastewater. The removal of fluoride capacity was 3.65 mg/g and it is strongly depended on initial pH of solution and the removal level is high as 95% of the removal occurs in acidic to neutral pH. FTIR measurements indicated the formation of inner-spherically bonded complexes and the removal of outer-sphere water molecules paved the way to appear isolated OH groups in the IR spectra upon the adsorption of fluoride and that the data helped us to predict the mechanism for the adsorption. Computational chemistry is very useful tool for studying surface reactions when used in combination with a variety of experimental techniques. Density Functional Theory (DFT) can reproduce the structure and vibration frequencies of bulk γ-Fe_2O_3 and these methods were also applied to predict the surface structure.

Competing interests
The authors declare that they have no competing interests.

Authors' contributions
LJ conducted experiments and manuscript preparation. AB and RW revised the manuscript and WJN facilitated the TEM analysis. All authors read and approved the final manuscript.

Author details
[1]Material Technology Section, Industrial Technology Institute, No 363, Bauddhaloka Mawatha, Colombo 07, Sri Lanka. [2]Chemical and Environmental System Modeling group, Institute of Fundamental Studies, Hanthana Road, Kandy, Sri Lanka. [3]Department of Chemistry, University of Peradeniya, Peradeniya, Sri Lanka. [4]Nanyang Environment and Water Research Institute, Singapore, Singapore. [5]Department of Soil Science, University of Peradeniya, Peradeniya, Sri Lanka.

References
1. Tripathy SS, Bersillon J-L, Gopal K. Removal of fluoride from drinking water by adsorption onto alum-impregnated activated alumina. Sep Purif Technol. 2006;50:310–7.
2. Clarkson WW, Collins AG, Sheehan PL. Effect of fluoride on nitrification of a concentrated industrial waste. Appl Environ Microbiol. 1989;55:240–5.
3. Rahmani A, Rahmani K, Dobaradaran S, Mahvi AH, Mohamadjani R, Rahmani H. Child Dental caries in relation to fluoride and some inorganic constituents in drinking water in arsanjan, iran. Fluoride. 2010;43:180–7.
4. Zhao X, Wang J, Wu F, Wang T, Cai Y, Shi Y, et al. Removal of fluoride from aqueous media by $Fe_3O_4@Al(OH)_3$ magnetic nanoparticles. J Hazard Mater. 2010;173:102–9.
5. Tor A. Removal of fluoride from an aqueous solution by using montmorillonite. Desalination. 2006;201:267–76.
6. Harrington LF, Cooper EM, Vasudevan D. Fluoride sorption and associated aluminum release in variable charge soils. J Colloid Interface Sci. 2003;267:302–13.
7. Srimurali M, Pragathi A, Karthikeyan J. A study on removal of fluorides from drinking water by adsorption onto low-cost materials. Environ Pollut. 1998;99:285–9.
8. Reardon EJ, Wang Y. A limestone reactor for fluoride removal from wastewaters. Environ Sci Technol. 2000;34:3247–53.
9. Zazouli MA, Mahvi AH, Dobaradaran S, Barafrashtehpour M, Mahdavi Y, Balarak D. Adsorption of fluoride from aqueous solution by modified *azolla filiculoides*. Fluoride. 2014;47:349–58.
10. Meenakshi S, Viswanathan N. Identification of selective ion-exchange resin for fluoride sorption. J Colloid Interface Sci. 2007;308:438–50.
11. Liu K, Chen Q, Hu H, Yin Z. Characterization and leaching behaviour of lizardite in Yuanjiang laterite ore. Appl Clay Sci. 2010;47:311–6.
12. Mohapatra M, Anand S, Mishra BK, Giles DE, Singh P. Review of fluoride removal from drinking water. J Environ Manage. 2009;91:67–77.
13. Shams M, Qasemi M, Dobaradaran S, Mahvi AH. Evaluation of waste aluminum filing in removal of fluoride from aqueous solution. Fresenius Environ Bull. 2013;22:2604–9.
14. Shams M, Nabipour I, Dobaradaran S, Ramavandi B, Qasemi M, Afsharnia M. An environmental friendly and cheap adsorbent (municipal solid waste compost ash) with high efficiency in removal of phosphorus from aqueous solution. Fresenius Environ Bull. 2013;22:723–7.
15. Dobaradaran S, Nabipour I, Mahvi AH, Keshtkar M, Elmi F, Amanollahzade F, et al. Flouride removal from aqueous solutions using shrimp shell waste as a cheap biosorbent. Fluoride. 2014;47:253–7.
16. Zhong LS, Hu JS, Liang HP, Cao AM, Song WG, Wan LJ. Self-assembled 3D flowerlike iron oxide nanostructures and their application in water treatment. Adv Mater (Weinheim, Ger). 2006;18:2426–31.
17. Chang Y-C, Chang S-W, Chen D-H. Magnetic chitosan nanoparticles: Studies on chitosan binding and adsorption of Co(II) ions. React Funct Polym. 2006;66:335–41.
18. Ma W, Ya F-Q, Han M, Wang R. Characteristics of equilibrium, kinetics studies for adsorption of fluoride on magnetic-chitosan particle. J Hazard Mater. 2007;143:296–302.
19. Ninjbadgar T, Yamamoto S, Fukuda T. Synthesis and magnetic properties of the γ-Fe_2O_3/poly-(methyl methacrylate)-core/shell nanoparticles. Solid State Sci. 2004;6:879–85.
20. Yu S, Chow GM. Carboxyl group (-CO_2H) functionalized ferrimagnetic iron oxide nanoparticles for potential bio-applications. J Mater Chem. 2004;14:2781–6.
21. Hiemstra T, Rahnemaie R, van Riemsdijk WH. Surface complexation of carbonate on goethite: IR spectroscopy, structure and charge distribution. J Colloid Interface Sci. 2004;278:282–90.
22. Meeussen JCL, Scheidegger A, Hiemstra T, van Riemsdijk WH, Borkovec M. Predicting multicomponent adsorption and transport of fluoride at variable pH in a goethite-silica sand system. Environ Sci Technol. 1996;30:481–8.
23. Löwdin P-O. Some properties of linear functionals and adjoint operators. Adv quantum Chem. 1996;27:371–97.
24. Madejova J, Palkova H, Komadel P. Behaviour of Li^+ and Cu^{2+} in heated montmorillonite: Evidence from far-, mid-, and near-IR regions. Vib Spectrosc. 2006;40:80–8.
25. Lewars EG. Computational chemistry: introduction to the theory and applications of molecular and quantum mechanics. 1st ed. Boston: Springer; 2003.
26. Foresman JB, Frisch A. Exploring chemistry with electronic structure methods: a guide using Gaussian. 2nd ed. Wallingford: Gaussian; 1996.
27. Hwang YS, Liu J, Lenhart JJ, Hadad CM. Surface complexes of phthalic acid at the hematite/water interface. J Colloid Interface Sci. 2007;307:124–34.
28. Frisch MJ, Trucks GW, Schlegel HB, Scuseria GE, Robb MA, Cheeseman JR, et al. Gaussian 09. Wallingford: Gaussian, Inc.; 2009.
29. Durães L, Costa BFO, Vasques J, Campos J, Portugal A. Phase investigation of as-prepared iron oxide/hydroxide produced by sol–gel synthesis. Mater Lett. 2005;59:859–63.
30. Jayarathne L, Ng WJ, Bandara A, Vitanage M, Dissanayake CB, Weerasooriya R. Fabrication of succinic acid-γ-Fe_2O_3 nano core-shells. Colloids Surf A Physicochem Eng Asp. 2012;403:96–102.

31. Bronstein LM, Huang X, Retrum J, Schmucker A, Pink M, Stein BD, et al. Influence of iron oleate complex structure on iron oxide nanoparticle formation. Chem Mater. 2007;19:3624–32.

32. Glisenti A. Interaction of formic acid with Fe_2O_3 powders under different atmospheres: an XPS and FTIR study. J Chem Soc, Faraday Trans. 1998;94:3671–6.

33. Cheng F-Y, Su C-H, Yang Y-S, Yeh C-S, Tsai C-Y, Wu C-L, et al. Characterization of aqueous dispersions of Fe_3O_4 nanoparticles and their biomedical applications. Biomaterials. 2005;26:729–38.

34. Vithanage M, Jayarathna L, Rajapaksha AU, Dissanayake CB, Bootharaju MS, Pradeep T. Modeling sorption of fluoride on to iron rich laterite. Colloids Surf A Physicochem Eng Asp. 2012;398:69–75.

35. Tang Y, Guan X, Wang J, Gao N, McPhail MR, Chusuei CC. Fluoride adsorption onto granular ferric hydroxide: Effects of ionic strength, pH, surface loading, and major co-existing anions. J Hazard Mater. 2009;171:774–9.

36. Sarkar M, Banerjee A, Pramanick PP, Sarkar AR. Use of laterite for the removal of fluoride from contaminated drinking water. J Colloid Interface Sci. 2006;302:432–41.

37. Coessens V, Schacht E, Domurado D. Synthesis of polyglutamine and dextran conjugates of streptomycin with an acid-sensitive drug-carrier linkage. J Controlled Release. 1996;38:141–50.

38. Rao KV, Purushottam D, Vaidyanadham D. Uptake of fluoride by serpentine. Geochim Cosmochim Acta. 1975;39:1403–11.

39. Kubicki J, Paul K, Sparks D. Periodic density functional theory calculations of bulk and the (010) surface of goethite. Geochem Trans. 2008;9:4.

40. Stachowicz M, Hiemstra T, van Riemsdijk WH. Surface speciation of As(III) and As(V) in relation to charge distribution. J Colloid Interface Sci. 2006;302:62–75.

41. Kubicki JD, Kwon KD, Paul KW, Sparks DL. Surface complex structures modelled with quantum chemical calculations: carbonate, phosphate, sulphate, arsenate and arsenite. Eur J Soil Sci. 2007;58:932–44.

11

Diesel biodegradation capacities of indigenous bacterial species isolated from diesel contaminated soil

Nandhini Palanisamy[1], Jayaprakash Ramya[1], Srilakshman Kumar[1], NS Vasanthi[1], Preethy Chandran[2] and Sudheer Khan[1,2*]

Abstract

Petroleum based products are the major source of energy for industries and daily life. Leaks and accidental spills occur regularly during the exploration, production, refining, transport, and storage of petroleum and petroleum products. In the present study we isolated the bacteria from diesel contaminated soil and screened them for diesel biodegradation capacity. One monoculture isolate identified by 16S rRNA gene sequence analysis to be *Acinetobacter baumannii* was further studied for diesel oil biodegradation. The effects of various culture parameters (pH, temperature, NaCl concentrations, initial hydrocarbon concentration, initial inoculum size, role of chemical surfactant, and role of carbon and nitrogen sources) on biodegradation of diesel oil were evaluated. Optimal diesel oil biodegradation by *A. baumanii* occurred at initial pH 7, 35°C and initial hydrocarbon concentration at 4%. The biodegradation products under optimal cultural conditions were analyzed by GC-MS. The present study suggests that *A. baumannii* can be used for effective degradation of diesel oil from industrial effluents contaminated with diesel oil.

Keywords: *Acinetobacter baumannii*, Diesel oil, Parameter optimization, Biodegradation, GC-MS analysis

Introduction

Increasing industrial activities and technological progress have demanded a drastic increase in the use of petroleum hydrocarbons. There is considerable risk of environmental contamination during extraction, transportation, refining, storage, usage and ultimate disposal of these non-aqueous phase liquids composed of a large number of hazardous and toxic constituents [1]. Marine and subsurface environment contamination has been reported due to accidental leakage of ships, cars, trucks, etc., leaks and spills from underground storage tank, pipelines and illegal disposals.

Diesel oil, a complex hydrocarbon pollutant is a mixture of alkanes and aromatic compounds that are reported frequently as soil contaminants [2]. One of the best approaches to restoring contaminated soil is to make use of microorganisms able to degrade those toxic compounds in a bioremediation process [3]. Bioremediation is a cost effective approach for cleaning up petroleum hydrocarbons from contaminated area because it is simple to maintain, applicable over large areas and leads to the complete destruction of the contaminant [4]. Bioremediation of these pollutants are significantly affected by the inherent capabilities of the microorganisms to overcome the bioavailability limitations in multiphase environmental scenarios (oil–water–soil) and environmental factors such as temperature, pH, nutrients and electron acceptor availability [5].

The present study employed bacteria isolated from diesel contaminated soil for the biodegradation of diesel oil. Culture parameters for efficient degradation of diesel oil by a bacterial isolate were optimized.

* Correspondence: ssudheerkhan@gmail.com
[1]Department of Biotechnology, Bannari Amman Institute of Technology, Sathyamangalam, Tamil Nadu, India
[2]CeNTAB, School of Chemical and Biotechnology, SASTRA University, Thanjavur 613401, Tamil Nadu, India

Materials and methods

Materials

Diesel oil used in this study was obtained from local petrol bunk, Tamil Nadu, India and sterilized by filter sterilization. All other chemicals used in the present study were of highest purity in grade.

Isolation, screening and identification

Bacterial species were isolated from diesel contaminated soil (petrol bunk, lorry shed, bus shed and automobile garage) from Sathyamangalam, Tamil Nadu, India [6,7]. Soil samples were serially diluted, 100 μL of the diluted samples spread on the surface of nutrient agar plates and the plates incubated at 37°C for 24 h. The colonies obtained from the agar plates were further sub-cultured to obtain the pure colonies. The bacterial species were screened based on the ability of the bacterial species to degrade diesel oil. The degradation studies were performed in Mineral Salt Medium (MSM) that contains diesel oil as the sole source of carbon. The MSM consists of $NH_4 NO_3$ – 3 g/L, KH_2PO_4 - 0.5 g/L, $K_2HPO_4.3H_2O$ - 0.5 g/L and trace amounts of $MgSO_4.7H_2O$ - 0.008 g/L, $CuSO_4.4H_2O$ - 0.002 g/L, $MnSO_4.H_2O$ - 0.002 g/L, $FeSO_4.7H_2O$ - 0.002 g/L and $CaCl_2.2H_2O$ - 0.002 g/L. The isolated bacterial species from different sites were cultured in Mineral Salt Medium containing diesel oil and incubated for 3 days at 37°C. The growths of the isolated bacterial species were monitored at regular intervals by measuring the optical density at 600 nm. Based on the growth of bacterial species on diesel oil degradation, the best degrader of diesel oil was selected and identified by 16S rRNA gene sequence analysis and used for further studies. Polymerase chain reaction (PCR) was conducted using 16S rRNA forward primer: AGA GTT TGA TCC TGG CTC AG and 16S rRNA reverse primer: ACG GCT ACC TTG TTA CGA CTT. The sequence of the PCR amplicon was submitted to GenBank to obtain accession number. NCBI BLASTN 2.2.26+ comparison software was used to reveal the identities of the isolate.

Effect of pH

The bacterial species were cultured in Luria Bertani (LB) broth and incubated at 37°C for 24 h. Then the culture was centrifuged at 5000 × g for 10 min, the pellets were collected and washed twice with saline to remove the trace amount of LB medium. To study the effect of pH on degradation, MSM supplemented with 1% of diesel oil was adjusted to different initial pH 4-11 using HCl or NaOH. Two hundred and fifty milliliter Erlenmeyer conical flasks containing 100 mL MSM were inoculated with 2 mL of 1.5 OD inoculum size of strain and incubated at 37°C under shaking condition (120 rpm). In the Control, without the inoculation of bacterial species was kept under similar condition. Samples were collected at regular intervals of time (6 h) up to 120 h.

Effect of temperature

The bacterial species was cultured in LB broth and was incubated at 37°C for 24 h. Then the culture was centrifuged at 5000 × g for 10 min, the pellets were collected and washed twice with saline to remove the trace amount of LB medium. To study the effect of temperature on degradation, MSM supplemented with 1% of diesel oil was incubated to different temperature such as 25, 30, 35, 40 and 45°C for optimizing the temperature. Two hundred and fifty milliliter Erlenmeyer conical flasks were inoculated with 1.5 optical density inoculum size which was prepared by diluting the bacterial culture with 0.9% NaCl and optical density value was adjusted to 1.5 using colorimeter. Flasks were kept under shaking condition (120 rpm). At regular intervals, the growth of the bacterial species was measured by recording the absorbance at 600 nm. In the control experiments, the flask containing inoculum devoid of diesel oil and the flasks containing diesel oil devoid of inoculum were run parallel at similar conditions.

Effect of initial hydrocarbon concentration

To study the effect of initial hydrocarbon concentration on degradation, MSM supplemented with different concentrations of diesel oil such as 1-5% was used as substrate. The flasks were inoculated with strain and incubated at 37°C under shaking condition (120 rpm). In the controls, without inoculation of strain were kept under similar conditions with inoculated flasks. Samples were collected at regular intervals of time (6 h) up to 120 h and analyzed for residual diesel oil.

Effect of initial inoculum size

To study the effect of initial inoculum concentration on degradation, MSM supplemented with 1% of diesel oil was inoculated with different inoculum quantity such as 0.1, 0.6, 1 and 1.5 OD (Absorbance at 600 nm). Two hundred and fifty milliliter Erlenmeyer conical flasks were incubated at 37°C under shaking condition (120 rpm). In the controls, without inoculation of strain were kept under similar conditions with inoculated flasks. Samples were collected at regular intervals of time (6 h) up to 120 h.

Role of chemical surfactants

To study the role of chemical surfactants on degradation, MSM supplemented with 4% of diesel oil along with 0.02% (w/v)/ (v/v) of different surfactants such as sodium dodecyl sulfate (SDS) and Tween 80 for enhancing the degradation. Two hundred and fifty milliliter Erlenmeyer conical flasks were inoculated with 1.5 OD inoculum size of strain flasks were inoculated with strain and incubated at 37°C under shaking condition (120 rpm). In the controls, without inoculation of strain, were kept under similar conditions with inoculated flasks. Samples were collected at regular intervals of time (6 h) up to 120 h.

Role of NaCl concentration

To study the role of NaCl concentration on degradation, MSM supplemented with 4% of diesel oil along with different concentrations of NaCl such as 1 mM, 100 mM, 500 mM, 1 M, 2 M and 5 M for checking the salinity condition. Two hundred and fifty milliliter Erlenmeyer conical flasks were inoculated with 1.5 OD inoculum size of bacterial culture. Flasks were inoculated with the bacterial strain and incubated at 37°C under shaking condition (120 rpm). In the controls, without inoculation of strain, were kept under similar conditions with inoculated flasks. Samples were collected at regular intervals of time (6 h) up to 120 h.

Effect of carbon and nitrogen sources

To study the effect of carbon and nitrogen sources on degradation, MSM supplemented with 4% of diesel oil along with 0.1% of carbon source such as Dextrose, Maltose and 0.1% of nitrogen source such Peptone and Yeast extract. Two hundred and fifty milliliter Erlenmeyer conical flasks were inoculated with 1.5 optical density inoculum size of strain was incubated at 37°C under shaking condition (120 rpm). In the controls, without inoculation of strain, were kept under similar condition with inoculated flasks. Samples were collected at regular intervals of time (6 h) up to 120 h.

Gas Chromatography- Mass Spectrometry (GC-MS) analysis

GC-MS analysis was done for detecting the degradation effect of diesel oil [8]. After the incubation period, 5 mL of the cultures were extracted with two 20 mL volumes of n-hexane as a solvent by using separating funnels to remove cellular material. The residues were transferred to tarred vials and the volume of each extract was adjusted to 100 mL by adding further n-hexane. The vials were kept at 4°C until the gas chromatographic analysis. Uninoculated control was incubated in parallel to monitor abiotic losses of the substrate.

The degradation effect of diesel oil was detected by GC-MS (Thermo GC- Trace Ultra ver: 5.0, Thermo MS DSQ II), which was equipped with a DB 35- MS Capillary Standard Non-polar column (30 m × 0.25 mm × 0.25 μm). 1 microliter of the organic phase was analyzed by GC-MS. The gas chromatograph equipped with a split–split less injector (split ratios of 50:1) was used for the GC-MS analysis. The oven temperature was initially at 40°C and then programmed to 270°C at a rate of 8°C/min where it was held for 5 min. The temperatures of injector, transfer line and ionization source were all 250°C. The electron impact ionization was tuned at 70 eV and Helium was used as carrier gas with an average linear velocity of 1.0 mL/min.

Results and discussion

Biodegradation capacity can be evaluated by performing a laboratory study or extensive waste characterization are put together with the bioremediation potential which depends on biodegradability of a specific type of hydrocarbon compound. In this section, isolation of pure cultures and effect of various parameters on hydrocarbon degradation, such as pH, temperature, NaCl concentration, initial hydrocarbon concentration, initial inoculums size, different carbon and nitrogen sources, and chemical surfactant were reported. The roles of biosurfactant effect in contaminant solubilization and biodegradation experiments also have been documented.

Isolation and identification of bacterial species

The isolated bacterial strain was screened based on the ability to utilize diesel oil. The isolate was identified by 16s rRNA analysis and the organism was *Acinetobacter baumannii* (Accession No. JQ975035). It utilizes hydrocarbons as the sole source of carbon and degrades it to maximum extent. Haritash and Kaushik [9] demonstrated that microorganisms are the main degraders of hydrocarbons. There are various abiotic factors which were optimized for maximum degradation of diesel oil.

Effect of pH

Effect of pH for growth of *A. baumannii* with diesel oil was evaluated. The *A. baumannii* showed maximum growth in pH 7. The growth of the bacterial species was decreased while decreasing or increasing the pH. Hence it is understood that neutral pH is required for optimum growth of bacteria, acidic and basic conditions did not favour the growth of *A. baumannii*. Figure 1 shows the growth of bacterial species at different pH in MSM containing 1% diesel oil. According to Whang et al. [10] microbial growth and diesel biodegradation was found to be at a pH 7.2, while decreasing or increasing the pH reduced the degradation efficiency considerably. Xia et al. [11] studied the effect of pH on diesel degradation through the response surface methodology (RSM) using a central composite design and they found the and the optimal biodegradation conditions of diesel oil was pH 7.4. According to Luo et al. [12] at pH level of 7 *Pseudomonas* sp. strain F4 showed efficient diesel oil degradation potential. Hence, the optimization of pH is very important for the enhanced growth of bacteria and also for selection of effective bioremediation strategy. Sathishkumar et al. [13] reported that the optimum pH for the degradation of crude oil by individual bacterial strains and a mixed bacterial consortium was found to be 7.

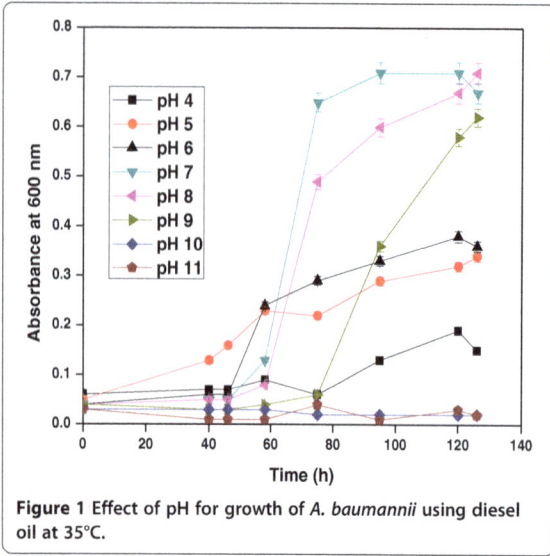

Figure 1 Effect of pH for growth of *A. baumannii* using diesel oil at 35°C.

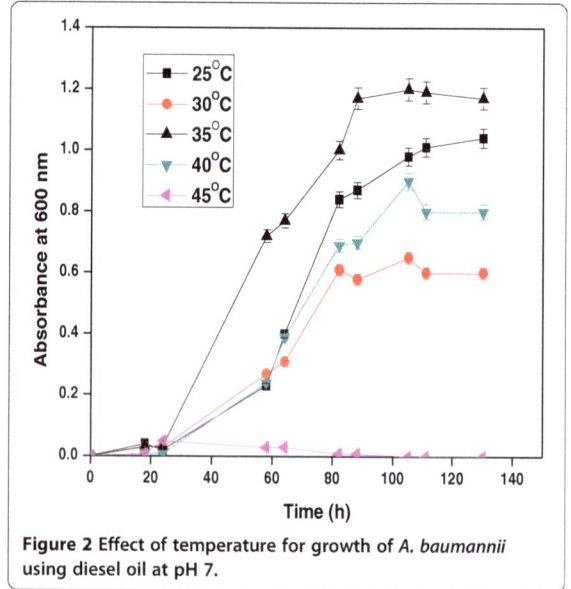

Figure 2 Effect of temperature for growth of *A. baumannii* using diesel oil at pH 7.

Effect of temperature

Effect of temperature for growth of *A. baumannii* with diesel oil was evaluated. We found that temperature is an important factor that affects the diesel degradation potential by bacteria. Mnif et al. [14] found that 30°C was the optimum condition for the degradation of diesel by *Bacillus subtilis* SPB1. Same time, the diesel oil-degrading ability of *Pseudomonas* sp. strain F4 was reported to be 37°C [12]. The present study analyzed the optimum temperature for the degradation of diesel oil and it was found maximum at 35°C. The minimum growth was observed at 45°C. Maintenance of temperature is as important as pH which strongly affects bacterial growth. The growth of *A. baumannii* was directly proportional to diesel oil degradation, since the medium with diesel oil as the sole source of carbon. The degradation efficiency decreased greatly with the increase of temperature. Figure 2 shows the growth of bacterial species at different temperature in MSM containing 1% diesel oil.

Effect of initial hydrocarbon concentration

Effect of initial hydrocarbon concentration for growth of *A. baumannii* with diesel oil was evaluated. Initial hydrocarbon concentrations of 1-5% were used for this degradation studies in which MSM containing 4% of diesel oil showed maximum growth of *A. baumannii*, as shown in Figure 3. Minimum growth of bacterial species was observed in 1% concentration of initial hydrocarbon. At high diesel oil concentration, diesel oil provided a better carbon source for the growth of bacteria Luo et al. [15]. The growth of the bacterial species was increased with increase in diesel concentration. The bacterial species did not possess higher growth at above 4%

diesel oil concentration. The reason for decreased consumption of diesel oil at high concentration may be attributed due to stress of hydrocarbons on bacterial species.

Effect of initial inoculum size

Effect of initial inoculum size for growth of *A. baumannii* with diesel oil was evaluated. MSM medium was inoculated with initial inoculum concentration of 0.1, 0.6, 1.0 and 1.5 optical density measured at 600 nm. Figure 4 shows the effect of inoculum quantity of bacterial species

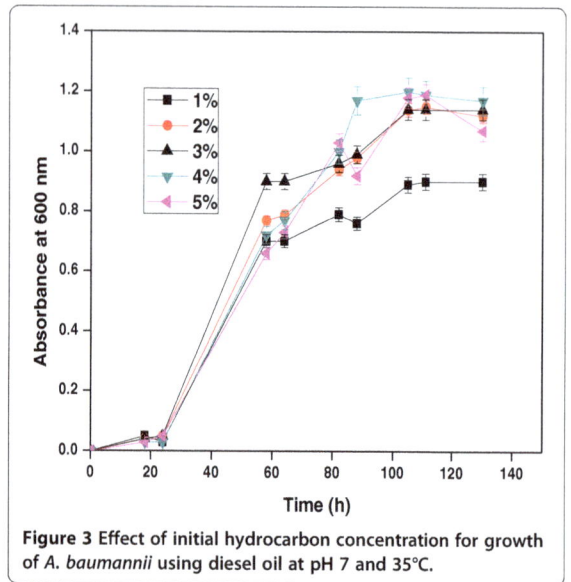

Figure 3 Effect of initial hydrocarbon concentration for growth of *A. baumannii* using diesel oil at pH 7 and 35°C.

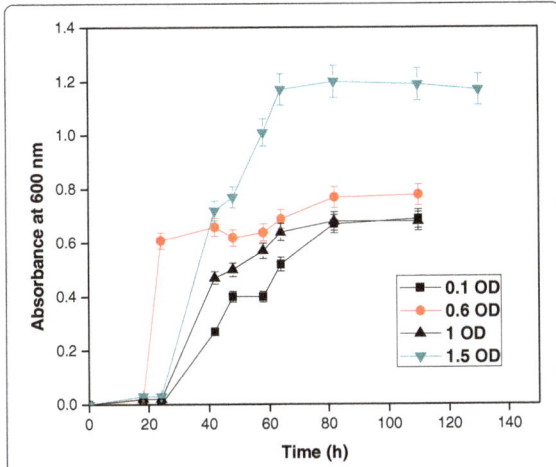

Figure 4 Effect of initial inoculum concentration for growth of *A. baumannii* using diesel oil at pH 7 and 35°C.

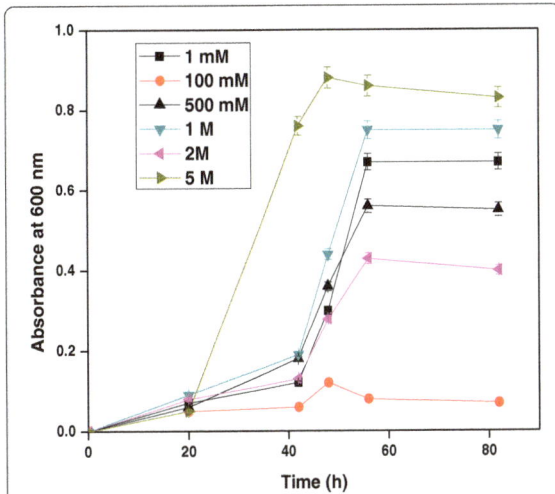

Figure 6 Effect of carbon and nitrogen sources for growth of *A. baumannii* using diesel oil at pH 7 and 35°C.

on diesel oil degradation. The removal of hydrocarbon mainly depends on the capabilities of the microorganisms [16]. When inoculum of *A. baumannii* increased up to 1.5 optical density, lag period for growth of strain decreased and subsequently resulted in higher growth indicates higher diesel oil degradation as depicted in Figure 4. *Pseudomonas* sp. showed efficient diesel degradation efficiency when the inoculums size was adjusted to 2%(v/v) [12]. Luo et al. [15] shows that when the bacteria concentration was 4×10^7 cells/mL, the cell density and biodegradation rate were all the highest after seven days incubation. The removal of hydrocarbon mainly depends on the capabilities of the microorganisms. Lakshmi et al. [17] reported that immobilized (18.6×10^6 cells/mL) and free cells of *Mycoplana* sp. MVMB2, a soil isolate was able

to degrade 95% phenanthrene within 72 h and 120 h respectively.

Effect of NaCl concentrations

Effect of NaCl concentration for growth of *A. baumannii* with diesel oil was evaluated. Metabolic capacities of the bacterial species with respect to salinity were tested for degradation of diesel oil. *A. baumannii* was degraded in the presence of 1 mM, 100 mM, 500 mM, 1 M, 2 M and 5 M NaCl concentration. Increase in NaCl concentration increases the growth of the bacteria in MSM containing 4% diesel oil. Figure 5 shows that variations in salinity had a strong influence on biodegradation which progressively increased when salinity increased. This would be particularly appropriate for removal of stranded oil on beaches or intertidal areas where bacterial species would have to

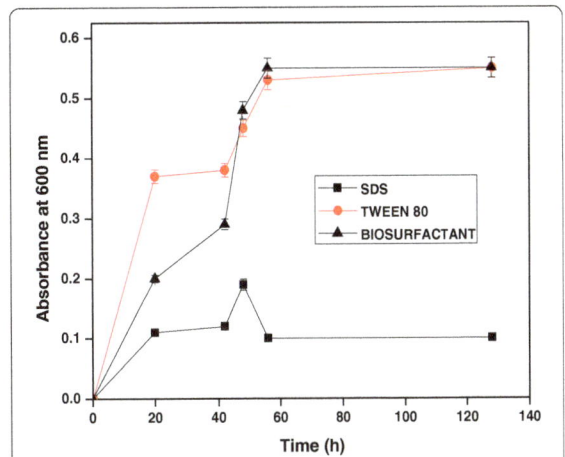

Figure 5 Role of NaCl concentration for growth of *A. baumannii* using diesel oil at pH 7 and 35°C.

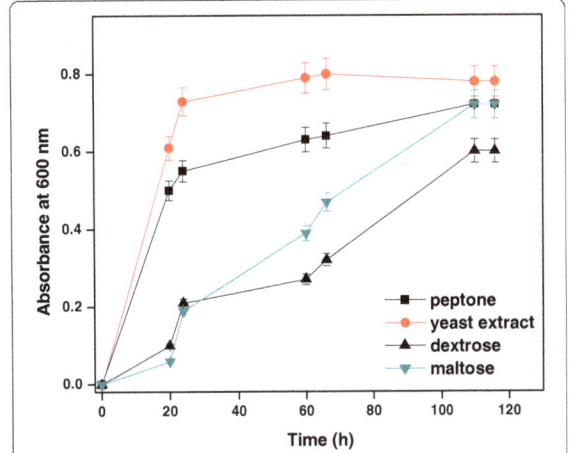

Figure 7 Role of surfactants for growth of *A. baumannii* using diesel oil at pH 7 and 35°C.

survive exposure to high concentrations of sodium chloride, or in estuarine areas where salinity gradients occur. Mnif et al. [18] isolated a novel aromatic-degrading bacterium from a geothermal oil field under saline and thermophilic conditions and the optimum NaCl concentration for degradation was found to be 10 g/L.

Role of carbon and nitrogen sources

Role of carbon and nitrogen sources for growth of *A. baumannii* with diesel oil was evaluated. Various carbon and nitrogen sources such as dextrose, maltose, yeast extract and peptone (0.1%) were added to the MSM medium containing 4% diesel oil as nutrient addition for diesel oil degradation, amongst them peptone and yeast extract were found to enhance diesel oil degradation). Prathima Devi et al [19] reported that addition of external nutrients enhance the degradation of crude petroleum sludge. As shown in Figure 6, growth of *A. baumannii* was increased in the presence of peptone within 120 h. Consequently, peptone was selected as the

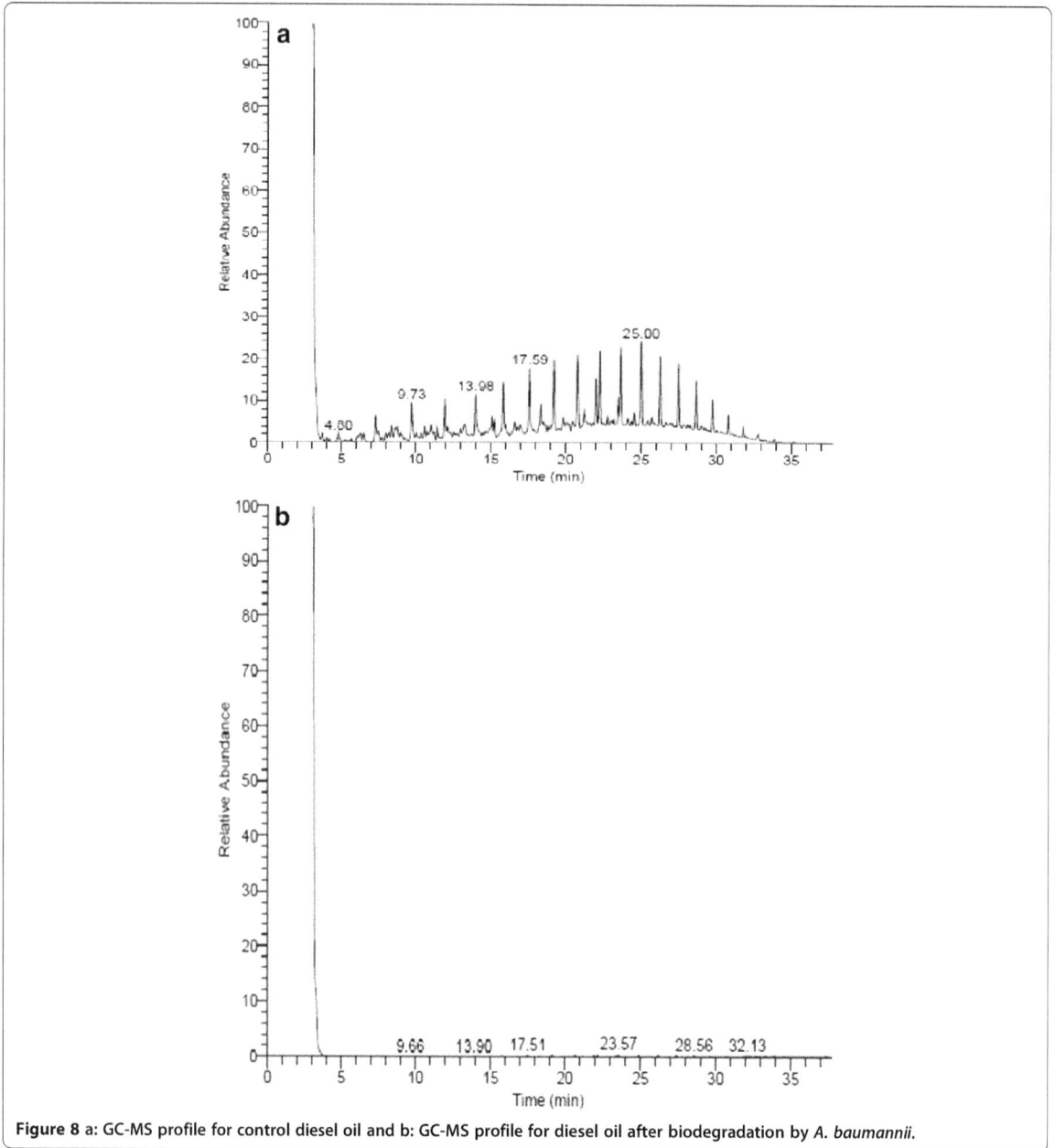

Figure 8 a: GC-MS profile for control diesel oil and b: GC-MS profile for diesel oil after biodegradation by *A. baumannii*.

nitrogen source for further studies. The addition of glucose caused a significant difference in the ability of the diesel degraders to break down diesel of up to 84% [20]. Zahed at al. [21] reported that crude oil removal of 69.5% was observed in presence of 16.05 mg/L nitrogen. Liu et al. [22] reported that addition of carbon source increased the rate of degradation of hydrocarbon by *Rhodococcus* sp. demonstrated that minerals or supplementary carbon substrates increase the rate of biodegradation. NH_3 may be used by microbes as a source of N which undergo nitrification and volatilize. $NO_3^- + NO_2^-$ may be used by microbes as a source of N, undergo denitrification [23]. Carbon sources were not utilized by the bacterial species at earlier stage and there was minimum growth in the presence of carbon sources.

Effect of chemical surfactants

The effect of surfactants (chemical and biological) on diesel oil degradation by bacterial strain is depicted in Figure 7. Tween 80, SDS and biosurfactant enhanced the growth of bacterial species within 72 h compared to the growth of bacterial species without surfactants. However, addition of non-ionic surfactants did not show any significant increase in diesel oil degradation in comparison to control (without surfactants). Bautista et al. [24] had also reported that Tween 80 was the best amongst non-ionic surfactants in improving the degradation of PAHs.

Possibly, *A. baumannii* produced biosurfactants, which increased pseudo solubilization of diesel oil in the medium.

GC-MS analysis

The biodegradation of diesel oil by *A. baumannii* was confirmed by GC-MS analysis. GC-MS chromatogram showed reduction in the intensity of diesel oil peaks after the degradation with *A. baumannii* when compared with control diesel oil (Figure 8). The study shows that *A. baumannii* was able to degrade >99% of the diesel oil within 5 days of incubation at initial pH 7 and 37°C. The previous study achieved degradation of 95.01% diesel by *Trichosporon asahii* [25]. Similarly Chandran and Das [26] reported that free cells of *C. tropicalis* able to degrade 80% of the diesel oil over a period of one week.

Conclusion

The present study focused on the degradation of diesel oil by a bacterial species isolated from diesel contaminated site. The success of oil bioremediation depends on our ability to optimize various physical, chemical, and biological conditions in the contaminated environment. Here, *A. baumannii* was able to degrade more than 99% of diesel oil at pH 7, 35°C and initial hydrocarbon concentration of 4%. The present study may be applied for the efficient removal of diesel oil containing industrial effluents released from petroleum refineries.

Competing interests

The authors declare that they have no competing interests.

Authors' contributions

The overall implementation of this study including experimental design, data analysis and manuscript preparation were done by NP, JR, SK and NSV. PC & Khan critically reviewed and revised the article. All authors read and approved the final manuscript.

Acknowledgement

Authors thank Bannari Amman Institute of Technology, Sathyamangalam and SASTRA University, Thanjavur for providing facility to carry out the research work.

References

1. Kingston PF: Long-term environmental impact of oil spills. *Spill Sci Technol Bull* 2007, **7**:53–61.
2. Gallego JLR, Loredo J, Llamas JF, Vázquez F, Sánchez J: Bioremediation of diesel-contaminated soils: evaluation of potential in situ techniques by study of bacterial degradation. *Biodegradation* 2001, **12**:325–335.
3. Pala DM, De Carvalho DD, Pinto JC, Sant Anna Jr GL: Suitable model to describe bioremediation of a petroleum contaminated soil. *Int Biodeterior Biodegrad* 2006, **58**:254–260.
4. Makkar RS, Rockne KJ: Comparison of synthetic surfactants and biosurfactants in enhancing biodegradation of polycyclic aromatic hydrocarbons. *Environ Toxicol Chem* 2003, **22**:2280–2292.
5. Mukherji S, Vijay A: Critical issues in bioremediation of oil and tar contaminated sites, In: Proceedings of the International Conference on Advances in Civil Engineering. *Civil Eng Dept* 2002, 507-516
6. Holt JG, Krieg NR, Sneath PHA: *Bergey's Manual of Determinative Bacteriology*. 9th edition. Baltimore: Williamsand Wilkins; 1994.
7. Buchanan RE, Gibbons NE: *Bergey's Manual of Determinative Bacteriology*, (8th edition, translated in Chinese) [M]. Beijing: Science Press; 1984.
8. Zhenle T, Lihua Z, Yinghui W, Heqing T: Evalution of biodegradation of petroleum hydrocarbons pollutions by gas chromatography and spectrophotometry. *Chin J Analyt Chem* 2006, **34**:343–346.
9. Haritash AK, Kaushik CP: Biodegradation aspects of Polycyclic Aromatic Hydrocarbons (PAHs): a review. *J Hazard Mater* 2009, **169**:1–15.
10. Whang L-M, Liu P-WG, Ma C-C, Cheng S-S: Application of rhamnolipid and surfactin for enhanced diesel biodegradation-Effects of pH and ammonium addition. *J Hazard Mater* 2009, **164**:1045–1050.
11. Xia W, Li J, Xia Y, Song Z, Zhou J: Optimization of diesel oil biodegradation in seawater using statistical experimental methodology. *Water Sci Technol* 2012, **66**:1301–1309.
12. Luo Q, Zhang J-G, Shen X-R, Sui X, Fan Z-Q: Characterization of a novel diesel oil-degrading pseudomonas sp. strain F4. *Fresenius Environ Bullet* 2013, **22**:689–697.
13. Sathishkumar M, Binupriya AR, Baik S-H, Yun S-E: Biodegradation of crude oil by individual bacterial strains and a mixed bacterial consortium isolated from hydrocarbon contaminated areas. *Clean-Soil Air Water* 2008, **36**:92–96.
14. Mnif I, Sahnoun R, Ellouze-Chaabouni S, Ghribi D: Evaluation of *B. subtilis* SPB1 biosurfactants' potency for diesel-contaminated soil washing: optimization of oil desorption using Taguchi design. *Environ Sci Pollut Res* 2014, **21**:851–861.
15. Luo Q, Xian-Rong S, Jian-Guo Z, Zheng-Qiu F, Ying H: Isolation, identification and biodegradation ability of diesel oil degrading *Pseudomonas* sp. strain C7 from bilge water. *Afr J Microbiol* 2012, **6**:1033–1040.
16. Greenwood PF, Wibrow S, George SJ, Tibbett M: Sequential hydro- carbon biodegradation in a soil from arid coastal Australia treated with oil under laboratory controlled conditions. *Org Geochem* 2008, **39**:1336–1346.
17. Lakshmi MB, Muthukumar K, Velan M: Immobilization of mycoplana sp. mvmb2 isolated from petroleum contaminated soil onto papaya stem (*carica papaya* l.) and its application on degradation of phenanthrene. *Clean-Soil Air Water* 2012, **40**:870–877.
18. Mnif S, Sayadi S, Chamkha M: Biodegradative potential and characterization of a novel aromatic-degrading bacterium isolated from a geothermal oil

field under saline and thermophilic conditions. *Int Biodeterior Biodegrad* 2014, **86**:258–264.

19. Prathima Devi M, Venkateswar Reddy M, Juwarkar A, Nageswara Sarma P, Venkata Mohan SR: **Effect of co-culture and nutrients supplementation on bioremediation of crude petroleum sludg.** *Clean-Soil Air Water* 2011, **39**:900–907.

20. Ganesh A, Lin B: **Diesel degradation and biosurfactant production by Gram- positive isolate.** *Afr J Biotechnol* 2009, **8**:5847–5854.

21. Zahed MA, Aziz HA, Isa MH, Mohajeri L: **Response surface analysis to improve dispersed crude oil biodegradation.** *Clean-Soil Air Water* 2012, **40**:262–267.

22. Liu CW, Chang WN, Liu H-S: *Speedy Hydrocarbon Pollutants Treatment Through the Cell Interaction by a Novel Strain Rhodococcus: Its Fundamental Characteristics and Applications.* Washington, DC; United States: 238th National Meeting and Exposition of the American Chemical Society, ACS; 2009.

23. Delgado JA, Follett RF: **Nitrogen fate and transport in agricultural systems.** *J Soil Water Conserv* 2002, **57**:402–407.

24. Bautista LF, Sanz R, Molina MC, González N, Sánchez D: **Effect of different non-ionic surfactants on the biodegradation of PAHs by diverse bacteria.** *Int Biodeter Biodegr* 2009, **63**:913–922.

25. Ilori MO, Adebusoye SA, Ojo AC: **Isolation and characterization of hydrocarbon degrading and biosurfactant producing yeast strains obtained from lagoon water.** *World J Microbiol Biotechnol* 2008, **24**:2539–2545.

26. Chandran P, Das N: **Degradation of diesel oil by immobilized** *Candida tropicalis* **and biofilm formed on gravels.** *Biodegradation* 2011, **22**:1181–1189.

12

Dietary exposure to tetracycline residues through milk consumption in Iran

Fathollah Aalipour[1], Maryam Mirlohi[2*], Mohammad Jalali[2] and Leila Azadbakht[2]

Abstract

Background: The abundant use of tetracycline antibiotics in veterinary medicine may result in the presence of their residues in milk at unsafe concentrations that can adversely affect public health. The aim of the current study was to evaluate the risk of tetracycline residue (TET) intake via milk consumption amongst different age groups of human consumers in Iran.

Methods: To quantify the drug residues, HPLC analysis was performed under isocratic conditions using UV detection at 355 nm. Milk consumption patterns were determined using a recent nutrition survey in Iran.

Results: The average concentration of total TETs in milk was determined to be 252.41 μg/kg, which is approximately 2.5 times greater than the maximum residue limit (MRL) set by codex. Of the four different tetracycline antibiotics analyzed, oxytetracyclin had the highest share (86 %) of the determined contamination.

Discussion: Daily exposure to TETs through milk using an average data on milk consumption was estimated to range from 58–62 μg. but, distribution based exposure to TETs in milk appeared as 0–99.3 μg per day.

Conclusions: Risk characterization of dietary exposure to TETs residue via milk intake in different age groups showed that considering the standard dietary recommendation that advices on two servings of milk per day (480 ml), consumers may receive
7–30 % of the determined ADI via bovine milk consumption.

Keyword: Risk assessment, Tetracycline, Milk, Exposure, Iran

Background

Tetracyclines (TETs) are a group of broad-spectrum antibacterial agents indicated for use against a large variety of microorganisms including aerobic, anaerobic, Gram-positive and Gram-negative pathogenic bacteria, as well as some types of protozoa. They are widely used in veterinary medicine for prevention and treatment of infectious diseases and also serve as growth promoter additives in animal feed [1]. Although the use of TETs in the treatment of human infectious disease was more common in the past, they are still routinely used, especially in patients susceptible to allergic reactions to ß-lactams and macrolides. Four types of tetracycline antibiotics are routinely used in animal husbandry practice: oxytetracycline (OXT), tetracycline (TET), chlortetracycline (CHT) and doxycyclin (DOT). In cattle, oxytetracycline is frequently applied to treat enteritis, endometritis, septicemia, and mastitis, as well as other bacterial infections [2]. They may be introduced orally through feed or drinking water, or parenterally via intra-mammary infusion [3]. In either case, upon administration, the drug is partially excreted through maternal milk. Its concentration in milk is approximately 70 % of that of the concentration within maternal serum [4–6]. Improper veterinary use of tetracyclines, as well as inadequate knowledge of the necessary withdrawal time, can easily make the tetracyclines or their derivatives appear in the marketed milk.

Intake of contaminated milk can result in adverse human health effects including allergic reactions, the development of bacterial resistance, and the risk of teratogenicity when administered during the first trimester of pregnancy. Moreover, primary and permanent teeth discoloration often occurs when milk

* Correspondence: m_mirlohi@hlth.mui.ac.ir
[2]Food Security Research Center, School of Nutrition and Food Sciences, Isfahan University of Medical Sciences, Hezargrib Street, Isfahan, Iran
Full list of author information is available at the end of the article

contaminated with tetracycline residues is consumed by infants or children less than 12 years old [7].

Regulations have been established by food safety authorities in order to prevent or reduce the negative health effects of tetracyclines on consumers; the Food and Agricultural Organization and World Health Organization (FAO/WHO) and the European Union (EU) have recommended a maximum residue limit (MRL) of 100 µg/kg for tetracycline, oxytetracycline and/or chlortetracycline (singly or in combination) in milk [8–10]. Meanwhile, the U.S. Food and Drug Administration (FDA) has set an upper legal level of 300 µg/kg for the combined residues of TET, OXT and CHT [11]. The Joint Expert Committee on Food Additives has recommended an acceptable daily intake (ADI) for TETs residues at a concentration of 0–30 µg/kg $_{bw}$/day [8].

In Iran, several studies have been published examining the presence of antibiotic residues in milk. Chronologically, the levels of antibiotic residues in milk found to exceed recommended levels have increased from 5 % in the first study completed in 2009 to more than 20 % in the latest one in 2012 [12–15], indicating that in Iran the risk of antibiotic intake through milk consumption must be assessed using both qualitative and quantitative methods. In this regard, previously published information has been insufficient in determining the presence of antibiotic residues in milk as they have primarily focused on the general antibacterial activity, while determination of specific antibiotics was rarely carried out.

The objective of this study was to evaluate the risk associated with the presence of antibiotic residues in marketed milk in Iran. A chemical risk assessment is defined as the likelihood of a chemical having adverse health effects and is divided into four steps: 1) hazard identification; 2) hazard characterization/dose-response assessment; 3) exposure assessment; and 4) risk characterization [16]. In the first component of the present study, the utilization pattern of antibiotic drugs in livestock and poultry farms was surveyed. It was indicated that TETs, in particular oxytetracycline, were in abundant use on both animal and poultry farms [17]. Secondly, a market study was carried out and the contamination rate of pasteurized and sterilized milk (n = 180) from 25 commercial brands offered in retail was investigated. Approximately 20 % of the samples tested were found to have antibiotic residues at levels greater than the legally permitted levels [18].

In the current study, the analytical determination of four types of tetracyclines within the previously identified contaminated milk samples was planned, and then, based on the existing and representative data on the average milk consumption for adults and children, the risk of TETs intake via milk intake in Iran was evaluated.

Method

Milk samples

A total of 187 commercial cows' milk samples, consisting of 33 sterilized and 154 pasteurized products, from 24 Iranian dairy brands were randomly collected from the Shahre-kourd city market from January to July 2012. Sample size was determined based on the previously reported values of antibiotic contaminated milk samples in Iran (5–20 %) [12–15]. Also availability of the commercial brands in the market was considered in sampling. Samples were sent to the Food and Drink Quality Control laboratory of the Shahre-kourd University of Medical Science, and were examined for the presence of antimicrobials.

Screening test

Screening tests were conducted using a commercial kit (Eclipse 100-kit (Zeu. Inmunotec, Spain)) by which the presence of antimicrobial residues could be detected based on the inhibition of microbial activity. According to the manufacturing company, the sensitivity of the test for TETs is between 50–150 µg/kg. Positively detected milk samples were stored at −20°c until HPLC analysis.

Chemical and reagents

Analytical standards of tetracycline antibiotics, including tetracycline hydrochloride, oxytetracyclin hydrochloride, chlortetracyclin hydrochloride and doxycyclin hydrochloride, were provided from Alderck Company. Analytical grade chemicals including acetic acid, sodium chloride, oxalic acid, citric acid, ethylene-diaminetetraacetic acid (EDTA) and disodium hydrogen phosphate (Na2HPO4) were provided from Merck, Germany. Methanol and Acetonitril of HPLC grade were also from Merck-Germany. Solid phase extraction cartridges (500 mg, 6 ml) were provided from Macherey-Nagel Company (Germany).

Preparation of standard curves

Stock solutions of tetracyclines in methanol (Merck-Germany) at the concentration of 100 µg/ml were prepared using pharmaceutical standards; working standards in methanol were made in the concentration rage of 50–400 ng/ml. The working standards were kept in refrigerator for less than one month [19].

Preparation milk samples

Two milliliters of 20 % TCA (trichloroacetic acid-Merck-Germany) and 20 ml of EDTA-McIlvaine buffer were added to 5 ml of milk sample. EDTA-McIlvaine was prepared using 11.8 g of citric acid monohydrate, 13.72 g of disodium hydrogen phosphate dehydrate, and 33.62 g of ethylene-diaminetetraacetic acid disodium salt dissolved in 1 l of double-distilled water (0.01 M). The mixture was then mixed thoroughly and centrifuged at 4000 rpm for 20 min. The resultant supernatant was removed and then applied to the solid phase extraction (SPE HLB C18) cartridge that had been previously activated with 3 ml methanol followed by rinsing with 2 ml double-distilled water. After loading each sample, the cartridge was washed with 2 ml methanol solution (5 %) in double-distilled water, the analytics was eluted with 3 ml of pure methanol, instantly removed under a nitrogen stream, and the residue was resolved in 1 ml mobile phase. In each batch of injection, 10 µl of the given aliquot or standard was filtered through a 0.45-µm micro-filter and injected in to system [20].

Instruments and HPLC condition
The UHPLC-KUNAER system was used (model A69420, Germany) which was equipped with double pumps (model A60015, Germany), UV-vis detector (model MW-1 A61031, Germany) set at 355 nm and auto sampler (model AS-1 A63500, Germany). Separation was carried out under isocratic conditions using a C_{18} Column, 250 mm × 4.6 mm I.D., containing 5 µm particles. The mobile phase consisted of 0.01 M oxalic acid solution, methanol and acetonitril (60:25:15 v/v/v), and was filtered through a 0.45-µm micro-filter at an adjusted flow rate of 1 ml/min.

Validity parameters
Obtained validity parameters, including linearity, limit of detection (LOD), limit of qualification (LOQ), entry-day precision, and intra-day precision, are listed in Table 1. Calibration curves of mixed standard TETs were provided for five concentration levels (50, 100, 200, 300 and 400 ng/ml) in a blank milk sample.

The limit of detection (LOD) and the limit of quantification (LOQ) were determined based on signal to noise ratios. For the entry-day precision, the relative standard deviation (RSD) of four repeated measurements of a blank milk sample containing 500 ng/ml mix standard through a one-day experiment run was considered. For the determination of intra-day precision, the relative standard deviation of the repeated measurements of a milk sample containing 100 ng/ml standard mix in blank milk sample through four subsequent days was taken in to account. A recovery test was performed using the spiked blank milk in three

concentration levels (100, 200, 300 ng/ml) of mix standard. The obtained recovery percentages are presented in Table 2. The highest and lowest recovery rates were obtained for tetracycline (88.78 %) and chlortetracyclin (64.68 %) respectively. Based on the European Commission, Regulation 2002/657/EC, the RSD should be lower than 15 % [21].

Exposure assessment of TETs residues in milk
Exposure assessment was performed using deterministic (point estimate) and probabilistic distribution-based or population-related) approaches for three target groups; infants, children and adults were included.

Per capita milk consumption was obtained from the official report released by the deputy head of Iran Dairy Industries Society (IDIS) [22] as 85–90 kg and used in the deterministic exposure assessment. The reported results of three recent Iranian nutrition surveys were used to derive the lowest, the highest and average milk intake among adults and children [23–25] in the probabilistic exposure estimation. The following equations were used to determine the risk associated with the TETs residue through milk intake:

$$\text{Intake of TETs residue(µg/day)} = \text{Mean residue concentration(µg/kg)}$$
$$\times \text{ daily milk consumption(kg/day)} \tag{1}$$

$$\%ADI = 100 \times \text{Intake (µg/day)}$$
$$\div \left[\left(ADI \left(\text{µg/kg}_{bw}/\text{day}\right) \times \text{body weight (kg)}\right] \right. \tag{2}$$

Statistical analysis
The Statistical Package for Social Science (SPSS) software Version-19 was used for data analysis using descriptive statistics.

Results
The screening test revealed that 19.78 % of the samples were positive for antibacterial residues at levels above

Table 1 Validity parameters in analytical determination of tetracycline residues in milk

Parameter	Oxytetracyclin	Tetracycline	Chlortetracycline	Doxycycline
LOD ng/ml	1.18	1.12	1.7	1
LOQ ng/ml	4	4	5	3
Regression coefficient	0.991	0.994	0.966	0.915
Retention Time (min)	4.56	5.01	7.90	11.78
Inter day precision $N = 3$ day	0.78	1.06	3.09	8.07
Entry day precision	0.59	1.33	3.19	6.13
Linearity equation	Y = 0.001376x	Y = 0.0157643x	Y = 0.0308564x	Y = 0.019085x

Table 2 Recovery (%) of tetracyclines spiked in different concentration levels

Spiked samples	Oxytetracyclin	Tetracycline	Chlortetracycline	Doxycyclin	Total
100 ng/ml	68.67 ± 8.6[a]	90.73 ± 3.4	69.1 ± 3.1	62.4 ± 10.4	72.73 ± 19.1
Precision	0.41[b]	0.69	5.29	19.04	1.02
200 ng/ml	63.67 ± 7.9	84.68 ± 2.3	59.76 ± 2.8	65.64 ± 3.1	67.99 ± 6.1
Precision	0.62	0.84	2.21`	2.34	0.16
300 ng/ml	68.14 ± 13.2	72.93 ± 11.2	65.17 ± 3.5	71.47 ± 5.9	$69.16 \pm 17.$
Precision	0.99	1.63	1.78	2.84	0.46
Mean	66.83 ± 9.9	82.78 ± 5.6	64.68 ± 3.1	66.52 ± 6.4	$69.96 \pm 14.$
Precision	0.78	1.06	3.09	8.07	0.67
n = 20					

[a]Means of recovery data ± standard deviation for 20 measurements
[b]Precision is expressed as the percentage of relative standard deviation (RSD)

the defined MRL. Twenty-eight of the total 178 milk samples (14.97 %) were detected to have residues below the regulatory limit and 122 milk samples negatively responded to the test.

The results of the HPLC determination of TETs residues in the positive samples are presented in Table 3. Amongst the four investigated types of TETs, oxytetracycline distribution frequency and concentration levels were most prevalent, as it was found to be present in all of the tested samples ($n = 37$) and that it covered 86.71 % of the total concentration of measured TET residues. Considering that the standard limit of TET residues in milk defined by MRL-EU is 100 µg/kg [9, 10], the average concentration of the four examined drugs among the positive samples (1090 µg/kg) and in the whole sample population (252.4 µg/kg) were approximately 11 and 2.5 times greater than the standard level, respectively. Oxytetracycline in particular had the highest share in this contamination and its mean concentrations in the positive and total milk samples were 30 and 7.2 times greater than the mentioned permitted limit, respectively. The other TET drug types were detected in low frequency; chlortetracycline and doxycycline residues were found in just five and two out of the 37 samples. Their mean

values were also measured at concentrations below the legislated level.

Using a deterministic approach and considering 233–246 g milk intake per day, the daily intake of TETs residues through long term milk consumption by every Iranian was roughly estimated to be 58–62 µg. While, probabilistic exposure estimation using cross sectional studies revealed that such an intake could be between 8.5–99.3 µg for adults (more than 18 years old) and 0–28 µg for children under 12 years old during a short term exposure. For younger children and infants, since exact data on regular cows' milk consumption was not available, frequency distributed exposure could not be estimated.

In Table 4, the result of risk characterization of TETs residue via milk consumption is presented. Estimated daily exposure data for adults and children is also depicted in this table. The highest level of acceptable daily intake (ADI) of TETs established by JECFA (30 µg/kg $_{bw}$ /day) was taken into account for risk characterization.

Based on the presented results, estimated daily exposure for the high-intake consumers could reach up to about 6 % of the defined ADI at short term exposure. While, considering per capita milk consumption in long time, children are exposed to more possible risk due to their less body weight.

Table 3 Tetracycline residues in commercial milk samples(µg/kg)

	Number of positive samples	Minimum	Maximum	Mean[a]	Each drug ratio to the total TETs concentration
Oxytetracyclin	37	197.65	2137.27	945.90	86.71
Tetracycline	2	<LOD	241.45	28.71	2.63
Chlortetracycline	5	<LOD	288.19	71.93	6.59
Doxycycline	2	<LOD	206.13	44.283	4.05
Total residues in positive samples	37	220.93	2452.06	1090.83	100
Total residues in total samples	187			252.41[b]	

[a]The average concentration in the positive samples
[b]Mean concentration of TET residue in the whole sampled population was calculated as an average of 1090.83 µg/kg for positive samples (20 % of the total share), 100 µg/kg (maximum residue level) for 15 % of the total share and 25 µg/kg for the rest (65 %) which appeared negative in the screening test

Table 4 Risk characterization of dietary exposure to tetracycline residues[a] via milk intake in different age groups

Age group	Body weight (Kg)[b]	Probabilistic								Deterministic	
		Low consumers[c]			Average Consumers[d]			High Consumers[e]		Per capita consumption[f]	
		EDI[g] (μg/Kg$_{bw/d}$)	%ADI		EDI (μg/Kg$_{bw/d}$)	%ADI		EDI (μg/Kg$_{bw/d}$)	%ADI	EDI (μg/Kg$_{bw/d}$)	%ADI
Adults (>18y)	60	0.141	0.47		0.7	2.3		1.65	5.5	0.96	3.2
Children (6 < y < 12)	30	-	-		0.44	1.46		0.92	3.16	4.13	13.7
Children (2 < y < 3)	13	ND	ND		ND	ND		ND	ND	4.43	14.7

The average concentration of tetracycline in the total milk samples (252.4 μg/kg) was taken in to consideration in all cases
Body weights for different age groups were derived from FAO/WHO guideline and relevant nutritional studies
The lowest amount of milk intake among Iranian adults and children were adapted from [24] as 33.6 ml/day and [25] as negligible, respectively
The average amount of milk intake among Iranian adults and children were adapted from [23] as 165 ml/day and [25] as 52 ml/day, respectively
The highest amount of milk intake among Iranian adults and children were obtained from [24] as 393 ml/day and [25] as 113 ml/day, respectively
Per capita milk consumption data is considered as 246 g/d (the upper side of the rage announced by IDIS)
[g]Estimated daily intake (EDI):[Milk consumption X 252.41(mean concentration of TETs in milk)/1000]/body weight

Discussion

Today, in addition to the adverse effects that can occur as a result of the use of veterinary drugs, antibiotic resistance is considered to be a major threat to human health. The presence of antibiotic residues in foods of animal origin have attracted attentions as to whether their long-term intake could be contributing to the development of antibiotic resistance in humans. The present study demonstrated that tetracycline residues in the commercial milk brands marketed in Iran can easily exceed the safety limits recommended by the EU. The average concentration of oxytetracycline, tetracycline, chlortetracycline and, doxycycline residues in the milk samples studied ($n = 187$) were found to be 218.86, 6.63, 16.66, and 10.22 μg/kg, respectively. However, all of the tested samples in the present study were compliant with the standards set by the FDA as the sum of concentration of tetracycline, oxytetracycline, chlortetracycline and doxycycline were less than the MRL set by FDA (300 μg/kg). This difference may be in part due to the different dairy consumption patterns in the US and European zones versus those in Iran, and could be indicative of the necessity to establish national standards based on the dietary intake patterns in Iran.

Recently, several descriptive studies in Iran have been carried out on the presence of tetracycline residues in milk. However, the reported contamination levels were less than those found in the current study. In one of the previous studies which examined 90 pasteurized milk samples, the researchers showed that 7.8 % of the tested samples were contaminated with oxytetracycline and tetracycline residues at concentrations far below the safe limit [26]. Another study investigated the residue levels of three drugs of the tetracycline family including tetracycline, oxytetracycline and chlortetracycline (TCs) in different types of thermally-processed bovine milk collected from Ardebil. The violation rate of 24.4 %, 30 % and 28.6 % for pasteurized, sterilized and raw milk samples were reported respectively, while the overall concentration of tested TETs in milk samples averaged as 97.6 μg/kg

[27]. In a similar study conducted in the U.S, on average, 62.2 μg/kg of TETs were identified in the marketed milk samples, in which 57.2 % of the total detected contamination was due to oxytetracycline [5]. In another study carried out in Ethiopia, each kilogram of raw marketed milk was revealed to contain 142 μg of oxytetracycline residue [28]. In Yugoslavia, Vragović et al. reported that, on average, the examined milk samples were contaminated with 1.5 μg/kg of tetracycline [29]. This was significantly lower than the levels identified in the present study (6.63 μg/kg). This suggests that tetracycline-derivative residues in milk in Iran are more controversial than in the aforementioned countries. However, in a report from Romania, TET contamination of milk averaged 272.5 μg/kg, which is comparable to the results of the current study [30]. Beta-lactam drugs such as penicillin were previously found as a commonly consumed veterinary antibiotic through food in US [31].

In the present study, with respect to the exposure estimation, for children group, the deterministic approach resulted in a higher estimated daily exposure to TETs than the probabilistic model. This could be due to the low level of milk intake in a group of children studied in the used nutritional survey. Although none of the investigated groups in the present study received the TETs residues at a level greater than the higher board of the ADI (30 μg/kg $_{bw}$/day), considering the standard dietary recommendation that advise on 2 servings of milk per day (480 ml), consumers may get 7–30 % of the determined ADI via bovine milk consumption. The estimated risks for children passing their infancy could be several times more than the average of the whole society, when bovine milk is being introduced to them and being replaced by breast-milk.

Since TET residues may also be found in additional food sources, such as eggs and meat, it is possible that a societal group may receive residues of this drug type at a level greater than the defined ADI if several food sources are contaminated with tetracycline residues. A risk assessment study on TET residues through milk consumption carried out by the Environmental Protection Agency revealed that the estimated daily intake of oxytetracycline residues for Americans was 9.6 µg/person/day, which accounted for a low risk [32]. In Yugoslavia, a similar study showed that the hazards associated with tetracycline residue intake via milk was negligible [29].

It should be mentioned that in Iran, the average consumption of milk is lower than the recommended daily allowance (RDA) of two servings (or 500 ml) per day [23]. Therefore, increases in milk and dairy product intake at the advice of nutritionists can increase the risk of TET uptake through the food chain. Until recently, no national or administrative program has been introduced to monitor the presence of drug residues in milk in Iran. The results of the present study can be informative for the safety authorities to instigate new policies to restrict the present potential risk.

Conclusion

In the present study, the average concentration of total TETs in milk was determined as 2.5 times greater than the maximum residue limit (MRL) set by codex and oxytetracyclin had the highest share in this contamination. Daily exposure to Tetracycline residues intake through milk considering long term and short term milk consumption were estimated to range from 58–62 µg and 0–99.3 µg per person.

Competing interests
The authors declare that they have no competing interest.

Authors' contributions
FA carried out laboratory works and drafted the manuscript. MM supervised the work regarding design, experiments and manuscript preparation. MJ participated in manuscript edition and LA contributed in data collection. All authors read and approved the final manuscript.

Acknowledgment
This article is the result of project approved in the Isfahan University of Medical Sciences (IUMS). The authors wish to acknowledge the Vice Chancellery of Research of IUMS for the financial support, Research Project No. 390579.

Author details
[1]Food and Drug Administration, Sharekord University of Medical Sciences, Shahkord, Iran. [2]Food Security Research Center, School of Nutrition and Food Sciences, Isfahan University of Medical Sciences, Hezargrib Street, Isfahan, Iran.

References
1. Kapusnik-Uner JE, Sande MA, Chambers HF. Tetracyclines, chloramphenicol, erythromycin and miscellaneous antibacterial agents. The Pharmacological Basis of Therapeutics, vol. 1. 9th ed. New York (NY): McGraw-Hill Companies; 1996. p. 1123–53.
2. Furusawa F. Rapid liquid chromatographic determination of oxytetracycline in milk. J Chromato A. 1999;839(1–2):247–51.
3. Botsoglou NA, Fletuvris DJ. Drug residue in food; Pharmacology,Food Safety, and Analysis. Chapter 3. New York: Marcel Dekker; 2001. p. 27–117.
4. Elmund GK, Morrison SM, Grant DW, Nevins MP. Role of excreted chlorotetracycline in modifying the decomposition process in feedlot waste. Bulletin of Environmental Contamination and Toxicology. 1971;6:129–32.
5. Fritz JW, Zuo Y. Simultaneous determination of tetracycline, oxytetracycline, and 4-epitetracycline in milk by high-performance liquid chromatography. Food Chemistry. 2007;105(3):1297–301.
6. Kathleen P. Martindale: The Complete Drug Reference. 32th ed, 1 Chapter Antibacterials, UK, London: The Pharmaceutical press. 115–132.
7. Olatoye, IO, Ehinmowo, AA. Oxytetracycline residues in edible tissues of cattle slaughtered in Akure, Nigeria. Nigerian. Nigerian Veterinary Journal. 2010;31(2):93-103.
8. Codexalimentarius. Maximum Residue Limits for Veterinary Drugs in Foods. Updated as at the 35th Session of the Codex Alimentarius Commission 2012, 1–40. http://www.codexalimentarius.net/vetdrugs/data/vetdrugs/index.html?lang=en.
9. European Commission. On pharmacologically active substances and their classification regarding maximum residue limits in foodstuffs of animal origin. Commission Regulation (EU). No 37/2010 of 22 December 2009, 1–76. http://ec.europa.eu/health/files/eudralex/vol-5/reg_2010_37/reg_2010_37_en.pdf.
10. Applegren L, Arnold D, Boisseau J, Boobis A, Ellis R, Livingston R, et al. Evaluation of certain veterinary drug residues in food. Fiftieth report of the Joint FAO/WHO Expert Committee on Food Additives. 1999;888:1–95.
11. US-Food and Drug Administration (FDA): Tolerance for residues of new animal drugs in food. Code of Federal Regulations 2012. http://www.accessdata.fda.gov/scripts/cdrh/cfdocs/cfCFR/CFRSearch.cfm?CFRPart=556&showFR=1
12. Movassagh MH, Karami AR. Determination of antibiotic residues in bovine milk in Tabriz. Iran Global Veterinaria. 2010;5(3):195–7.
13. Sani AM, Nikpooyan H, Moshiri R. Aflatoxin M 1 contamination and antibiotic residue in milk in Khorasan province, Iran. Food and Chemical Toxicology. 2010;48(8):2130–2.
14. Movassagh MH. Study of antibiotics residues in Cow Raw milk by Copan milk test in parsabad region, Ardabil province, Iran. Annals Biol Res. 2011; 2(4):355–9.
15. Movassagh MH. Detection of antibiotics residues in cow raw milk in Bostanabad Region, Iran. Res Opin Anim Vet Sci. 2012;2(1):1–3.
16. Joint FAO/WHO Expert Committee (JECFA). On Dietary Exposure Assessment Methodologies for Residues of Veterinary Drugs. Rome, Italy: Joint FAO/WHO Expert Committee; 2012.
17. Aalipour F, Maryam M, Jalali M. Determination of antibiotic consumption index for animal originated foods produced in animal husbandry in Iran, 2010. J Env Health Sci Eng. 2014;12:42–7.
18. Aalipour F, Mirlohi M, Jalali M. Prevalence of contamination with antibiotic residues in commercial milk as affected by the season and different thermal processing. Int J Env Health Eng. 2013;2(1):41–7.
19. Eliangiringa K, Chambuso M, Kitwala J. Analysis of residual oxytetracycline in fresh milk using polymer reversed-phase column. Food Chem. 2008;107(3): 1289–93.
20. Cinquina AL, Longo F, Anastasi G, Giannetti L, Cozzani R. Validation of a high-performance liquid chromatography method for the determination of oxytetracycline, tetracycline, chlortetracycline and doxycycline in bovine milk and muscle. J Chromato A. 2003;987(1–2):227–33.
21. Bratinova S, Raffael B, Simoneau C: Guidelines for performance criteria and validation procedures of analytical methods used in controls of food contact materials. EUR 24105 EN 2009, 1st ed. 1–74. http://www.irna.ir/en/News/2716240/Social/Iran%E2%80%99s_per_capita_milk_consumption.
22. Azadbakht L, Mirmiran P, Esmaillzadeh A, Azizi F. Dairy consumption is inversely associated with the prevalence of the metabolic syndrome in Tehranian adults. The Am J Clin Nut. 2005;82(3):523–30.
23. Rafraf M, Bazyun B. Food Habits Related To Osteoporosis in Women in Iran. Health promotion perspectives. 2011;1(2):111.
24. Ramezani Tehrani F, Moslehi N, Asghari G, Gholami R, Mirmiran P, Azizi F. Intake of dairy products, calcium, magnesium, and phosphorus in childhood and age at menarche in the Tehran lipid and glucose study. PloS one. 2013; 8(2):e57696.
25. Rassouli A, Abdolmaleki Z, Kamkar A, Shams GHR. A cross- sectional study on oxytetracyclin and tetracyclin residues in pasteurized milk supplied in tehran by an HPLC method. International J Vet Res. 2010;4(1):1–3.
26. Abbasi M, Mesgari B, Hossein A, Masoud N, Ashraf-o S, Nemati M. Simultaneous Determination of Tetracyclines Residues in Bovine Milk

Samples by Solid Phase Extraction and HPLC-FL Method. Ad Pharmac Bulletin. 2011;1(1):34–9.

27. 28- Abebew DS: Detertion and determination of oxytetracycline and penicillin g antibiotic reidue levels in bovine bulk milk from Debrezeit and Nazareth dariy farms. Proceedings of the 1 International Technology, Education and Environment Conference, African Society for Scientific Research 2008. http://www.hrmars.com/admin/pics/241.pdf

28. Vragović N, Bažulić D, Njari B. Risk assessment of streptomycin and tetracycline residues in meat and milk on Croatian market. Food Chem Toxicol. 2011;49(2):352–5.

29. Grădinaru AC, Popescu O, Solcan G. Antibiotic residues in milk from Moldavia. Romania HVM Bioflux. 2011;3(2):133–41.

30. Molina MP, Althaus RL, Balasch S, Torres A, Peris C, Fernandez N. Evaluation of Screening Test for Detection of Antimicrobial Residues in Ewe Milk. J Dairy Sci. 2003;86(6):1947–52.

31. Vragovic N, Bazulic D, Jakupovic E, Zdolec N. Dietary exposure assessment of streptomycin and tetracycline in food of animal origin on the Croatian market. Food Addit Contam: Part B. 2012;5(no):336–240.

13

Biodegradation kinetics and interactions of styrene and ethylbenzene as single and dual substrates for a mixed bacterial culture

Hossein Hazrati, Jalal Shayegan* and Seyed Mojtaba Seyedi

Abstract

This study examined biodegradation kinetics of styrene and ethylbenzene as representatives of alkenylbenzenes and mono-alkylbenzenes, respectively. The compounds were studied independently and as binary mixtures using a series of aerobic batch degradation experiments introduced by acclimatized mix culture. Initial concentration of styrene and ethylbenzene in the liquid phase vacillated from 0 to 220 mg/l. The Andrew model was applied for the biodegradation of individual substrates and the estimated constants of the equation for styrene and ethylbenzene were $\mu_{max} = 0.1581$, 0.2090 (1/h), $K_S = 25.91$, 37.77 (mg/L), $K_I = 13.15$, 62.62 (mg/L), respectively. The accomplished parameters from single substrate degradation tests were used to predict possible interaction factors achieved from dual substrate experiments. The Sum Kinetics with Interaction Parameters (SKIP) model and the purely competitive enzyme kinetics model were employed to evaluate any interactions. The SKIP model was found to accurately describe these interactions. Moreover, it was revealed that ethylbenzene plays an influential role on styrene consumption (e.g. $I_{E,S} = 1.64$) compared to styrene which has insignificant inhibitory effect on ethylbenzene usage (e.g. $I_{S,E} = 0.4$). The active site differences for styrene and ethylbenzene biodegradation and the pathway variations for biodegradation are among the major potential reasons for failure of the estimation that occurred in purely competitive kinetics model. This study is the first to calculate the interactions between styrene and ethylbenzene.

Keywords: Styrene, Ethylbenzene, Mixed culture, Andrews model, SKIP model

Introduction

Nowadays, the elimination of Volatile Organic Compounds (VOCs) from contaminated airstreams and groundwater has become one of the main issues facing the industrialized world [1].

Various methods exist for the treatment of the wastewaters containing VOCs such as physical (e.g. adsorption with the activated carbon) or chemical processes (e.g. Advanced Oxidation Processes). Meanwhile, the biological treatments are gradually becoming popular due to the complete destruction of contaminants [2, 3]. Compared to the physical and chemical methods, biological processes have excessive green benefits and are potentially cost saving either for capital or operating costs. These benefits made biological processes appropriate for the treatment of wastewaters containing

* Correspondence: shayegan@sharif.edu
Department of Chemical and Petroleum Engineering, Sharif University of Technology, Tehran, Iran

various organic pollutants [4]. Two main VOCs that are widely used in many petrochemical and polymer-processing industries are styrene and ethylbenzene. In the chemical industry, styrene monomer plays a pivotal role and it is mainly used in the production of polystyrene and several copolymers [5]. This significant component is produced in large scales through dehydrogenation of ethylbenzene. Despite various advantages, the low conversion rate of styrene and ethylbenzene are considered as residues present in most solvents [6]. These VOCs are known to be hazardous to human health and the environment. The negative effects of their short-term and long-term exposure to human health and environment have been validated in several studies [7, 8]. The Maximum Contaminant Level (MCL) for ethylbenzene and styrene is 0.7 mg/L and 0.1 mg/L, respectively [9]. Hence, to achieve the standard concentration (e.g. below MCL) of styrene and ethylbenzene levels, the liquid and gaseous effluents from petrochemical complexes, polystyrene factories, and plastic industries need to undergo an appropriate treatment before being discharged to the environment.

One of the major steps to forecast and optimize the biodegradation procedures at commercial scales is to specify the degradation kinetics of these contaminants by bacterial populations. Therefore, the employment of an appropriate kinetic model is necessary. For instance, Monod derived models are employed for population growth studies during the microbial growth kinetics [10]. The Monod kinetic model is commonly employed in previous studies where there were pure culture, restricted substrate, and non-inhibitory biomass growth [11, 12]. However, the modified Monod models have been used to investigate the effects of substrate inhibition on biomass growth at large quantity of substrates [13–17].

The styrene biodegradation kinetics have been previously studied and modeled for special isolated microorganisms as well as mixed bacterial cultures. The fungus *E. oligosperma* [18] and some bacteria such as *Rhodococcus pyridinovorans PYJ-1* [19] constitute small part of the microorganisms which are able to treat synthetic wastewater and gases containing styrene. R. Babaee et al. [16] used the industrial activated sludge as a mixed culture to biodegrade styrene in synthetic wastewater. The authors have been successfully modeled and fit the Andrews kinetics to their data [16]. Ethylbenzene biodegradation kinetics were modeled either by special isolated microorganisms or through the mixed bacterial culture as a sole source of carbon and energy. Bacterial strains, including *Pseudomonas fluorescens-CS2* [20], several strains of *Pseudomonas putida* [21], and *Alcaligenes xylosoxidans* [22] were used to biodegrade ethylbenzene. Another research study by H. Shim et al. [23] also evaluated the biodegradation of ethylbenzene in coculture. The study concluded that there is a direct relationship among biodegradation rates of BTEX, BTEX concentration, and the reactor loading rate [23].

The results from previous studies indicate that other substituents of the mixture can be intensely affected by the microbial degradation of a compound [24]. Such interactions can involve the enhancement (positive effect) or inhibition (negative effect) of degradation of substrates in mixtures. However, negative interactions are reported more frequently [25, 26]. This indicates that taking the metabolic influence of each compound into account is vital for further understanding of the mixture effects in microorganisms.

The analytical outcomes assist to develop the applications of the biological systems for efficient VOCs degradation which enhances the energy savings. In addition, there are limited number of studies which model the styrene and ethylbenzene biodegradation mixtures in water. Although in the petrochemical industry these materials coexist with each other, the interaction between

the two substrates (styrene and ethylbenzene as a binary mixture) and the development of the mixed culture in this circumstance have not been modeled to date.

Overall, this study aims to determine the biodegradation kinetic constants under well-defined conditions in the laboratory and quantify the interactions that emerge during the degradation of styrene and ethylbenzene by making use of the interaction equations. It also endeavors to devise a method to calculate the approximate value of biomass growing for a bacterial group.

Materials and methods
Assessment of model adequacy and parameters
Due to suitable and long-term maintenance of the cultures under strong substrate, this study chose the batch cultures to attain biodegradation kinetics, to estimate models, and to define model parameters. Biomass growing can be defined by Eq. (1) in batch situation. This equation is suitable for biomass growth due to different number of substrates [27].

$$\frac{dX}{dt} = \mu X \tag{1}$$

For a low volatility chemical, in a batch degradation for a given substrate, i, the substrate depletion equation was:

$$\frac{dS_i}{dt} = -\frac{\mu_i X}{Y_{X/S_i}} \tag{2}$$

In these expressions, X is biomass concentration (mg/l), t is time (h), $\mu_{(i)}$ is specific growth rate (h^{-1}), S_i is substrate concentration (mg/l), and Y_{X/S_i} is the observed yield coefficient (mg/mg), defined as the proportion of the biomass mass generated to the mass of substrate consumed.

The specific growth rate in Eq. (1) and Eq. (2) could be described by various models. Monod initially suggested the idea that the microbial growth kinetics has been controlled by an empirical model (Eq. 3) [28].

$$\mu_i = \frac{\mu_{\max_i} S_i}{K_{si} + S_i} \tag{3}$$

Where specific growth rate of biomass is μ_i (h^{-1}), μ_{max_i} is the maximum specific growth rate of biomass (h^{-1}), S_i is substrate concentration (mg/l), and K_{si} is substrate half-saturation constant (i.e. substrate concentration at half μ_{max_i}).

A modified version of the Monod model is hired to deliver an improved fit for the achieved data from the sole substrate tests. In this case, Monod derivative (e.g. the Andrews model) shown as Eq. (4) was used for substrate inhibition [28].

$$\mu_i = \frac{\mu_{max_i} S_i}{K_{si} + S_i + \frac{S^2_i}{K_I}} \tag{4}$$

In Eq. (4), μ_{max_i} is the maximum specific growth rate (h^{-1}), K_{si} is the half velocity constant, and K_I is the substrate inhibition constant which quantifies the influence of a toxic compound on its biodegradation. It is important to note that the Andrews model considered as a nonlinear equation. Thus, it is hard to reliably estimate parameters such as linier equations. To overcome this barrier, nonlinear least-squares regression was hired to minimize possible errors during the prediction of mentioned kinetic parameters.

The growing number of substrates led to further complexity of degradation models. Moreover, kinetic parameters for a single substrate are not able to describe the phenomena observed during the degradation of mixtures. Uncompetitive inhibition, non-competitive inhibition and competitive inhibition are some interactions that can take place once multiple substrates are present. One of the most common types of these models is obtained through the summation of specific growth rates on each substrate. In this environment, a sum kinetics model which incorporates purely competitive substrate kinetics could be useful (Eq. 5) [29].

$$
\begin{aligned}
\mu_{tot} &= \mu_1 + \mu_2 \\
&= \frac{\mu_{max_1} S_1}{K_{s1} + S_1 + \left(\frac{K_{s1}}{K_{s2}}\right)S_2} \\
&+ \frac{\mu_{max_2} S_2}{K_{s2} + S_2 + \left(\frac{K_{s2}}{K_{s1}}\right)S_1}
\end{aligned} \tag{5}
$$

Nevertheless, when the ways of interactions among substrates are not thoroughly necessary, use of model which deal with different interactions deprived of their specifics might be the best choice. This model is known as Sum Kinetics with Interaction Parameters (SKIP) and formulated by incorporating $I_{i,j}$ as an unidentified interaction factor.

$$
\begin{aligned}
\mu_{tot} &= \mu_1 + \mu_2 \\
&= \frac{\mu_{max_1} S_1}{K_{s1} + S_1 + I_{2,1}S_2} + \frac{\mu_{max_2} S_2}{K_{s2} + S_2 + I_{1,2}S_1}
\end{aligned} \tag{6}
$$

In Eq. 6, $I_{i,j}$ specifies the degree of impact which the "i" substrate enforces to the "j" substrate biodegradation. According to the model as the amount inhibitory increases, the $I_{i,j}$ (interaction factor) enhances gradually [26]. The value of $I_{i,j}$ is calculated by fitting the SKIP model to a dual of mixture of containments. This is performed by defining some basic parameters related to each substrate such as μ_m, K_s, and $Y_{X/S}$ and replacing them in the cited equation.

Culture and media

A suitable culture with a reliable background plays significant role to properly evaluate the kinetic parameters on growth and substrate removal of any compound [30]. In this study, the industrial mix culture is supplied from a petrochemical complex[1] located in Tabriz (North Western Iran). Tabriz petrochemical complex is a producer of raw polymers such as polystyrene as well as consumed raw materials such as styrene and ethylbenzene. In order to improve the capability of the bacteria and to modify their macromolecular composition (e.g. protein, RNA, and DNA) in response to their environment, the bacterium culture was grown under aeration in the synthetic wastewater. Table 1, shows the components of the carbon-free growth medium formulation. All nutrients used in the growth medium were obtained from Merck Ltd. The mineral media was supplemented with styrene and the ethylbenzene were obtained from Merck Ltd. The concentration ranges of styrene varied from 0 to 220 mg/l while ethylbenzene ranged from 0 to 220 mg/l in single substrate experimental. Both nutrients as carbon sources were equivalent to a COD of 675 ± 10 mg/l and were prepared in test flasks. In dual substrates tests various concentration of styrene and ethylbenzene were used as a binary mixture to set a COD equivalent to 200 ± 10.

Since ammonia was employed as a nitrogen source in biomass cells during the production of main elements, Ammonium chloride (NH_4Cl) was used to supply nitrogen for the medium. The phosphate salts were added to the synthetic medium to provide a buffer capacity and to acts as a source of phosphorus for the microorganisms. EDTA was used in low level of concentration as a chelating agent.

Experiments

To distinctively evaluate the values of each parameter and identify the parameters correctly, kinetic experiments

Table 1 Mineral concentrations in the bioreactor at the beginning and the end of the acclimation step

Constituent	Concentration (mg/L)
NH_4Cl	560
K2HPO4	35
KH_2PO_4	45
$MgSO_4.7H_2O$	13
$CaCl_2.2H_2O$	7
$FeCl_3$	5
$ZnSO_4$	2
$NaHCO_3$	500
EDTA ($C_{10}H_{16}N_2O_8$)	7

need to be performed in a condition with minimum error levels. Therefore, instabilities must be small and/or the test session that leads to the kinetic measurements needs to be short. In addition, another element that intensely impact the estimated value of each parameter in kinetic experiments is the ratio of the initial substrate concentration, S_0, to the initial biomass concentration, X_0, [30]. In our experiments, since inflexibility exists during continuous tests (as mentioned previously in this report), batch kinetic experiments were carried out in 250-mL amber-colored serum bottles. Separate tests were performed for styrene and/or ethylbenzene and the biodegradation of these compounds were examined individually and together.

The first set of experiments involved the use of sole substrate (e.g. styrene or ethylbenzene, separately) to become biodegraded by the mix culture. Therefore, kinetics experiments on nine original concentrations of substrate, from 8 up to 220 mg/L which is the maximum dosage of petrochemical plant (8, 21, 28, 37, 60, 80, 103, 122, 162, 220 mg/L for styrene as well as 12, 23, 32, 42, 64, 83, 98, 130, 158, 220 mg/L for ethylbenzene). Concurrently, the early concentration of biomass for the all bottles is kept static at 25 mg/L. Besides, the biomass concentration has been experimentally observed over time in order to obtain the specific growth rate and to combine this parameter into a model. The concentration of biomass and substrate at different time pauses were detected using the technique cited in the following section (e.g. Section 2.4). The gas phase was also observed randomly during the experiments. 25 mg/L of the biomass in addition to the 100 mL mineral medium were in 250 mL sample bottles (sterile amber-colored serum bottles sealed with Teflon-coated silicone septa and Paraffin layer to prevent volatilization) in the rotary incubator shaker at 160 rpm. Besides, the experiments in which styrene and ethylbenzene were simultaneously biodegraded as mixed substrates, the initial biomass concentration range was 8–12 mg/L. In fact, the initial substrate concentration to the initial biomass concentration ratio ranged from 22.0 to 27.5 on COD basis. This range tolerates determination of the inherent growth related kinetic factors. It characterizes the abilities of the members who belong to the activated sludge with the rapid growth kinetics [30]. In addition, the biomass quantities were selected to diminish the possible errors caused by leaks, to reduce the required time for complete biodegradation, and to eliminate any changes in the characteristics of the biomass caused by long-term contact to the VOCs or some probable by products. The temperature and pH of the aqueous solution were kept stable at 32 °C and 7, correspondingly, until the tests were finalized. It is worth noting that all batch experiments were achieved by concurrent incorporation of similar batches which are free from substrates.

Analytical methods

The GC (Agilent 6890) was set with a Flame Ionizing Detector (FID) and attached to a silica HP-Innowax capillary column (30 m × 0.32 mm × 0.5 lm, J&W Scientific, USA) that was aimed for an appropriate analysis of volatile elements. High-purity helium poured through the column at 1.5 ml/min and 45 psi as a carrier gas. The injector and detector temperatures were fixed at 220 °C and 280 °C. The initial temperature was programmed at 60 °C for 6 min long and it remained constant for 10 min after it was increased to 150 °C at heating rate of 30 °C/min. Once several test time intervals are given, a 25 µL gas-tight syringe employed to extract suitable amount of gas (10 µL) from each serum bottles. The achieved outcome from the gas chromatograph device was documented on a computer fitted out with Agilent data analysis chemstation[2] software to execute peak integration and the related exploration. The achieved output was also compared with the calibration curves of individual components and consequently the VOC concentrations were attained. In order to assess the concentration of carbon source(s) (styrene, ethylbenzene) in the aqueous solutions, partition coefficient of carbon source(s) as well as other equations accosiated with Vapor-Laquid Equilibrium (VLE) was hired. To gain the amount of partition coefficient for each components, specific quantity of styrene and ethylbenzene (0.2, 0.5, 0.75, 2.5, 4.5, and 5 µL) were added independently to 250 mL amber-colored serum bottles holding 100 mL mineral medium. Moreover, to avoid any volatilization of styrene and ethylbenzene, the bottles were sealed by Teflon-coated silicone septa and with Paraffin. Once the proper time was allowed and the VLE circumstance is attain, the styrene or ethylbenzene concentration in the gas phase was evaluated with GC. The partition coefficient of styrene and ethylbenzene was calculated using the mathematical relations and considering the entire volume of VOCs added to serum bottle [19]. To estimate the biomass concentration in the liquid medium the cell concentration suspending in the liquid was examined. The results from fresh culture medium were used as an index in order to compare and evaluate the obtained results from samples with various amount of biomasses. The Optical Density (OD) was determined at 600 nm using a spectrophotometer (Spectro Direct 712000; Lovibond®, Tintometer® Ltd, England). The output was plotted in a graph as a standard curve for desiccated mass of biomass per volume, mg/l, against different quantities of the ODs' resulted at 600 nm. As a consequence, with the acceptable total regression ($R^2 = 0.9979$) a linear curve was established up to concentration of 250 mg/l. To discover other crucial factors such as TSS, VSS, MLSS, MLVSS, and the COD analysis were accomplished according to the standard methods [31].

During the tests time the pH value was fixed and measured by a pH-meter (SCHOTT® CG825). All quantities were prepared using the same method if necessary.

Results and discussions
Single substrate experiments and biodegradation kinetics
The effect of single substrates concentration
To handle kinetic experiments, a specific amount of styrene and/or ethylbenzene (deeply discuss in section 2.4) were added to the serum bottles via a 10 µL syringe inserted through the Paraffin protective layer into Teflon cap. The concentration of gaseous phase was monitored during incubation until complete degradation of styrene was attained.

Figure 1a and b show the variation of styrene and ethylbenzene concentration in the liquid phase during the batch growths over different time intervals. Various dosages (e.g. 8 up to 220 mg/L for styrene and 12 up to 220 mg/L for ethylbenzene mg/L) were degraded by the microbial community presented in the industrial activated sludge from 2 up to almost 127 h for styrene and from 2 up to 106 h for ethylbenzene. The temperature

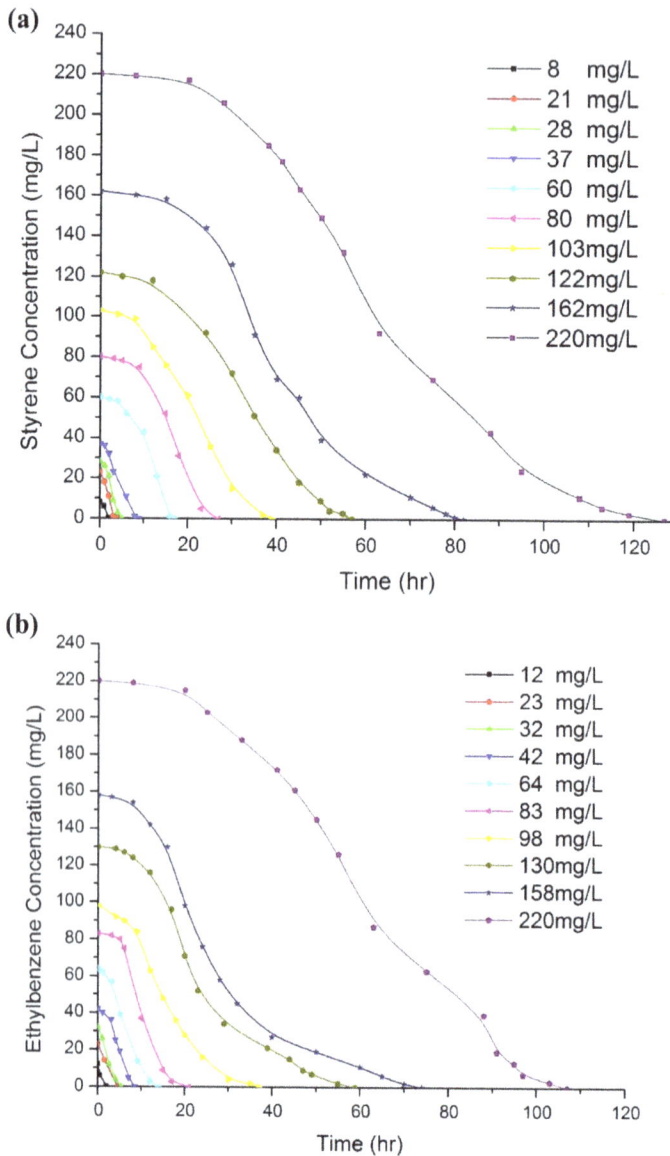

Fig. 1 a The effect of various initial concentrations of styrene in the liquid phase on degradation period, **b** The effect of different preliminary concentrations of ethylbenzene in the liquid phase on depletion period

was fixed at 32 °C and the pH level was 7 for styrene and ethylbenzene. That degradation rate depends on the initial concentration. In addition, inhibition at high concentration diminishes the biodegradation rate [16, 19]. A similar explanation could be drawn from Fig. 1. The figure shows that as the initial concentration grows, the slope of the curve constantly starts to decrease. Moreover, due to the enrichment of inhibition by substrate, the amount of substrate intake declines with the increment of original concentration.

The kinetics of biodegradation
Generally, the kinetic model factors are obtained by observing the biomass growth rate over time at different

initial substrate concentrations during batch experiments. Consequently, if endogenous decay was abandoned, Eq. 7 in exponential growth phase could be used to calculate the specific growth rate (μ_g) values [32].

$$\ln\left(\frac{X}{X_0}\right) = \mu_g.t \qquad (7)$$

Where X_0 and X indicate the biomass level at the beginning and at time t; (mg/l) respectively. μ_g is the specific growth rate (h^{-1}) and t represents the specific time (h). Hence, to calculate the value of the specific growth rate (μ_g), the biomass concentration has been experimentally observed during the batch experiments with

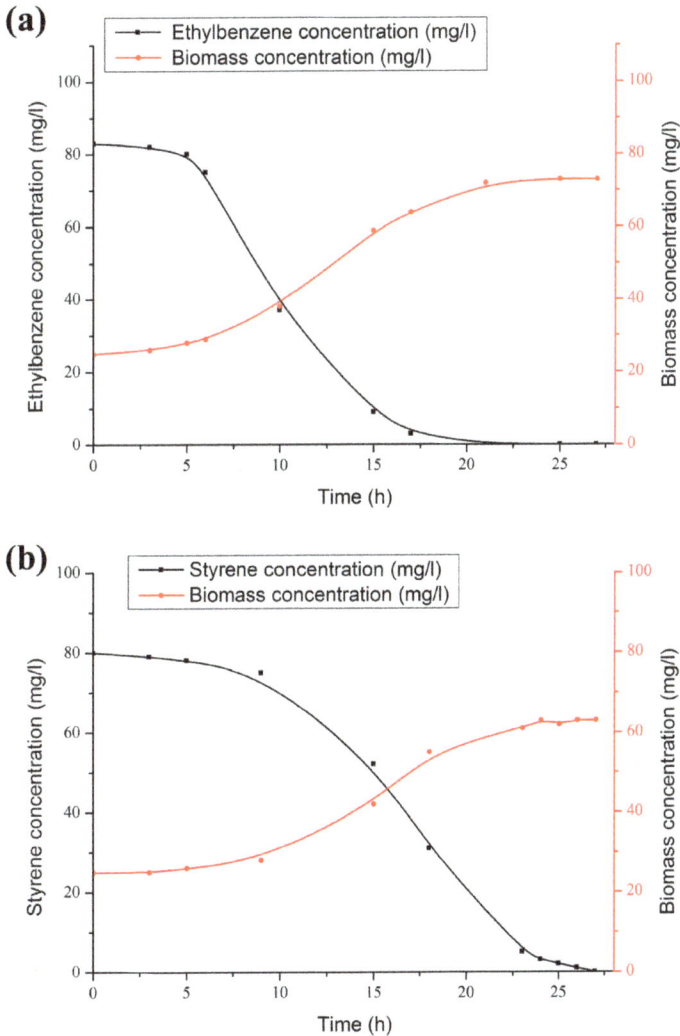

Fig. 2 a The experimental data for ethylbenzene and biomass concentrations in liquid phase taken from single substrate tests; **b** The experimental data for styrene and biomass concentrations in liquid phase taken from single substrate tests

different substrate concentration and fixed initial concentration of biomass.

The kinetics of styrene or ethylbenzene consumption as sole carbon sources by the mixed bacterial cultures which exist in industrial activated sludge is revealed in Fig. 2. The ethylbenzene degradation movements and the parallel biomass progress during the time variation is illustrated clearly in Fig. 2a. The figure specifies that ethylbenzene is fully consumed by nearly 21 h and the sludge growth was imitated by the ethylbenzene consumption. Besides, the sludge growth was apparent after a short lag period (less than 5 h). Unlike the ethylbenzene biodegradation, the data achieved from the styrene biodegradation experiment (Fig. 2b) reveals longer lag period (nearly 9 h) prior to biomass development (Fig. 2a). Also, the styrene consumption was considerably slower (27 h), and similar to the ethylbenzene its growth stopped while styrene concentration was depleted.

Styrene

Figure 3 depicts the specific growth rates (μ_g) plotted against various initial concentrations of styrene as an individual source of carbon. As can be seen, the overall trend shows a similar pattern for the obtained results. The specific growth rate continues to reach the peak value before starting to decline. This indicated the inhibitory effect of substrate above a certain concentration for bacterial activity. Similar to previous researches, trends of graphs and slope of the curves in the second part of plots (after peaks) revealed that as the concentrations of substrates increase, styrene start to make barriers for the bacterial activity [25, 33]. According to the figure, the value of maximum specific growth rates for

styrene is 0.042 h^{-1}. In addition, the initial concentration of styrene is about 21 mg/l when the peaks achieved in the plots.

The parameters of Andrews model evaluated by curve fitting toolbox provided in Matlab 7.14 software to minimize the least-square error. An acceptable fitness occurs between the parameters of Andrews equation and the experimental data for the mix culture. Details of parameters obtained for styrene are shown in Table 2.

The R^2 and RMSE parameters approved that the Andrews model has appropriate qualification for our experimental data. Also, former studies of batch operation situations exhibited that the Andrews model has further superior substrate inhibition effect on the cell growth compared to the Monod model when the initial concentration of toxic substrates is increasing (typically S_0 of beyond 30 mg/L^{-1}) [34]. The μ_{max} value shows the capability of microbial culture to use the special pollutant as a source of carbon and energy. As can be seen in Table 3 μ_{max} obtained for styrene is relatively at low level compared to the prior researches that considered the various pure and mix cultures to evaluate the kinetic of biodegradation [35, 36]. This appears to be reasonable enough since the utilized activated sludge constitutes a mixture of several other microorganisms. The abundance of the microorganisms leads to a competition between the bacterium cultures for the common substrate.

"K_s" or the value of half-saturation constant depends on the affinity of a bacterium for a substrate. In other words, bacterium activity with a lower K_s value could be further efficient to eliminate the pollutant compared to the upper K_s value. According to Table 3, the obtained values for K_s are relatively high compared to the

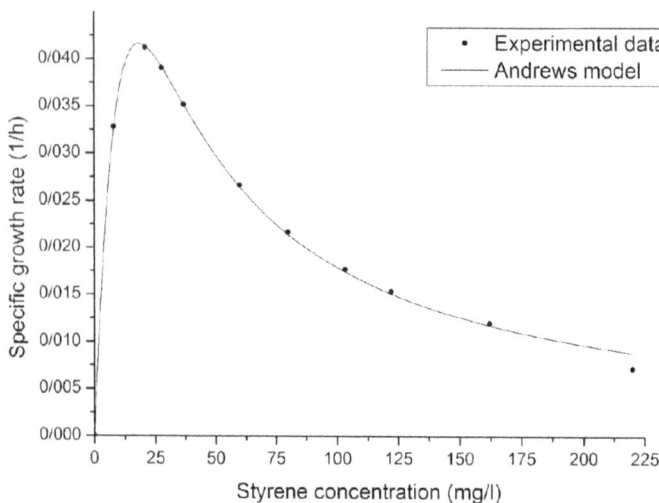

Fig. 3 Specific growth rate of the mixed culture at various concentrations of styrene as a sole substrate. Observed experimental data (shapes) and simulation outcomes acquired via Andrews model (lines)

Table 2 Estimated parameter values for Andrews kinetic models to biodegrade styrene as a single substrate

Compound	μ_{max} (1/h)	K_S (mg/L)	K_i (mg/L)	$Y_{X/S}$ (g/g)	R^2	RMSE
Styrene	0.1581	25.91	13.15	1.19 ± 0.24	0.9917	0.0041

previous reports which used pure culture to remove styrene as a substrate [35, 36]. This could be attributed to the activated sludge usage as a bacterial culture as well as to the amount of various microbial species that are probably incapable or poorly accomplished for metabolizing specific carbon sources. This is contrary to the pure culture which included appropriate microbial species to develop growth rate at several concentrations of substrates.

The inhibition coefficient (K_i), illustrates the impact of component toxicity during treatment process. When bacterial culture has high sensitivity to the component toxic effect, K_i quantity is small. However, the large values of K_i indicate that the culture is less delicate to substrate inhibition and therefore, the Andrews equation is simplified to the Monod equation. As shown in Table 3, the inhibition coefficients obtained in this study for styrene are fairly high. Thus, it can be concluded that inhibitory effects of styrene is relatively at low level. This is expected since the biodegradation in this study is handled by the mixed culture which covers a wide spectrum of microorganisms and can sustain high tolerances of toxic components compared to the pure microbial culture.

Ethylbenzene

Data achieved from various initial ethylbenzene concentrations versus specific growth rate (μ_g) (Fig. 4) reveal that ethylbenzene follows a similar pattern to styrene.

The correlation between the specific growth rate and the initial concentration of the substrate was described by the Andrew kinetic equation with the expected constants (Table 4): μ_{max} (1/h) = 2090, half velocity constant (mg/L) = 37.77, and K_I (mg/L) = 62.62. The value of the maximum growth rate for ethylbenzene achieved by the authors is higher than styrene. However, the inhibition

coefficient gained for ethylbenzene demonstrates lower toxicity compared to styrene.

The R^2 value for ethylbenzene was 0.988. This demonstrates an acceptable correlation between the experimental and predicted values obtained from the Andrews model.

Table 5 shows the comparison of kinetic parameters for ethylbenzene in bacterial pure and mixed cultures. The maximum specific growth rates, μ_{max}, obtained from different Studies was vacillated from 0.006 to 0.26 h^{-1} [36, 37]. The value of μ_{max} in this study also shows the same pattern while the acclimatized mixed culture bacterium shows higher activity of degrading ethylbenzene with μ_{max} of 0.21 h^{-1}. The achieved results are close enough to the maximum range reported by Trigueros et al. [36]. Meanwhile, the circumstances in our research with a high amount of substrate concentration created more barriers for microbial culture to biodegrade. Trigueros et al. [36] examined the biodegradation kinetics of ethylbenzene by *R. pyridinovorans PYJ-1*. The specific ethylbenzene degradation rate followed the Andrew model in which S_0 is the initial ethylbenzene concentration during gas phase. For *pyridinovorans PYJ-1*, μ_{max} (1/h) = 0.26, K_S (mg/L) = 1.5, and the inhibition coefficient of 20 mg/L was estimated [36]. The inhibition coefficients K_I for ethylbenzene was estimated to be 62.62 mg/L. Compared with previous studies, the relatively high value of K_I for ethylbenzene indicated that the mix culture was less sensitive to substrate inhibition [38].

In addition, when the obtained values for styrene and ethylbenzene were assessed, the styrene showed lower K_s value than ethylbenzene. Hence, the estimated K_s value on styrene indicates that affiliation of mixed culture to styrene is higher than ethylbenzene. Besides, the difference between μ_{max} factors revealed limited styrene degrading capability (in contrast to ethylbenzene) in the

Table 3 Comparisons between kinetic parameters estimated for the biodegradation of styrene in batch culture in different studies

Strain	Maximum substrate concentration (mg/L)	μ_{max} (1/h)	K_s (mg/l)	K_i (mg/l)	T (°C)	pH	References
Mix culture adapted with petrochemical residue	220	0.1581	25.91	13.15	32	7	This study
Mix culture adapted with industrial residue	123.4	0.1601	13.8	21.57	32	7	[16]
exophiala jeanselmei	104.15	1.26	0.1	3.3	25	5.7	[35]
P. putida F1	43	0.86 ± 0.01	13.8 ± 0.9	———————	30	7	[25]
Pseudomonas sp. E-93486	90	0.1188	5.984	156.6	30	7	[36]
exophiala oligosperma	19.3	0.160	7.381	———————	32.2	5.75	[18]

The parameters' values are for the **Andrews** model if a value of K_i is given. If the K_i value is not given the parameters' values are for the **Monod** model

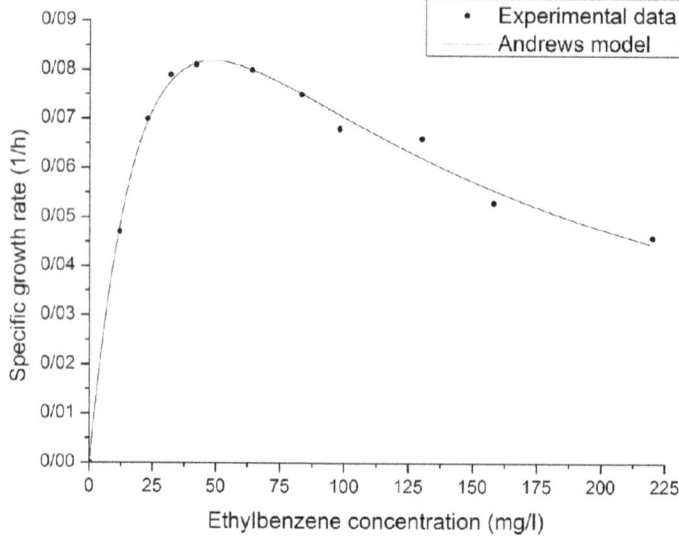

Fig. 4 Specific growth rate of the mixed culture at various concentrations of ethylbenzene as a sole substrate. Observed experimental data (shapes) and simulation outcomes acquired via Andrews model (lines)

environment of an activated sludge plant used for treating petrochemical wastewaters. In addition, different values of μ_{max} demonstrate different pathways in order to completely catabolize the selected components using the microbial species picked to attack and catabolized the carbon sources [39].

The achieved outcome through this part of study revealed that mixed culture has a good resistance to the substrate inhibition compared to other culture studies. Furthermore, the main reason for the differences in biodegradation kinetics of these substrates could be attributed to: a) metabolic pathways which according to former studies leads to the formation of the intermediate products caused by the reaction rate. b) Mass transfer process as an essential part of transforming the substrate into bio-cells, and c) Additional physicochemical conditions that might affect the biodegradation reaction rate [39, 40].

Dual substrate experiment and substrate interactions

Single substrate kinetic parameters cannot deal with the tough situation detected in biodegradation of the mixed toxic substrates. Thus, experiments with specific conditions were performed to evaluate the effects of interactions among substrates on styrene and ethylbenzene degradations (as a binary mixture) which have been nearly neglected by other authors. Since in low level of substrate concentrations the differences between the Andrews model and the Monod model are negligible and additional parameters in the Andrews model are ineffective on fitting the model to the experimental data, the Monod model was chosen to evaluate the mixture experiments parameters in this section. The results of a biodegradation experiments are shown in Fig. 4. Although the mixed bacterial culture used both substrates at the same time during most of the treatment period, ethylbenzene biodegradation as well as ethylbenzene depletion began before that of styrene.

This specifies that the styrene degradation was inhibited due to the existence of second substrate i.e. ethylbenzene, while the presence of styrene had little effect on ethylbenzene depletion. In addition, the prolonged lag and poor degradation at early times could be related to the competition of microorganism and its adaption to the new situation in order to predominate the special species which can degrade substrates easily. On the other hand, smooth and late biomass growth probably

Table 4 Estimated parameter values for Andrews kinetic models to biodegrade ethylbenzene as a single substrate

Compound	μ_{max} (1/h)	K_S (mg/L)	K_I (mg/L)	$Y_{X/S}$ (g/g)	R^2	RMSE
Ethylbenzene	0.2090	37.77	62.62	1.13 ± 0.17	0.9885	0.0030

Table 5 Comparisons between kinetic parameters estimated for the biodegradation of ethylbenzene in batch culture in different studies

Strain	Maximum substrate concentration (mg/L)	μ_{max} (1/h)	K_s (mg/l)	K_i (mg/l)	T (°C)	pH	References
Mix culture adapted with petrochemical residue	220	0.2090	37.77	62.62	32	7	This study
pseudomonas putida f1	80	0.26	1.5	20	35	7	[36]
Pseudomonas species	80	0.13	0.36	————	30	——	[24]
Aerobic bacterial consortium	100	0.05	0.11	100	30	6.2–6.9	[33]
Immobilized *Pseudomonas putida* and *Pseudomonas fluorescens*	150	0.012	236	429	25	6.4	[23]
Comamonas sp. JB	50	0.0064	49.35	————	30	7.5 to 8.0	[37]

The parameters' values are for the **Andrews** model if a value of K_i is given. If the K_i value is not given the parameters' values are for the **Monod** model

support the construction of the intermediates during the biodegradation reactions.

The SKIP model (Eq. 6) was used to define any interactions among the observed substrate. To obtain the values of the interaction parameters, kinetic parameters from single substrate experiments were used and substituted into substrate depletion and biomass growth equations (Eqs. 1 and 2). The results are shown in Table 6. As can be seen in Fig. 5 the fitted SKIP model perfectly explains the biodegradation data for styrene and ethylbenzene binary mixture. Although the biomass prediction depicts slightly higher yield than the experimental data. The large value of $I_{E,S}$ compared to $I_{S,E}$ approves the assertion made from experimental outcomes. It also represents a high degree for the inhibition of ethylbenzene on styrene biodegradation versus the insignificant impact of styrene on ethylbenzene consumption. Moreover, Fig. 5 demonstrates that the SKIP model can be used appropriately to fit unspecified types of inhibition between the two substrates.

To evaluate the similarity of metabolic pathway that was used in the catabolism of both components and to estimate the possible compete for the active site, specific growth rate models similar to the SKIP model were determined.

The models account for competitive inhibition among dual substrates (Eq. 5). Figure 6 depicts the output curves and the fitness quality of the purely competitive model gained from experimental data. It is obvious that this model does not deliver a correct fit to the

Table 6 Bio-kinetic interactive parameter estimated for styrene and ethylbenzene dual substrate experiments

Model type	Parameter	Value
SKIP	$I_{S,E}$	0.4
	$I_{E,S}$	1.64
Competitive inhibition	$K_{s\ S}/K_{s\ EB}$	0.16
	$K_{s\ EB}/K_{s\ S}$	6.4

experimental data for styrene and ethylbenzene combinations. Potential reasons for inappropriateness of such kinetics and experimental results are interactions among the substrates which are transported into the cytoplasm and enzymatic reaction complexity. Besides, the achieved results suggest that the active site for styrene biodegradation was not similar to ethylbenzene. This theory is supported by the research led by Deebet al. [41] performed for BTEX and MTBE biodegradation. Thus, the inhibition between styrene and ethylbenzene is not thoroughly competitive and due to various biodegradation pathways to metabolize the substrates a single interaction method does not explain the biodegradation kinetics easily. Supplementary studies of the metabolic pathways will make promise to deepen the understanding of interactions between mentioned substrates.

Conclusion

As outlined above, the biodegradation of styrene and ethylbenzene as a sole and binary source of carbon and energy in water was studied using industrial mixed culture. According to the biodegradation batch experiments and the attained data, it was observed that Andrews model successfully predicted kinetic biodegradation for wide and various ranges of styrene concentration (from 8 up to 220 mg/L) and ethylbenzene concentration (from 12 up to 220 mg/L) in a single substrates experiments. Yet, the simulation obtained from the Monod model in low concentration level for organic component is reliable. The comprehensive biodegradation of styrene and ethylbenzene occurred within various hours after being in contact with the microbial culture. Lag phase in styrene and ethylbenzene biodegradation increased efficiently as the organic concentration upraised. Nevertheless, lag phase time for styrene is slightly higher than that for ethylbenzene and the differences among the two lag phase time were extended at higher concentrations. Microorganism growth as well as organic component depletion for ethylbenzene was slightly faster than that

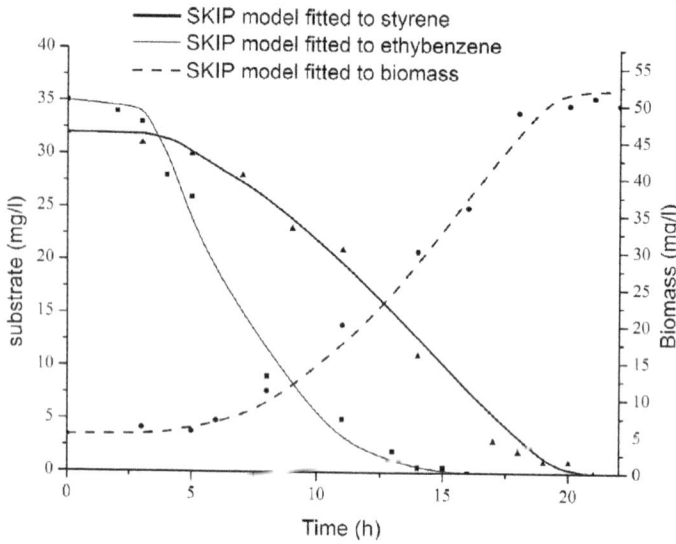

Fig. 5 Styrene and ethylbenzene dual degradation and biomass growth experimental data (shapes) and simulation outcomes acquired via SKIP model (lines)

of for styrene. Furthermore, the bio-kinetic factors obtained from single substrate biodegradation studies were used to assess the parameter for the interaction effect. This had a significant effect on biodegradation of a binary mixture. Hence, the styrene and ethylbenzene were used as substrates to evaluate the inhibition of toxic compounds and their interactions in a binary mixture. To the best of our knowledge the interaction between styrene and ethylbenzene has not been investigated until the current study. The outcomes of dual experiments were modeled using a purely competitive model and the SKIP model. Through using the corporate SKIP model, an accurate description for the biodegradation process for a dual mixture of styrene and ethylbenzene was attained. However, the purely competitive model had a poor estimation and fitness for experimental data. Finally, the difference between the biodegradation pathways was the main potential reason for the inaccurate description.

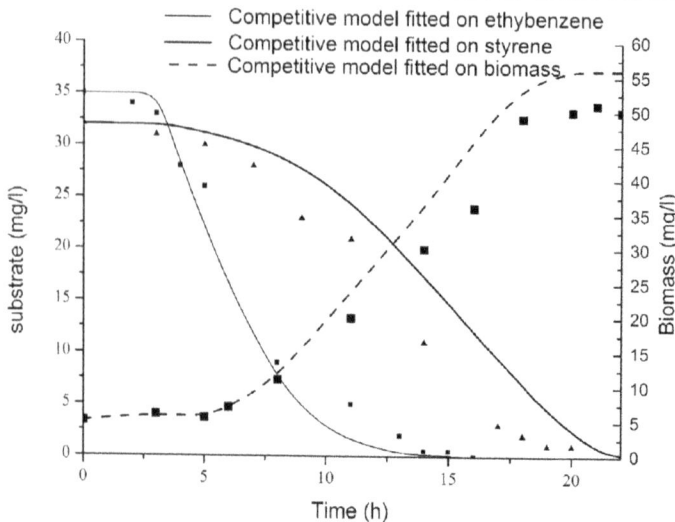

Fig. 6 Styrene and ethylbenzene dual degradation and biomass growth experimental data (shapes) and simulation outcomes acquired via purely competitive inhibition model (lines)

Endnotes

[1] *Web site address: http://www.tpco.ir/*
[2] Web site address: http://www.chem.agilent.com/

Acknowledgment
The authors would like to thank the support from the Iranian Ministry of Petroleum, National petrochemical Company.

References

1. Chen W-H, Yang W-B, Yuan C-S, Yang J-C, Zhao Q-L. Influences of aeration and biological treatment on the fates of aromatic VOCs in wastewater treatment processes. Aerosol Air Qual Res. 2013;13:225–36.

2. Ruel SM, Choubert J-M, Budzinski H, Miege C, Esperanza M, Coquery M. Occurrence and fate of relevant substances in wastewater treatment plants regarding Water Framework Directive and future legislations. Water Sci Technol. 2012;65:1179–89.

3. Malhautier L, Quijano G, Avezac M, Rocher J, Fanlo JL. Kinetic characterization of toluene biodegradation by Rhodococcus erythropolis: Towards a rationale for microflora enhancement in bioreactors devoted to air treatment. Chem Eng J. 2014;247:199–204.

4. Ahansazan B, Afrashteh H, Ahansazan N, Ahansazan Z. Activated sludge process overview, international journal of environmental science & development 5. 2014.

5. Sannino D, Vaiano V, Ciambelli P. A green route for selective synthesis of styrene from ethylbenzene by means of a photocatalytic system. Res Chem Intermed. 2013;39:4145–57.

6. Parreira FV, de Carvalho CR, Cardeal ZL. Evaluation of indoor exposition to benzene, toluene, ethylbenzene, xylene, and styrene by passive sampling with a solid-phase microextraction device. J Chromatogr Sci. 2002;40:122–6.

7. Zhou J, You Y, Bai Z, Hu Y, Zhang J, Zhang N. Health risk assessment of personal inhalation exposure to volatile organic compounds in Tianjin, China Sci Total Environ. 2011;409:452–9.

8. Sofuoglu SC, Aslan G, Inal F, Sofuoglu A. An assessment of indoor air concentrations and health risks of volatile organic compounds in three primary schools. Int J Hyg Environ Health. 2011;214:36–46.

9. US-EPA, List of contaminants and their maximum contaminant level (MCLs). http://water.epa.gov/drink/contaminants/index.cfm. (January 2013).

10. Liu Y. Overview of some theoretical approaches for derivation of the Monod equation. Appl Microbiol Biotechnol. 2007;73:1241–50.

11. De Lucas A, Rodriguez L, Villasenor J, Fernandez F. Biodegradation kinetics of stored wastewater substrates by a mixed microbial culture. Biochem Eng J. 2005;26:191–7.

12. Pala-Ozkok I, Rehman A, Yagci N, Ubay-Cokgor E, Jonas D, Orhon D. Characteristics of mixed microbial culture at different sludge ages: effect on variable kinetics for substrate utilization. Bioresour Technol. 2012;126:274–82.

13. Delhoménie M-C, Nikiema J, Bibeau L, Heitz M. A new method to determine the microbial kinetic parameters in biological air filters. Chem Eng Sci. 2008;63:4126–34.

14. Karamanev D, Margaritis A. Kinetic modeling of the biodegradation of the aqueous *p*-xylene in the immobilized soil bioreactor. Biochem Eng J. 2006;27:204–11.

15. Avalos Ramirez A, Bénard S, Giroir-Fendler A, Jones JP, Heitz M. Kinetics of microbial growth and biodegradation of methanol and toluene in biofilters and an analysis of the energetic indicators. J Biotechnol. 2008;138:88–95.

16. Babaee R, Bonakdarpour B, Nasernejad B, Fallah N. Kinetics of styrene biodegradation in synthetic wastewaters using an industrial activated sludge. J Hazard Mater. 2010;184:111–7.

17. Gąszczak A, Bartelmus G, Greń I. Kinetics of styrene biodegradation by Pseudomonas sp. E-93486. Appl Microbiol Biotechnol. 2012;93:565–73.

18. Rene ER, Bernat P, Długoński J, Veiga MC, Kennes C. Use of styrene as sole carbon source by the fungus: optimization and modeling of biodegradation. Pathway Elucidation Cell Membrane Compos Appl Biochem Biotechnol. 2012;168:1351–71.

19. Jung I-G, Park C-H. Characteristics of styrene degradation by *Rhodococcus pyridinovorans* isolated from a biofilter. Chemosphere. 2005;61:451–6.

20. Parameswarappa S, Karigar C, Nagenahalli M. Degradation of ethylbenzene by free and immobilized Pseudomonas fluorescens-CS2. Biodegradation. 2008;19:137–44.

21. Kim L-H, Lee S-S. Isolation and characterization of ethylbenzene degrading Pseudomonas putida E41. J Microbiol. 2011;49:575–84.

22. Yeom S-H, Yoo Y-J. Analysis of microbial adaptation at enzyme level for enhancing biodegradation rate of BTX. Korean J Chem Eng. 2002;19:780–2.

23. Shim H, Shin E, Yang S-T. A continuous fibrous-bed bioreactor for BTEX biodegradation by a co-culture of *Pseudomonas putida* and *Pseudomonas fluorescens*. Adv Environ Res. 2002;7:203–16.

24. Littlejohns JV, Daugulis AJ. Kinetics and interactions of BTEX compounds during degradation by a bacterial consortium. Process Biochem. 2008;43:1068–76.

25. Abuhamed T, Bayraktar E, Mehmetoğlu T, Mehmetoğlu Ü. Kinetics model for growth of *Pseudomonas putida* F1 during benzene, toluene and phenol biodegradation. Process Biochem. 2004;39:983–8.

26. Reardon KF, Mosteller DC, Bull Rogers JD. Biodegradation kinetics of benzene, toluene, and phenol as single and mixed substrates for Pseudomonas putida F 1. Biotechnol Bioeng. 2000;69:385–400.

27. Shuler ML KF. Bioprocess engineering. 2nd ed. New Jersey: Prentice Hall; 2002.

28. Andrews JF. A mathematical model for the continuous culture of microorganisms utilizing inhibitory substrates. Biotechnol Bioeng. 1968;10:707–23.

29. Okpokwasili G, Nweke C. Microbial growth and substrate utilization kinetics. 2006.

30. Grady Jr C, Smets BF, Barbeau DS. Variability in kinetic parameter estimates: a review of possible causes and a proposed terminology. Water Res. 1996;30:742–8.

31. APHA. Standard methods for the examination of water and wastewater. 19th ed. Washington, DC, USA: American Public Health Association; 1995.

32. Kim D-J, Choi J-W, Choi N-C, Mahendran B, Lee C-E. Modeling of growth kinetics for Pseudomonas spp. during benzene degradation. Appl Microbiol Biotechnol. 2005;69:456–62.

33. Datta A, Philip L, Bhallamudi SM. Modeling the biodegradation kinetics of aromatic and aliphatic volatile pollutant mixture in liquid phase. Chem Eng J. 2014;241:288–300.

34. Lin C-W, Cheng Y-W, Tsai S-L. Multi-substrate biodegradation kinetics of MTBE and BTEX mixtures by Pseudomonas aeruginosa. Process Biochem. 2007;42:1211–7.

35. Cox H, Moerman R, Van Baalen S, Van Heiningen W, Doddema H, Harder W. Performance of a styrene-degrading biofilter containing the yeast Exophiala jeanselmei. Biotechnol Bioeng. 1997;53:259–66.

36. Trigueros DE, Módenes AN, Kroumov AD, Espinoza-Quiñones FR. Modeling of biodegradation process of BTEX compounds: kinetic parameters estimation by using Particle Swarm Global Optimizer. Process Biochem. 2010;45:1355–61.

37. Jiang B, Zhou Z, Dong Y, Tao W, Wang B, Jiang J, et al. Biodegradation of benzene, toluene, ethylbenzene, and o-, m-, and p-xylenes by the newly isolated bacterium Comamonas sp. JB, Appl Biochem Biotechnol. 2014;1–9.

38. Chen D-Z, Ding Y-F, Zhou Y-Y, Ye J-X, Chen J-M. Biodegradation kinetics of tetrahydrofuran, benzene, toluene, and ethylbenzene as multi-substrate by pseudomonas oleovorans DT4. Int J Environ Res Public Health. 2014;12:371–84.

39. Smith MR. The biodegradation of aromatic hydrocarbons by bacteria. Physiology of Biodegradative Microorganisms: Springer; 1991. p. 191–206.

40. Ramos JL, Duque E, Gallegos M-T, Godoy P, Ramos-González MI, Rojas A, et al. Mechanisms of solvent tolerance in gram-negative bacteria. Ann Rev Microbiol. 2002;56:743–68.

41. Deeb RA, Hu H-Y, Hanson JR, Scow KM, Alvarez-Cohen L. Substrate interactions in BTEX and MTBE mixtures by an MTBE-degrading isolate. Environ Sci Technol. 2001;35:312–7.

14

Effects of ethanol on the electrochemical removal of *Bacillus subtilis* spores from water

Masuma Moghaddam Arjmand[1], Abbas Rezaee[1*], Simin Nasseri[2] and Said Eshraghi[3]

Abstract

This study aimed to characterize the effects of ethanol on the monopolar electrochemical process to remove *Bacillus subtilis* spores from drinking water. In particular, spores' destruction was tested by applying 20–100 mA current for 15–60 min to *B. subtilis* spores (10^2–10^4 CFU/mL density), with stainless steel electrodes. The experimental results showed electrochemical removal of spores in the presence of 0.4 M ethanol at 15, 45, and 60 min and 5 mA/cm^2 current density. However, the use of ethanol or the electrochemical process alone did not eliminate *B. subtilis* spores at these time points. Overall, this study suggests that adding ethanol to the electrochemical process successfully removes *B. subtilis* spores from drinking water.

Introduction

Cryptosporidium parvum is an important microbial contaminant found in drinking water and is associated with a waterborne disease in humans [1]. Recently, *Bacillus subtilis* spores were used to evaluate the inactivation of *C. parvum* during water treatment [2]. To date, several methods of water treatment have been proposed, including chlorine. Although chlorination represents an efficient method of water treatment, it presents several disadvantages such as unfavorable taste and odor and the generation of potentially toxic disinfection products. In ddition, chlorine is ineffective when used alone against resistant microorganisms such as *C. parvum* and *Giardia* spp. [3]. A number of alternatives to chlorination have been suggested, including chemical (e.g., ozone and electrochemical treatments), physical (e.g., ultraviolet irradiation), and microwave systems [4]. In recent years, increasing attention has been paid to the electrochemical process as an alternative method to chlorination in water disinfection [5]. This treatment has been proposed since the 1950s [4] and can be divided into two categories: direct electrolyzers and mixed oxidant generators [6, 7]. The electrochemical process has several advantages, including the simplicity of the equipment and the fact that no additional chemicals are required for this method, as they

can be generated during the process [8]. In the presence of iron and stainless steel electrodes, the general electrochemical mechanism for this process can be illustrated as follows:

$$\text{Fe (S)} \rightarrow \text{Fe}^{+3}{}_{\text{aq}} + 3\,\text{e (Anode)} \qquad (1)$$

$$3\,\text{H}_2\text{O} + 3\text{e}^- \rightarrow 3/2\,\text{H}_{2\,\text{g}} + 3\,\text{OH}^- \text{(Cathode)} \qquad (2)$$

Fe^{3+} and OH^- ions, generated at the electrodes surface, react to generate Fe(OH)_3 compounds that can remove pollutants from aqueous solutions [9, 10]. Ethanol has also been used in water treatment processes, even though it has not been found to be an efficient sporicidal agent [11]. According to previous studies, 875 mg/L ethanol is needed to reduce *B. subtilis* populations over 6 log10 [12]. Ethanol is a membrane disrupter that induces rapid release of intracellular components and membrane disorganization, most likely due to the penetration of solvents into the hydrophobic region of the membrane bilayer. The aim of this study was to evaluate the efficiency of an electrochemical process in the presence of ethanol in water treatment. The effects of the operative parameters on *Bacillus B. subtilis* spore removal were also studied.

Materials and methods

Bacterial strain and culture conditions

The *B. subtilis* ATCC 6633 strain was obtained from the culture collection at the Tehran University (Iran). The strain was maintained on slant nutrient agar at 4 °C.

* Correspondence: rezaee@modares.ac.ir
[1]Department of Environmental Health Engineering, Faculty of Medical Sciences, Tarbiat Modares University, Tehran, Iran
Full list of author information is available at the end of the article

Table 1 Analysis of the water quality

Parameter	Value (mg/L)	Parameter	Value (mg/L)	Parameter	Value
Ca	59	Cl$^-$	34.2	pH	7.5
Mg	20.2	SO$_4$	21.4	TDS (mg/l)	364
Na	86	F$^-$	0.25	Conductivity(μs)	638
K	1.25	NH$_4$- N	0.1	Nitrate (mg/l)	12

Stocks were stored in aliquots containing 10 % glycerinated nutrient broth at –18 °C. 0.5 McFarland standards (corresponding to ~1.5×10^8 CFU/mL) spores were reactivated by incubation in 100 mL Erlenmeyer flasks containing 50 mL fresh trypticase soy broth (Merck) at 37 °C for 24 h, under aerobic conditions. Next, spore suspension was poured into sterile Erlenmeyer flasks and placed in a water-bath at 80 °C for 15 min to eliminate vegetative cells. Sporulation was confirmed by optical microscopy using the Gram staining technique, and spores were diluted into water. Total counts of bacterial spore suspensions were made using the pour plate method. Briefly, after incubation at 37 °C for 48 h, spore-forming bacteria were counted, and the results were expressed as the mean number of spores/mL. Culture media and equipment were sterilized by autoclaving at 121 °C for 15 min. The pH was adjusted at 0.1 M by adding NaOH or HCl.

Minimal inhibitory concentration (MIC)

Minimal inhibitory concentration (MIC) is defined as the lowest concentration of an antimicrobial agent that prevents the visible growth of a microorganism under certain *in vitro* condition. In this study, MICs were tested using the dilution broth method, according to the National Committee for Clinical Laboratory Standards [13]. Briefly, 1 mL of the culture was transferred into a first sterile glass tube containing 10 mL TBS medium

and 0.2 to 4 M ethanol. Next, after stirring, 1 mL was transferred from the first to the second tube and so on, to obtain a total of 12 dilutions. Tubes were incubated at optimal temperature for 24 and 48 h, prior to determine the MICs.

Experimental set up and operation

Electrochemical treatments were conducted in a single electrochemical reactor. The electrochemical reactor was equipped with two sheets of stainless steel that were used as anode and cathode electrodes. The distance between the electrodes was adjusted to 2 cm and maintained by placing plastic spacers. Experimental runs were conducted by imposing current densities ranging from 1 to 5 mA/cm^2 (Atten APS 3005S-3, China). The electrochemical process operated in batch mode and was performed using a 500 mL capacity glass beaker with 300 mL water at room temperature. Commercially available steel plates (size $15 \times 4 \times 0.1$ cm) were applied as electrodes and dipped in water to a depth of 10 cm. The effects of the operating conditions on the efficacy of the process were evaluated, including the applied current (50–150 mA), ethanol concentrations (0.2–0.4 M), spore concentrations (10^2–10^4), retention time(60 min), and current density (1–5 mA/cm^2). The quality of water is presented in Table 1.

The electrodes were connected to a DC power supply (Atten APS 3005S-3, China) with operational options for controlling the constant voltage and current density. The current density was calculated through the following equation as follows:

$$CD = I/S \qquad (3)$$

where I is the current through the solutions (A) and S is the area of the electrode (cm^2). The pH level and

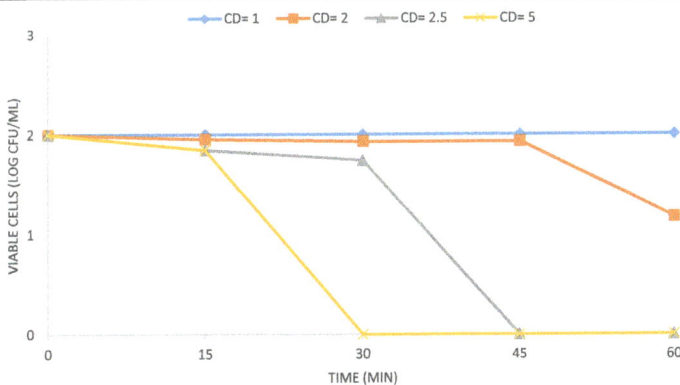

Fig. 1 Kinetic of electrochemical *B. subtilis* spore removal from water. Experimental conditions: current density CD = 1–5 mA/cm^2; spore density 10^2 CFU/mL; stainless steel as electrode anode; T = 25 °C; pH = 7.2; electrodes gap = 2 cm

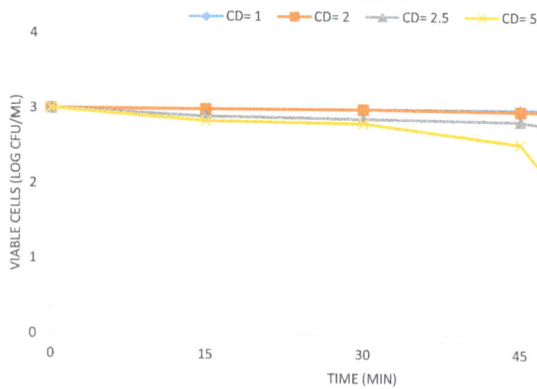

Fig. 2 Kinetic of electrochemical *B. subtilis* spore removal from water. Experimental conditions: current density CD = 1–5 mA/cm²; spores density 10³ CFU/mL; stainless steel as electrode anode; T = 25 °C; pH = 7.2; electrodes gap = 2 cm

conductivity of the solution were measured using a portable pH and EC meters (Eutech, Singapore). All the experiments were performed at a pH of 7.2. At the end of each experiment, the DC power source was switched off and the electrodes were removed from the water. During the experiments, samples were taken at 15 min interval and plated on TBS plates. All the experiments were repeated twice. The density of *B. subtilis* spore (10^2–10^4 m/l) removal was assessed in the electrochemical treatment using ethanol (0.2–0.4 M) at 20–100 mA current (1–5 mA/cm²).

Results

In this study, the experiments were carried out in laboratory scale to evaluate the sporicidal effect of ethanol (0.2–0.4 M) in combination with the electrochemical treatment, using *B. subtilis* ATCC 6633 strain as a surrogate microorganism. In order to evaluate the sporicidal efficacy of ethanol, MIC determination was performed. The results showed that ethanol alone is not an efficient sporicidal agent (Figs. 1, 2 and 3).

However, in combination with the electrochemical treatment, it acquires a high antimicrobial activity (Figs. 4, 5 and 6).

These results demonstrated that the sporicidal efficiency is inversely proportional to the initial number of spore in solution. In addition, the sporicidal activity of the electrochemical treatment was directly proportional to ethanol concentrations, as shown in Figs. 4, 5 and 6. Overall, the results obtained in this study demonstrated that ethanol in combination with the electrochemical treatment improves the sporicidal efficiency of water disinfection, suggesting a synergistic effect between these two agents. The best result was obtained on 10^3 *B. subtilis* spores, using 0.4 M ethanol for 45 min at 5 mA/cm² current density (Fig. 5).

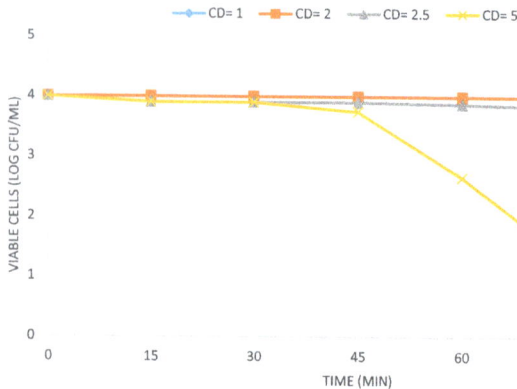

Fig. 3 Kinetic of electrochemical *B. subtilis* spore removal from water. Experimental conditions: current density CD = 1–5 mA/cm²; spore density 10⁴ CFU/mL; stainless steel as electrode anode; T = 25 ° C; pH = 7.2; electrodes gap = 2 cm

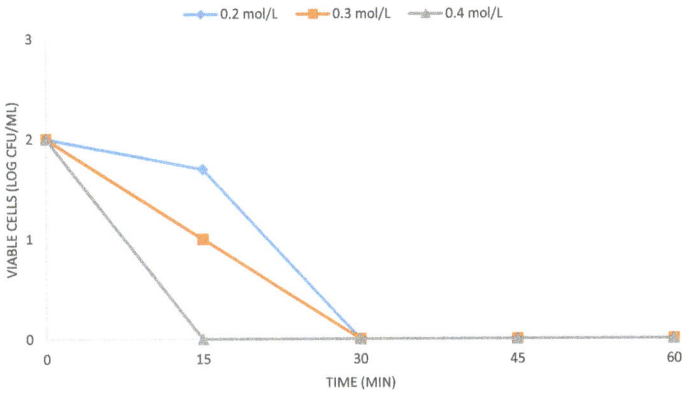

Fig. 4 Antimicrobial efficacy of ethanol and the electrochemical treatment on *B. subtilis* spore removal from water. Experimental conditions: current density CD = 5 mA/cm^2; spore density 10^2 CFU/mL; ethanol concentrations 0.2–0.4 M; stainless steel as electrode anode; T = 25 °C; pH = 7.2; electrodes gap = 2 cm

Discussion

Spores of *B. subtilis* are particularly resistant to conventional water disinfection treatments and, for this reason, they are used as surrogates for some waterborne pathogens such as *Cryptosporidium spp*, and as an indicator of hygienic quality of drinking water. Bacterial spores are more resistant to general sterilization and disinfection treatments such as heating, radiation, and the use of various chemicals than their vegetative cells [14]. Several parameters participate in spore resistance, including impermeability, low water content, high levels of pyridine-2,6-dicarboxylic acid and divalent cations, and outer membrane thickness [15]. In addition, spore DNA is protected against various types of damage [16]. In recent years, increasing attention has been paid to electrochemical oxidation as an efficient technology for water disinfection. During this process, free chlorine is produced. This chemical damages the bacterial outer membrane, penetrates into the periplasm, destroys the inner membrane and degenerates cytoplasmic proteins. Also, the process can oxidize the microbes on the electrode surfaces [17]. Usually, the oxidants of the electrochemical treatment are reactive oxygen species generated from the oxidation of water molecules [18]. It was shown that the electrochemical treatment efficiently removes bacteria spores but not their vegetative cells [19]. The aim of this study was to evaluate the efficiency of a combination of the electrochemical process and ethanol at low concentrations for disinfection of *B. subtilis* spores. Ethanol is a general bactericidal agent and has been widely applied for disinfection of human tissues and contaminated surfaces. According to the

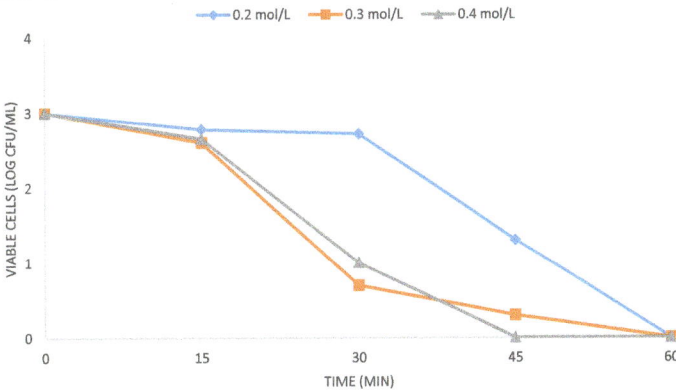

Fig. 5 Antimicrobial efficacy of ethanol and the electrochemical treatment on *B. subtilis* spore removal from water. Experimental conditions: current density CD = 5 mA/cm^2; spore density 10^3 CFU/mL; ethanol concentrations 0.2–0.4 M; stainless steel as electrode anode; T = 25 °C; pH = 7.2; electrodes gap = 2 cm

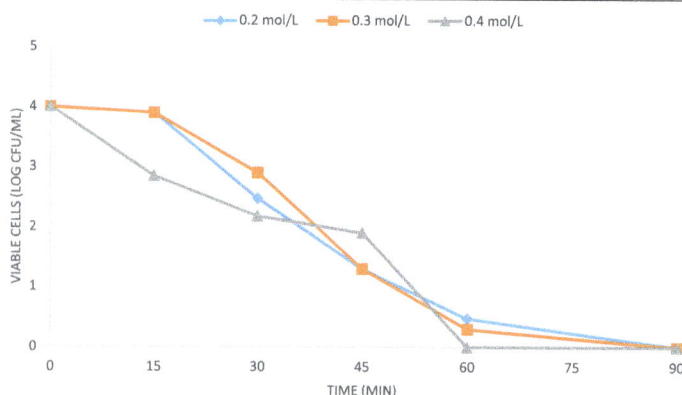

Fig. 6 Antimicrobial efficacy of ethanol and the electrochemical treatment on *B. subtilis* spore removal from water. Experimental conditions: current density CD = 5 mA/cm^2; spore density 10^4 CFU/mL; ethanol concentrations 0.2–0.4 M; stainless steel as electrode anode; T = 25 °C; pH = 7.2; electrodes gap = 2 cm

presented results, however, ethanol alone does not possess sporicidal activity [11]. It has been reported that ethanol causes microbial membrane damage and denaturation of proteins thus interfering with cell metabolism and inducing cell lysis. According to the obtained results, maximum sporicidal effects were obtained by adding 0.4 M ethanol to10^4 *B. subtilis* spores at 100 mA current. Lower ethanol concentration (0.2 M) increased the reaction time to 90 min. Ethanol alone does not possess high sporicidal efficiency. Some studies have reported that the combination of ethanol with ferric chloride and ethylenediaminetetraacetic acid can act as sporicidal agent [20]. Also, it has been shown that ethanol and anionic surfactants have sporicidal activity at low pH values. In this study, the sporicidal effects on *B. subtilis* spores was assessed by MIC. The results showed that ethanol alone has not sporicidal effect on *B. subtilis* spores. The electrochemical treatment exerted a low sporicidal effect on small numbers of *B. subtilis* spores, but it failed for large number of spores. These results indicated that the electrochemical treatment is an efficient method for water disinfection, but is not sufficient to remove disinfectant-resistant bacteria such as spore forming bacteria. This study showed that ethanol significantly increases the sporicidal efficiency of the electrochemical process.

Conclusion

The results obtained in this study show that *B. subtilis* spores were killed at 90 min by electrochemical water disinfection using ethanol. It was observed that increasing the operational time and adding ethanol to the electrochemical process improved the spore removal efficiency. Moreover, increasing the supporting electrolyte concentration in the solution reduces the specific electrical energy consumption.

Competing interests
The authors declare that they have no competing interests.

Authors' contributions
All authors read and approved the final manuscript.

Acknowledgements
The authors would like to thank the Tarbiat Modares University for funding and supporting this project (code: 1123803).

Author details
^{1}Department of Environmental Health Engineering, Faculty of Medical Sciences, Tarbiat Modares University, Tehran, Iran. ^{2}Department of Environmental Health Engineering, School of Public Health, and Center for Water Quality Research, Institute for Environmental Research, Tehran University of Medical Sciences, Tehran, Iran. ^{3}Department of Pathobiology, School of Public Health, Tehran University of Medical Sciences, Tehran, Iran.

References
1. Zhou P, Giovanni GDD, Meschke GS, Dodd MC. Enhanced Inactivation of *Cryptosporidium parvum* Oocysts during Solar Photolysis of Free Available Chlorine. Environ Sci Technol Lett. 2014;1(11):453–8.
2. Forsyth JE, Zhou P, Mao Q, Asato SS, Meschke JS, Dodd MC. Enhanced Inactivation of Bacillus subtilis Spores during Solar Photolysis of Free Available Chlorine. Environ Sci Technol. 2013;47(22):12976–84.
3. Rezaee A, Kashi G, Jonidi-Jafari A, Khataee AR, Nili-Ahmadabadi A. Effect of Hydrogen peroxide on Baciluss Subtilis spore removal in an electrophotocatalytic system. Fresenius Environ Bull. 2011;20(10a):2750–5.
4. Kerwick MI, Reddy SM, Chamberlain AHL, Holt DM. Electrochemical disinfection, an environmentally acceptable method of drinking water disinfection? Electrochim Acta. 2005;50:5270–7.
5. Kraft A. Electrochemical water disinfection: A short review. Platinum Metals Rev. 2008;52(3):177–85.
6. Cho J, Choi H, Kim IS, Amy S. Chemical aspects and byproducts of electrolyser. Water Sci Technol. 2001;1(4):159–64.
7. Qin GF, Li ZY, Chen XD, Russell AB. An experimental study of an NaClO generator for anti-microbial applications in the food industry. J Food Eng. 2002;54:111–7.
8. Mart'nez-Huitle CA, Brillas E. Electrochemical alternatives for drinking water disinfection. Angew Chem Int Ed. 2008;47:2–10.
9. Bazrafshan E, Mahvi AH, Nasseri S, Shaieghi M. Performance evaluation of electrocoagulation process for diazinon removal from aqueous environments by using iron electrodes. Iranian J Environ Health Sci Eng. 2007;4(2):127–32.

10. Bazrafshan E, Ownagh KA, Mahvi AH. Application of electrocoagulation process using iron and aluminum electrodes for fluoride removal from aqueous environment. E J Chem. 2012;9(4):2297–308.

11. Chambers ST, Peddie B, Pithie A. Ethanol disinfection of plastic-adherent micro-organisms. J Hospital Infect. 2006;63:193–6.

12. Priscila GM, Angela FJ, Leticia C, Patricia M, Thereza CVP. Minimal inhibitory concentration (MIC) determination of disinfectant and/or sterilizing agents. Brazilian J Pharmaceutical Sci. 2009;45(2):241–8.

13. Wayne PA, National Committee for Clinical Laboratory Standards: Methods for dilution antimicrobial susceptibility tests for bacteria that grow aerobically. Approved Standard, M7-A2, National Committee for Clinical Laboratory Standards: USA; 1990.

14. Setlow B, Loshon CA, Genest PC, Cowan AE, Setlow C, Setlow P. Mechanisms of killing of spores of Bacillus subtilis by acid, alkali and ethanol. J Appl Microbiol. 2002;92:362–75.

15. Setlow P. Spores of Bacillus subtilis: their resistance to and killing by radiation, heat and chemicals. J Appl Microbiol. 2006;101:514–25.

16. Tennen R, Setlow B, Davis KL, Loshon CA, Setlow P. Mechanisms of killing of spores of Bacillus subtilis by iodine, glutaraldehyde and nitrous acid. J Appl Microbiol. 2000;89:330–8.

17. Anglada A, Urtiaga A, Ortiz I. Contributions of electrochemical oxidation to waste-water treatment: Fundamentals and review of applications. J Chem Technol Biotechnol. 2009;84:1747–55.

18. Jeong J, Kim C, Yoon J. The effect of electrode material on the generation of oxidants and microbial inactivation in the electrochemical disinfection processes. Water Res. 2009;43:895–901.

19. Francisco V, Selma F. Electrochemical disinfection: An efficient treatment to inactivate Escherichia Coli 0157:H7 in process wash water containing organic matter. Microbiol. 2010;30:146–56.

20. Kida N, Mochizuki Y, Taguchi F. Effects on the sporicidal activity by using various metal ions in the formulation combining ferric chloride, ethylenediaminetetraacetic acid and ethanol. Biocontrol Sci. 2004;9:29–32.

15

Optimization of sonochemical degradation of tetracycline in aqueous solution using sono-activated persulfate process

Gholam Hossein Safari[1], Simin Nasseri[1,2*], Amir Hossein Mahvi[1,3], Kamyar Yaghmaeian[1,3], Ramin Nabizadeh[1,4] and Mahmood Alimohammadi[1]

Abstract

Background: In this study, a central composite design (CCD) was used for modeling and optimizing the operation parameters such as pH, initial tetracycline and persulfate concentration and reaction time on the tetracycline degradation using sono-activated persulfate process. The effect of temperature, degradation kinetics and mineralization, were also investigated.

Results: The results from CCD indicated that a quadratic model was appropriate to fit the experimental data ($p < 0.0001$) and maximum degradation of 95.01 % was predicted at pH = 10, persulfate concentration = 4 mM, initial tetracycline concentration = 30.05 mg/L, and reaction time = 119.99 min. Analysis of response surface plots revealed a significant positive effect of pH, persulfate concentration and reaction time, a negative effect of tetracycline concentration. The degradation process followed the pseudo-first-order kinetic. The activation energy value of 32.01 kJ/mol was obtained for $US/S_2O_8^{2-}$ process. Under the optimum condition, the removal efficiency of COD and TOC reached to 72.8 % and 59.7 %, respectively. The changes of UV–Vis spectra during the process was investigated. The possible degradation pathway of tetracycline based on loses of N-methyl, hydroxyl, and amino groups was proposed.

Conclusions: This study indicated that sono-activated persulfate process was found to be a promising method for the degradation of tetracycline.

Keywords: Tetracycline degradation, Persulfate, Response surface methodology, Central composite design, Optimization

Background

Tetracycline (TC) is extensively used for the prevention and treatment of infectious diseases in human and veterinary medicine and as feed additives for promote growth in agriculture [1, 2]. Because of their extensive usage, their strongly hydrophilic feature, low volatility [2] and relatively long half-life [3], TC antibiotic has been frequently detected in different environmental matrices: surface waters (0.07-1.34 µg/L) [4], soils (86.2-198.7 µg/kg) [5], liquid manures (0.05-5.36 µg/kg) [5] and in 90 % of farm lagoon samples (>3 µg/L) [6]. In addition to environmental contamination, the occurrence of TC in the aquatic environments would also increase antibiotic resistance genes [7]. However, due to the antibacterial nature of TC, they cannot effectively be removed by conventional biological processes [8]. In wastewater treatment plants, the TC removal efficiency varied in the range of 12 % to 80 % [9, 10]. For example, concentrations of TC residues have been detected in values of 0.97 to 2.37 µg/L in the final effluent from wastewater treatment plants [11]. Hence, the effort to develop new processes to minimize the tetracycline residues discharges into the environment is become essential. Physicochemical processes such as membrane filtration and adsorption using activated carbon have been used to removal of TC. These processes are not efficient enough, transfer the pollutant from one phase to another [12, 13]. Advanced oxidation processes (AOPs) such as (O_3/H_2O_2, US/O_3, UV/O_3, UV/H_2O_2, H_2O_2/Fe^{2+}, $US\text{-}TiO_2$ and

* Correspondence: naserise@tums.ac.ir
[1]Department of Environmental Health Engineering, School of Public Health, Tehran University of Medical Sciences, Tehran, Iran
[2]Center for Water Quality Research, Institute for Environmental Research, Tehran University of Medical Sciences, Tehran, Iran
Full list of author information is available at the end of the article

UV-TiO$_2$) have been proposed as very effective alternatives to degrade tetracycline antibiotics. The primary of AOPs is production of hydroxyl radical in water, a much powerful oxidant in the degradation of a wide range of organic pollutants [12–15]. Recently, the application of sulfate radical-based advanced oxidation processes (SR-AOPs) to oxidation of biorefractory organics have attracted great interest [16, 17]. Persulfate (PS, S$_2$O$_8^{2-}$) is a powerful and stable oxidizing agent (E$_0$ = 2.01 V vs. NHE), which has high aqueous solubility and high stability at room temperature as compared to hydrogen peroxide (H$_2$O$_2$, E$_0$ = 1.77 V vs. NHE) [18, 19].

Sulfate radicles could be produced through the activation of persulfate (PS, S$_2$O$_8^{2-}$) with ultraviolet [20], heat [21, 22], microwave [23], sonolysis [24], base [25], granular activated carbon [26], quinones [27], phenols [28], soil minerals [29], radiolysis [30] and transition metals [31, 32]. Sulfate radicals are more effective than hydroxyl radical in the oxidation of organic contaminants. They have higher redox potentials, longer half-life and higher selectivity in the oxidation of organic contaminants (SO$_4^-$•, E$_0$ = 2.5-3.1, half-life = 30–40 μs) than hydroxyl radical (HO•, E$_0$ = 1.89–2.72 V, half-life = 10^{-3} μs) [33–39]. Hence, the organic pollutants could be oxidized entirely by SO$_4^-$•, especially benzene derivatives compounds [18]. Generally, sulfate radical reacts with organic contaminants predominantly through selective electron transfer, while hydroxyl radical mainly reacts through hydrogen abstraction and addition. Therefore, the possibility of sulfate radical scavenging by nontarget compounds is lower than hydroxyl radical [39–42].

Sonochemical treatment is an emerging and efficient process that applied pyrolytic cleavages to degradation of organic compounds [42, 43]. This process is a cleaner and safe technique compared with UV, ozonation, and has the ability of operation under ambient conditions [43, 44]. However, combination of ultrasound with various processes has been detected as an economical and successful alternative for the degradation and mineralization of some recalcitrant organic compounds in aqueous solution [42]. The combination of ultrasound and persulfate (US/S$_2$O$_8^{2-}$) has been effective for the degradation of compounds such as; methyl tert-butyl ether (MTBE) [45], nitric oxide [18], 1,4-dioxane [46], arsenic(III) [44], amoxicillin [47], tetracycline [48] and dinitrotoluenes [24]. In aqueous solutions, acoustic cavitation leading to produce plasma in water and free radicals and other reactive species such as HO• and H• radicals due to the thermal degradation of water according to Reaction (1) and (2). The HO• and H• radicals can also react with PS to production of more reactive SO$_4^-$• radicals according to Reactions (3) to (7) [42, 44, 49, 50].

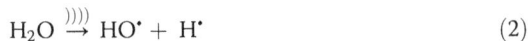

$$H_2O \overset{))))}{\to} H_2O \text{ plasma} \qquad (1)$$

$$H_2O \overset{))))}{\to} HO^\bullet + H^\bullet \qquad (2)$$

Where "))))))" refers to ultrasonication. In the presence of S$_2$O$_8^{2-}$:

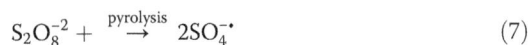

$$S_2O_8^{-2} + \overset{))))}{\to} 2SO_4^{-\bullet} \qquad (3)$$

$$SO_4^{-\bullet} + H_2O \to SO_4^{2-} + HO^\bullet + H^+ \qquad (4)$$

$$S_2O_8^{-2} + HO^\bullet \to HSO_4^- + SO_4^{-\bullet} + \frac{1}{2}O_2 \qquad (5)$$

$$S_2O_8^{-2} + H^\bullet \to HSO_4^- + SO_4^{-\bullet} \qquad (6)$$

$$S_2O_8^{-2} + \overset{pyrolysis}{\to} 2SO_4^{-\bullet} \qquad (7)$$

In aqueous solution, Hydroxyl radicals may be produced via the degradation of persulfate and/or ultrasonic irradiation. Ultrasonic irradiation could also lead to cavitation through the formation, growth and collapse of tiny gas bubbles in the water [51]. Moreover, during US irradiation, the collapse of cavitation bubbles leads to higher temperatures and pressures that produces free radicals and other reactive species and would also increase the number of collisions between free radicals and contaminants [42, 44, 49, 50].

The specific objectives of this study were to optimize the TC degradation in aqueous solution using US/S$_2$O$_8^{2-}$ process. Response surface methodology (RSM) is a reliable statistical technique for developing, improving and optimizing processes and can be used to assess the relative significance of several affecting factors with the least experiments [52–54]. Therefore, an experimental design methodology using RSM and CCD was used to evaluate the effect of operational parameters such as initial TC concentration, initial S$_2$O$_8^{2-}$ concentration, initial pH and reaction time on the sonochemical degradation of tetracycline. In addition, the effect of temperature, degradation kinetics, mineralization, changes of ultraviolet Visible (UV–Vis) spectra and the proposed degradation pathway of TC by the US/S$_2$O$_8^{2-}$ process were investigated. This study as part of a PhD dissertation of the first author was performed at Department of Environmental Health Engineering, School of Public Health, Tehran University of Medical Sciences in 2015.

Materials and methods
Materials
Tetracycline hydrochloride [C$_{22}$H$_{25}$N$_2$O$_8$Cl] (AR, 99 %), was provided from Sigma–Aldrich. Chemical properties of tetracycline hydrochloride are shown in Table 1 [2]. Sodium persulfate (Na$_2$S$_2$O$_8$, 98 %) was provided from Sigma–Aldrich. All other chemicals were of analytical

Table 1 Chemical properties of tetracycline hydrochloride

Molecule	Formula	Molecular weight (g/mol)	Solubility (mol/L)	pK_{a1}	pK_{a2}	pK_{a3}
TC	$C_{22}H_{24}O_8N_2.HCl$	480.9	0.041	3.2 ± 0.3	7.78 ± 0.05	9.6 ± 0.3

grade and were used without further purification. The water used in all experiments was purified by a Milli- Q system.

Procedure

Schematics of the experimental setup applied in this study is demonstrated in Fig. 1. A stock solution of tetracycline was daily prepared with distilled deionized water and diluted as required initial concentration. Sonochemical treatment was carried out with a fixed volume of 100 mL of TC solution in a glass vessel of 200 mL. The vessel was wrapped with tinfoil in order to avoid any photochemical effects. The pH adjustments were conducted with 1 m NaOH or 1 m HCL (Merck Co.) using a pH meter (E520, Metrohm, Tehran, Iran). Sonochemical treatment was performed with an ultrasonic generator at a frequency of 35 kHz and power of 500 W (Elma, Singen, Germany). The reactor was immersed into the ultrasonic bath and its location was always kept similarly. All experiments were conducted at constant temperature using cooling water and temperature controller. At pre-specified time intervals, 2 mL sample was withdrawn, filtered through 0.22 μm syringe filter and mixed with the same volume of methanol to quench the reaction before analysis [3].

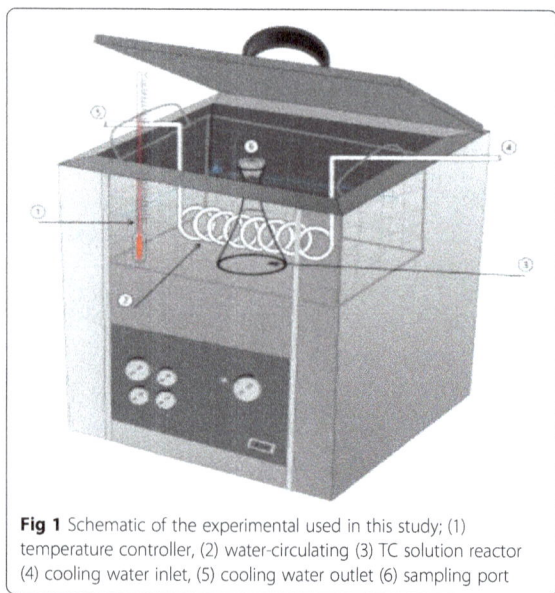

Fig 1 Schematic of the experimental used in this study; (1) temperature controller, (2) water-circulating (3) TC solution reactor (4) cooling water inlet, (5) cooling water outlet (6) sampling port

Analytical methodology

The pH was determined at room temperature using an S-20 pH meter, which was calibrated with pH 4.0 and 7 reference buffer solutions. The concentration of TC in aqueous solution was analyzed by HPLC, with a LC-20 AB pump, Shimadzu, Kyoto, Japan) with a reversed-phase column (VP-ODS-C18 4.6 mm × 250 mm, 5 μm, Shim-Pack, Kyoto, Japan), and UV detector (Shimadzu UV-1600 spectrophotometer). The injection volume was 20 μL; the mobile phase was acetonitrile 0.01 M, oxalic acid solution (31:69, v/v) with a flow rate of 1.0 mL min^{-1}. The detection wavelength and retention time of tetracycline were 360 nm and 2.38 min, respectively. In this study, limit of detection (LOD) were found to be 0.02-0.03 mg/L based on linear regression method.

Experimental design

A central composite statistical experiment design was used to evaluate the effects of four independent variables (initial solution pH (A), initial TC concentration (B), initial $S_2O_8^{-2}$ concentration (C) and reaction time (D)) on the TC degradation. The application of RSM provides a mathematical relationship between variables and experimental data can be fitted to an empirical second-order polynomial model as the following Eq. (8). [55–57].

$$Y = \beta_0 + \beta_1 A + \beta_2 B + \beta_3 C + \beta_4 D + \beta_{12} AB$$
$$+ \beta_{13} AC + \beta_{14} AD + \beta_{23} BC + \beta_{24} BD$$
$$+ \beta_{34} CD + \beta_{11} A^2 + \beta_{22} B^2 + \beta_{33} C^2$$
$$+ \beta_{44} D^2 \qquad (8)$$

Where, y (%) is the predicted response (TC degradation rate), β_0 is interception coefficient, β_1, β_2, β_3 and β_4 are the linear coefficients, β_{12}, β_{13}, β_{14}, β_{23}, β_{24} and β_{34} are interaction coefficients, β_{11}, β_{22}, β_{33} and β_{44} are the quadratic coefficients and A, B, C and D are the independent variables.

The natural and coded levels of independent variables based on the central composite design are shown in Table 2. The experimental values for each independent variables were chosen according to the results obtained from preliminary analysis. Table 3 indicates the four-factor, five-level CCD and the obtained and predicted values for the TC degradation rate (%) using the developed quadratic model. In RSM analysis, the approximation of y was proposed using the fitted second-order polynomial regression model which is called the quadratic model. A quadratic regression is the process of finding the equation of the parabola that fits best for a set of data [58].

Table 2 Natural and coded levels of independent variables based on the central composite design

Independent variable	Symbol	Coded levels				
		-2	-1	0	+1	+2
		Natural level				
pH	A	2.5	5	7.5	10	12.5
Tetracycline (mg/L)	B	10	30	50	70	90
Persulfate (Mm)	C	1	2	3	4	5
Reaction time (min)	D	30	60	90	120	150

Resulta and Disscusion

Analysis of variance (ANOVA)

The results of the analysis of variance test is summarized in Table 4. The probability > F for the model is less than 0.05 which implies that the model is significant and the terms in the model have significant effects on the response. In this case A, B, C, D, AB, AC, AD, BC, BD, CD, A2, B^2, C^2, D^2 are significant model terms at the 95 % confidence level (α =5 %). The model F-value of 1387.59 and P-value of < 0.0001 implies that the model is highly significant. Based on the ANOVA results, the values of R^2, Adjusted R^2 and Predicted R^2 were 0.9992, 0.9985 and 0.9971, respectively. This result suggests that the regression model is well interpreted the relationship

Table 3 Four-factor five-level central composite design for RSM

Run	Experimental conditions				TC degradation rate (%)	
	pH (A)	Tetracycline (mg/L) (B)	Persulfate (mM) (C)	Time (min) (D)	Observed (%)	Predicted (%)
1	7.5 (0)	50 (0)	3 (0)	90 (0)	51.06	49.82
2	2.5 (−2)	50 (0)	3 (0)	90 (0)	55.64	55.74
3	7.5 (0)	50 (0)	3 (0)	90 (0)	48.45	49.82
4	5 (−1)	70 (+1)	4 (+1)	60 (−1)	45.16	44.64
5	5 (−1)	30 (−1)	4 (+1)	120 (+1)	86.62	86.33
6	10 (+1)	70 (+1)	2 (−1)	120 (+1)	61.02	61.32
7	7.5 (0)	50 (0)	3 (0)	150 (+2)	81.85	81.17
8	5 (−1)	30 (−1)	2 (−1)	60 (−1)	34.55	35.25
9	10 (+1)	30 (−1)	2 (−1)	60 (−1)	47.25	46.91
10	10 (+1)	30 (−1)	4 (+1)	60 (−1)	70.44	69.95
11	5 (−1)	30 (−1)	2 (−1)	120 (+1)	61.85	61.79
12	7.5 (0)	50 (0)	1 (−2)	90 (0)	28.72	28.18
13	10 (+1)	70 (+1)	4 (+1)	120 (+1)	85.05	84.34
14	7.5 (0)	50 (0)	3 (0)	90 (0)	49.75	49.82
15	10 (+1)	30 (−1)	2 (−1)	120 (+1)	75.56	76.07
16	10 (+1)	30 (−1)	4 (+1)	120 (+1)	94.25	95.04
17	5 (−1)	70 (+1)	2 (−1)	60 (−1)	12.65	11.98
18	5 (−1)	30 (−1)	4 (+1)	60 (−1)	64.04	63.86
19	7.5 (0)	90 (+2)	3 (0)	90 (0)	41.15	41.31
20	10 (+1)	70 (+1)	4 (+1)	60 (−1)	54.88	55.05
21	7.5 (0)	50 (0)	3 (0)	90 (0)	50.55	49.82
22	7.5 (0)	50 (0)	3 (0)	90 (0)	49.75	49.82
23	7.5 (0)	50 (0)	5 (+2)	90 (0)	79.38	79.82
24	12.5 (+2)	50 (0)	3 (0)	90 (0)	80.62	80.42
25	7.5 (0)	10 (−2)	3 (0)	90 (0)	75.55	75.28
26	7.5 (0)	50 (0)	3 (0)	90 (0)	49.35	49.82
27	7.5 (0)	50 (0)	3 (0)	30 (−2)	24.75	25.33
28	5 (−1)	70 (+1)	2 (−1)	120 (+1)	42.25	42.72
29	10 (+1)	70 (+1)	2 (−1)	60 (−1)	27.68	27.96
30	5 (−1)	70 (+1)	4 (+1)	120 (+1)	70.85	71.31

Table 4 ANOVA results for Response Surface Quadratic Model

Source	Sum of squares	df	Mean square	F -Value	P- value
model	12060.48	1	861.46	1387.59	<0.0001
A-pH	914.15	1	914.15	1472.45	<0.0001
B-TCcon.	1730.94	1	1730.94	2788.08	<0.0001
C-PScon.	3999	1	3999	6441.32	<0.0001
D-Time	4676.04	1	4676.04	7531.85	<0.0001
AB	18.66	1	18.66	30.06	<0.0001
AC	30.97	1	30.97	49.88	<0.0001
AD	6.84	1	6.84	11.01	0.0047
BC	16.4	1	16.4	26.42	0.0001
BD	17.64	1	17.64	28.41	<0.0001
CD	16.61	1	16.61	26.75	0.0001
A^2	571.64	1	571.64	920.76	<0.0001
B^2	123.3	1	123.3	198.6	<0.0001
C^2	29.96	1	29.96	48.26	<0.0001
D^2	20.18	1	20.18	32.5	<0.0001
Residual	9.31	15	0.62		
Lack of Fit	5.13	10	0.51	0.61	0.7607
Pure Error	4.18	5	0.84		
Cor Total	12069.8	29			

$R^2 = 0.9992$; Adjusted $R^2 = 0.9985$ and Predicted $R^2 = 0.9971$

between the independent variables and the response. Furthermore, the adequate precision ratio of 149.08 in the study shows that this model could be applied to navigate the design space defined by the CCD.

Evaluation of model adequacy

There are many statistical techniques for the evaluation of model adequacy, but graphical residual analysis is the primary statistical method for assessment of model adequacy [59].

The normal probability plot indicates that the points on this plot are formed a nearly linear pattern (Fig. 2 (a)). Therefore, the normal distribution is a good model for this data set. Random scattering of the points of internally studentized residual (the residual divided by the estimated standard deviation of that residual) versus predicted values between −3 and +3 emphasizes highly accurate prediction of the experimental data through the derived quadratic model (Fig. 2 (b)).

The plot of predicted vs Actual values (Fig. 2 (c)) indicate a higher correlation and low differences between actual and predicted values. Hence, the predictions of the experimental data by developed quadratic models for the TC degradation is perfectly acceptable and this model fits the data better. Also, the random spread of

the residuals across the range of the data between −3 and +3 implies that there are no evident drift in this process and the model was a goodness fit (Fig. 2 (d)). The Box-cox plot is used for determine the suitability of a power low transformation for the selected data (Fig. 3 (a)). In this study, the best lambda values of 0.92 was obtained with low and high confidence interval 0.73 and 1.11, respectively. Therefore, recommend the standard transformation by the software is 'None'. The plot of points Leverage vs Run order is shown in Fig. 3 (b). The factorial and axial points have the most influence with a leverage of approximately 0.59, while the center points have the least effect with a leverage of 0.16.

Design matrix evaluation for response surface quadratic model

Design matrix evaluation implies that there are no aliases for the quadratic model. In general, a minimum of degrees of freedom 3 and 4 has been recommended for lack-of-fit and pure error, respectively. Therefore, degrees of freedom obtained in this study ensured a valid lack of fit test (Table 5).

The standard error (SE) used to measure the precision of the estimate of the coefficient. The smaller standard error implies the more accurate the estimate. The variables of A, B, C and D have a standard errors = 0.16. The interceptions of AB, AC, AD, BC, BD and CD have slightly high standard errors = 0.2, while A^2, B^2, C^2 and D^2 have standard errors = 0.15. An approximate 95 % confidence interval for the coefficient is given by the estimate plus and minus 2 times the standard error. For example, with 95 % confidence can be said that the value of the regression coefficient A is between 6.49 and 5.85 ($6.17 \pm 2 \times 0.16$).

The quadratic model coefficients for the CCD are shown in Table 6. This results suggested that the variables coefficients and their interactions are estimated adequately without multicollinearity. The low Ri-squared for independent variables and their interactions imply that the model is a good fit. In general, power should be approximately 80 % for detecting an effect [60]. In this study, there are more than 99 % chance of detecting a main effect while it is twice the background sigma.

Final equation and model graphs

The values of regression coefficients were determined and the experimental results of CCD were fitted with second order polynomial equation. The quadratic model for TC degradation rate in terms of coded were determined using as following Eq. (9):

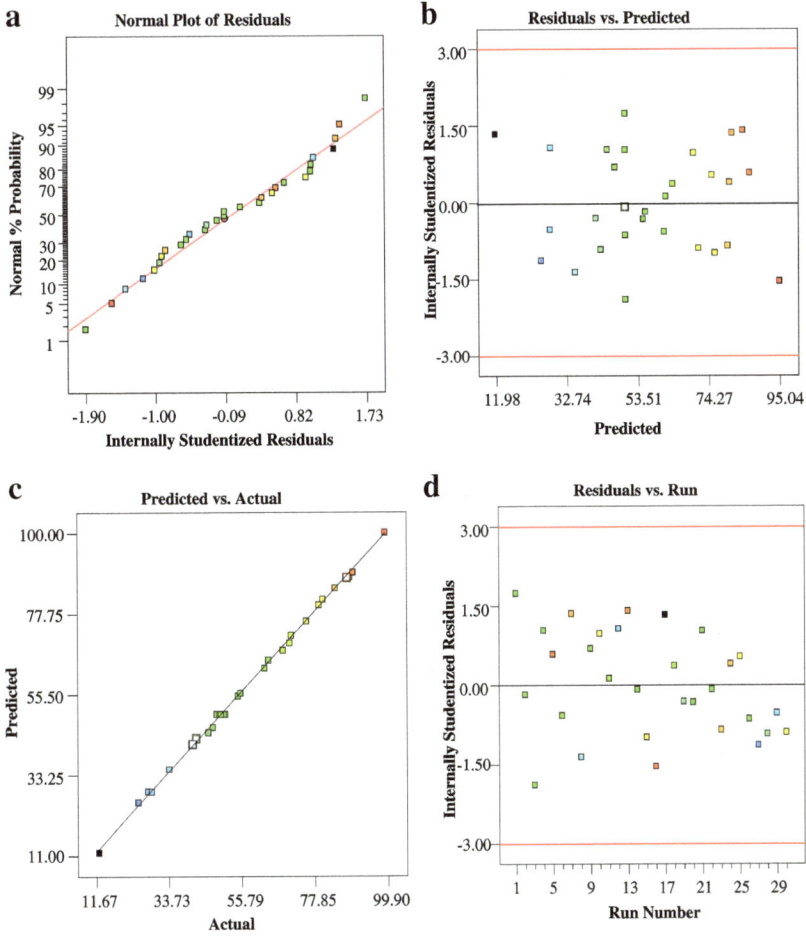

Fig 2 The plot of: (**a**) Normal Plot of Residuals; (**b**) residuals vs Predicted Response; (**c**) predicted vs Actual values; (**d**) Residuals vs Run Order

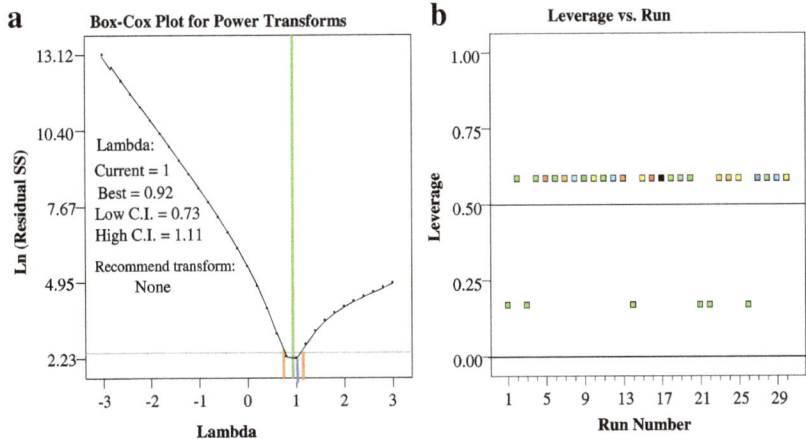

Fig 3 The plot of: (**a**) Box-Cox Plot for Power Transforms; (**b**) The points Leverage vs Run order for the CCD Design

Table 5 Degrees of freedom for evaluation

Model	14
Residuals	15
Lack of Fit	10
Pure Error	5

Final equation in terms of coded factors

$$Y = + 49.82 + 6.17 * A - 8.49 * B + 12.91 * C \\ + 13.96 * D + 1.08 * A * B - 1.39 * A * C \\ + 0.65 * A * D + 1.01 * B * C + 1.05 * B \\ * D - 1.02 * C * D + 4.57 * A^2 + 2.12 * B^2 \\ + 1.05 * C^2 + 0.86 * D^2 \quad (9)$$

The factors in the quadratic equation were coded to produce the response surface with limiting the responses into a range of −1 to +1. The ramp function graph for the maximum TC degradation rate is shown in Fig. 4. The optimization of experimental conditions was conducted for maximize the TC degradation at defined criteria of the variable. The developed quadratic model for the TC degradation (Eq. (8)) was applied as an objective function to the optimization of operating conditions. Consequently, the optimum parameters were achieved using the numerical technology based on the predicted model and the variable in their critical range. The maximum degradation of 95.01 % was achieved at pH = 9.9, TC concentration = 30.19 mg/L, PS concentration = 3.97 mM and reaction time = 119.98 min. in order to evaluation of the model validity, the experiments were carried out under the optimal operating conditions. 93.45 % TC degradation was

obtained under the optimum operating conditions, which supported the results of the developed model.

The perturbation Plot of independent variables implies that reaction time (D) has the most significant effect (steepest slope) on the TC degradation rate, followed by $S_2O_8^{-2}$ concentration (C) and TC concentration (B), whereas pH (A) has the lowest effect on the TC degradation. (Fig. 5).

Interactive effect of independent variables on the TC degradation

Three-dimensional surfaces and contour plots are graphical representation of regression equation for the optimization of reaction Status. The results of the interactions between four independent variables and dependent variable are indicated in Figs. 6 and 7.

Figure 6(a) indicates the interaction effect of TC concentration and PS concentration on the TC degradation rate with reaction time of 120 min. with the increasing PS concentration, the TC degradation rate significantly enhanced. With increasing PS concentration from 2 to 4 mM, the TC degradation rate increased from 75.56 % to 94.25 % at TC concentration of 30 mg/L. These results suggest that with increasing PS concentration, more sulfate radicals are produced which leads to more quickly TC degradation [32].

Figure 6(b) indicates the interaction effect of initial TC concentration and reaction time on the TC degradation rate. The TC degradation rate strongly increased with increase of sonication time from 60 to 120 min. with increasing reaction time from 60 to 120 min, TC concentration of 30 and 70 mg/L, the TC degradation rate

Table 6 The Quadratic model coefficients for the CCD

Term	StdErr**	VIF	Ri-Squared	Power at 5 % SN = 0.5	Power at 5 % SN = 1	Power at 5 % SN = 2
A	0.16	1	0	20.90 %	63.00 %	99.50 %
B	0.16	1	0	20.90 %	63.00 %	99.50 %
C	0.16	1	0	20.90 %	63.00 %	99.50 %
D	0.16	1	0	20.90 %	63.00 %	99.50 %
AB	0.2	1	0	15.50 %	46.50 %	96.20 %
AC	0.2	1	0	15.50 %	46.50 %	96.20 %
AD	0.2	1	0	15.50 %	46.50 %	96.20 %
BC	0.2	1	0	15.50 %	46.50 %	96.20 %
BD	0.2	1	0	15.50 %	46.50 %	96.20 %
CD	0.2	1	0	15.50 %	46.50 %	96.20 %
A^2	0.15	1.05	0.0476	68.70 %	99.80 %	99.90 %
B^2	0.15	1.05	0.0476	68.70 %	99.980 %	99.90 %
C^2	0.15	1.05	0.0476	68.70 %	99.80 %	99.90 %
D^2	0.15	1.05	0.0476	68.70 %	99.80 %	99.90 %

**Basis Std. Dev. = 1.0

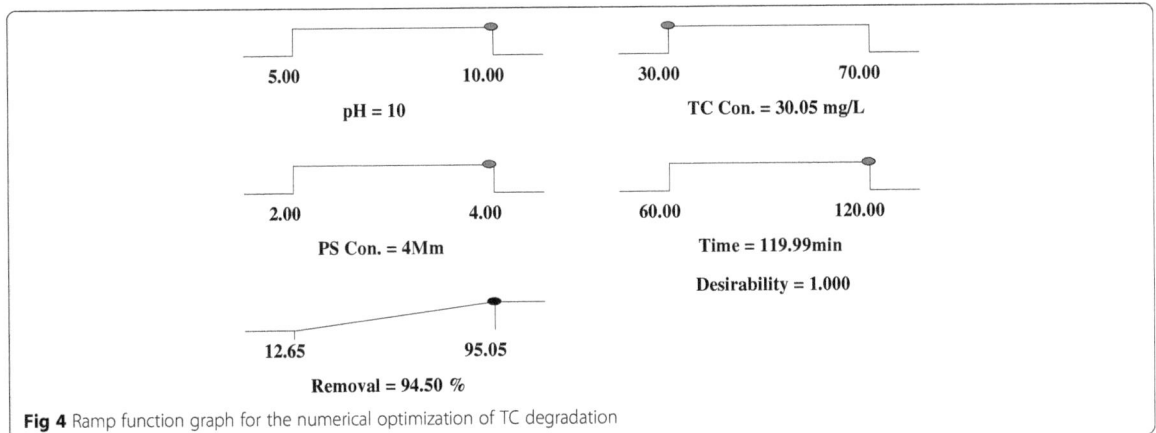

Fig 4 Ramp function graph for the numerical optimization of TC degradation

increased from 70.44 % to 94.25 % at TC concentration of 30 mg/L. With increasing the TC concentration from 30 to 70 mg/L, the TC degradation rate decreased from 94.25 % to 85.05 %. In the constant conditions, with the increasing TC concentration, possibility of reaction between TC molecules and reactive species were declined. Moreover, the higher concentration of TC may lead to the creation of resistant byproducts and consequently decreases the degradation rate of TC [14, 61]. However, the total amount of degraded TC increased with the increasing initial TC concentration. This results are in agreement with the results obtained by other researchers [50].

Figure 7 indicates the interaction influence of pH value and initial TC concentration on the TC degradation rate. With increasing pH from acidic (5) to natural (7.5), the degradation rate slightly decreased, whereas with increasing pH from neutral (7.5) to alkaline (10), the degradation rate significantly enhanced. The TC degradation rate increased from 86.62 % to 94.25 % with increasing pH from 5 to 10, at TC concentrations of 30 mg/L. Under alkaline conditions (pH $\geq$10), alkaline-activated persulfate is the primary responsible for the production of SO4$^{-\bullet}$, O$_2^{\bullet}$ and HO$^{\bullet}$ radicals as following equations: [62, 63].

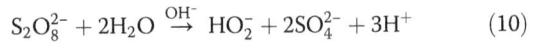

$$S_2O_8^{2-} + 2H_2O \overset{OH^-}{\to} HO_2^- + 2SO_4^{2-} + 3H^+ \qquad (10)$$

$$HO_2^- + S_2O_8^{2-} \to SO_4^{-\bullet} + SO_4^{2-} + H^+ + O_2^{-\bullet} \qquad (11)$$

$$SO_4^{-\bullet} + OH^- \to SO_4^{2-} + HO^{\bullet} \qquad (12)$$

Also, at alkaline pH, sulfate radicals can react with hydroxyl anions to generate hydroxyl radicals (HO$^{\bullet}$) according to Eq. (3). In addition, a theory was introduced by other researchers that with increasing pH, the PS degradation into HO$^{\bullet}$ and SO$_4^{-\bullet}$ increased [64].

The SO$_4^{-\bullet}$ is the predominant radical responsible for TC degradation at acidic pH, whereas both SO$_4^{-\bullet}$ and OH$^{\bullet}$ are contributing in TC degradation at natural pH. Thus, three reactions compete with each other in natural pH: the reaction between SO$_4^{-\bullet}$ and HO$^{\bullet}$, the reaction between SO$_4^{-\bullet}$ and TC, and the reaction between HO$^{\bullet}$ and TC, the simultaneous occurrence of these reactions may reduce the TC degradation rate [37, 65].

Kinetics of tetracycline degradation

The sonochemical degradation process typically follows pseudo first-order kinetics as shown in the following Eqs. (13) and (14). Many studies have suggested that oxidation of organic pollutants by ultrasound follows pseudo first-order kinetics [42, 47, 52].

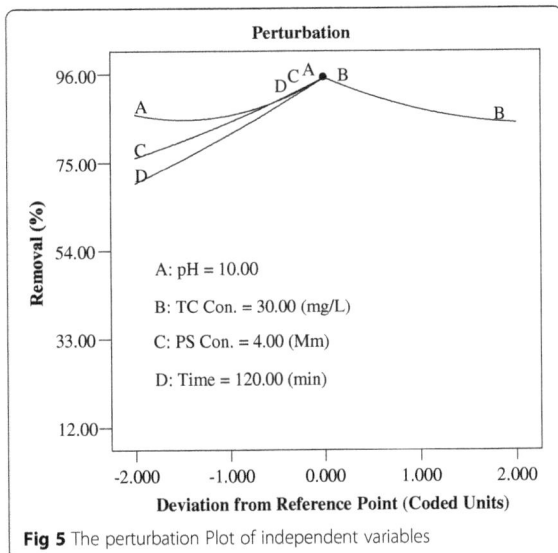

Fig 5 The perturbation Plot of independent variables

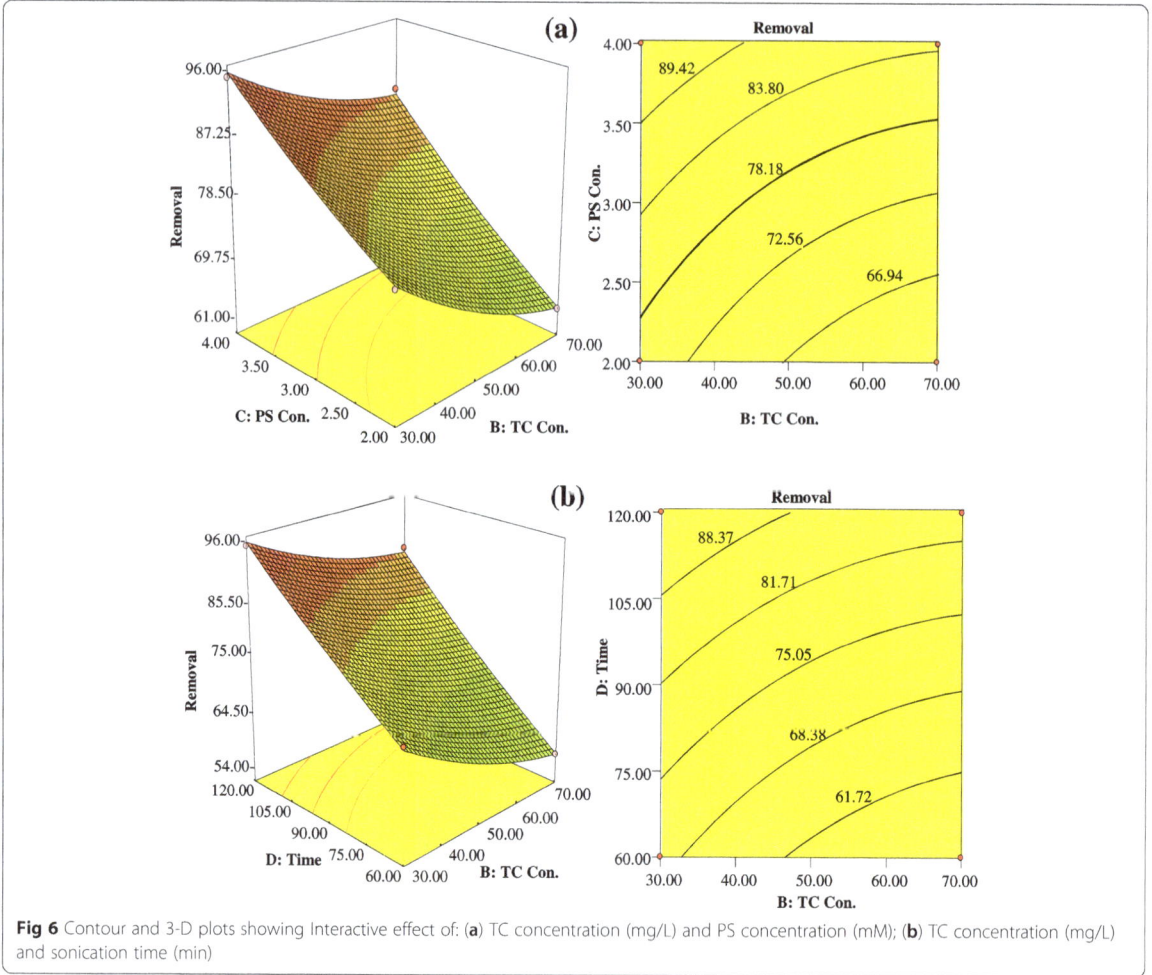

Fig 6 Contour and 3-D plots showing Interactive effect of: (**a**) TC concentration (mg/L) and PS concentration (mM); (**b**) TC concentration (mg/L) and sonication time (min)

$$-d[TC]/dt = k[TC] \tag{13}$$

Eq. (13) can be rewritten as:

$$\ln C_i/C_t = kt \tag{14}$$

Where C_i is the initial TC concentration, C_t is the TC concentration at time t, k is the pseudo first order reaction rate t is constant (min^{-1}) and the reaction time (min). To study the TC degradation by US/$S_2O_8^{2-}$ process, the data obtained was investigated using the pseudo first order kinetics. The effect of different parameters such as initial TC concentration, initial PS concentration, pH and temperature on the kinetic of TC degradation was evaluated. In all the experiments, TC degradation well-fitted to the using the pseudo first order kinetics with higher correlation coefficients (R^2). The values of kinetic rate constants

Table 7 Effect of operation parameters on the kinetics degradation of TC

parameter	Value	k_0 (min^{-1}) $\times 10^{-2}$	R^2	$t_{1/2}$ (min)
TC concentration (mg/L)	25	2.29	0.9973	30.2
	50	1.75	0.9952	39.6
	75	1.23	0.9956	56.3
PS concentration (mM)	2	1.15	0.9816	60.6
	3	1.52	0.9946	45.9
	4	2.29	0.9973	30.2
pH	5	1.62	0.9937	42.7
	7.5	1.12	0.9942	62.8
	10	2.29	0.9973	30.2
Temperature (°C)	25	2.29	0.9973	30.2
	45	5.70	0.9127	12.1
	55	7.87	0.921	8.8
	65	10.42	0.9824	6.6

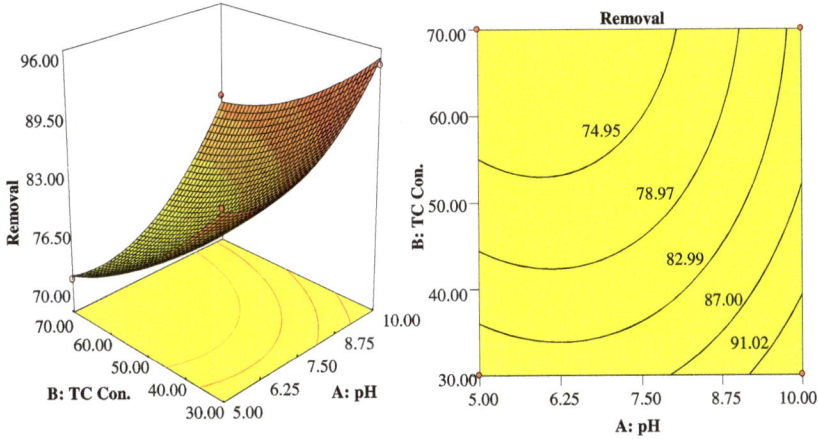

Fig 7 Contour and 3-D plots showing Interactive effect of pH and TC concentration (mg/L)

(k) related to the different parameters, with their regression coefficients R^2 are shown in Table 7.

The effect of temperature on the degradation of tetracycline

To investigate the effect of temperature on the TC degradation rate, experiments were done with various temperature varying from 25 to 65 °C. With increasing temperature from 25 to 65 °C, the degradation rate constant increased from 0.0229 to 0.1042 min $^{-1}$. Complete TC degradation occurs after 40, 60 and 75 min of reaction at 65, 55 and 45 °C respectively. The activation of $S_2O_8^{2-}$ can be done under heat to form $SO_4^{-\bullet}$ radical as following Eq. (15). Therefore, complete removal of TC by high temperature could be as a result of thermally activated

$S_2O_8^{2-}$ oxidation. Moreover, the increase of temperature significantly enhanced the cavitation activity and chemical effects, resulting in greater degradation rate of TC by US/$S_2O_8^{2-}$ process [22, 60].

$$S_2O_8^{-2} + \xrightarrow{\text{Termal-activation}} 2SO_4^{-\bullet} \qquad (30^{\circ\circ}C < T < 99^{\circ\circ}C) \tag{15}$$

To investigate the effect of ultrasound on the process kinetics, significant parameters such as activation energy (Ea) play a remarkable role. The effect of temperature on the rate of the reaction and rate constant (k) is obtained by Arrhenius equation according with Eq. (16) [66].

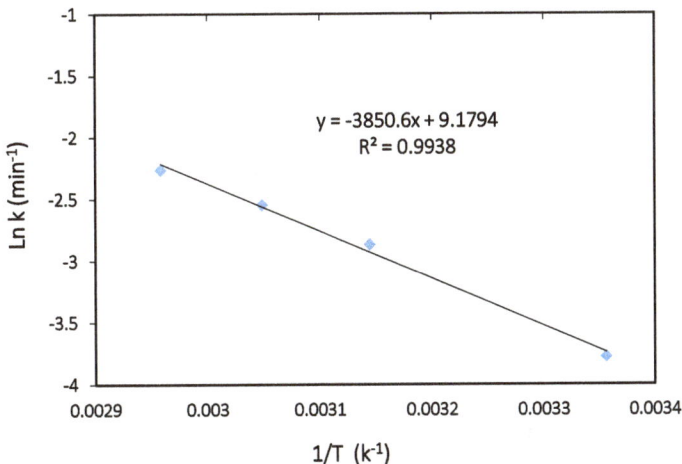

Fig 8 Arrhenius equation graph representation the temperature dependence on chemical reaction rate

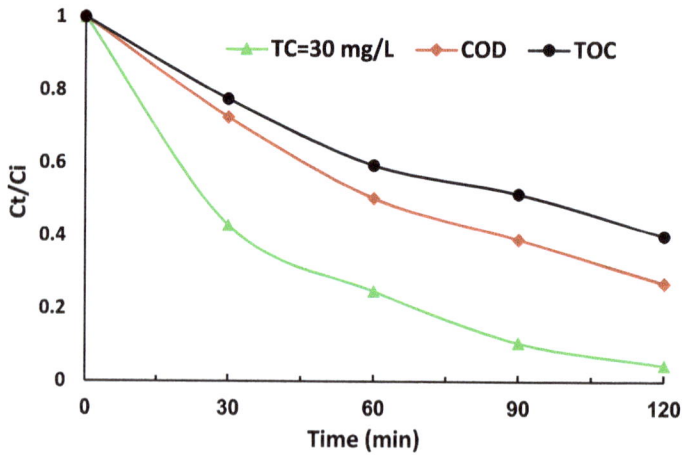

Fig 9 Removal of TC, COD and TOC by US/$S_2O_8^{2-}$ process; [$S_2O_8^{2-}$] = 4 mM; US: 500 W, 35 KHz; pH=10; T=25 ^{0}C

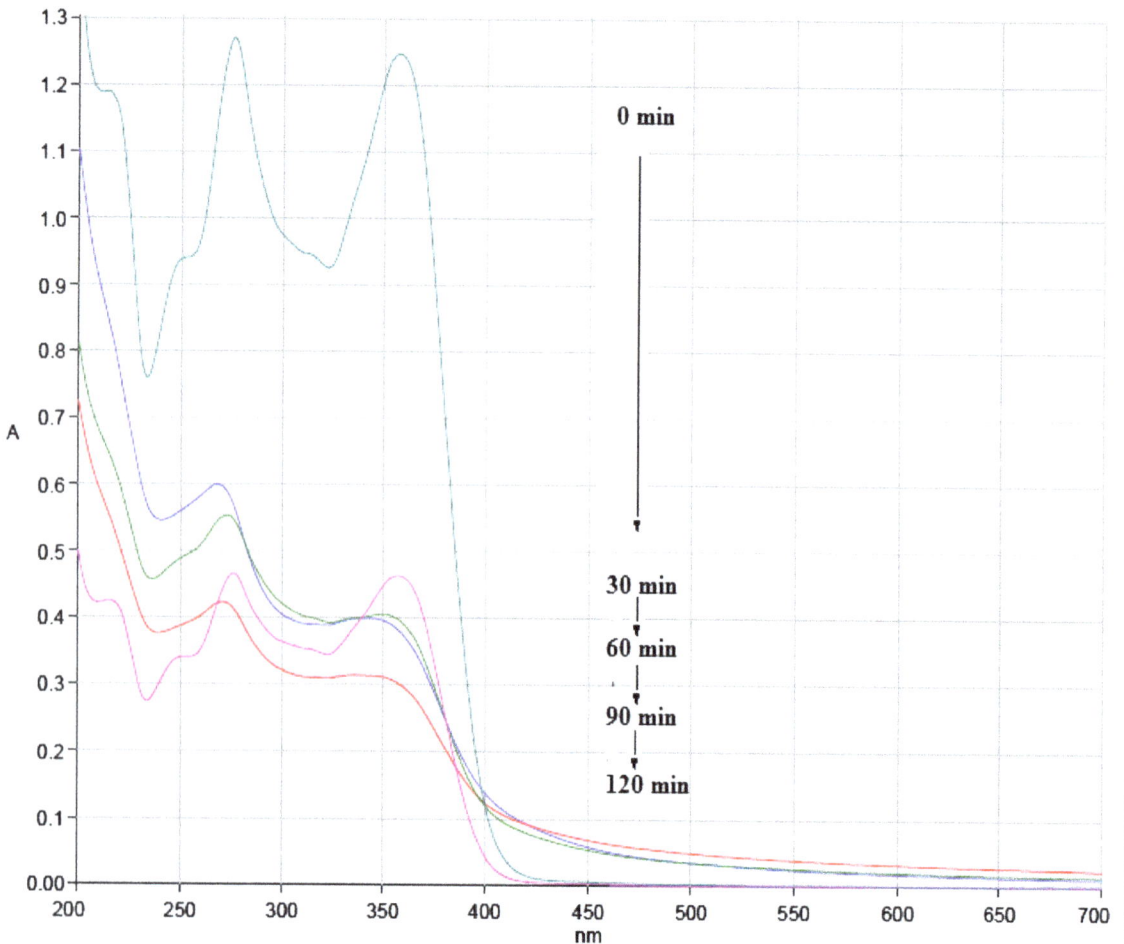

Fig 10 Changes of UV–Vis spectra of 50 mg/L aqueous solution of TC during process of US/$S_2O_8^{2-}$

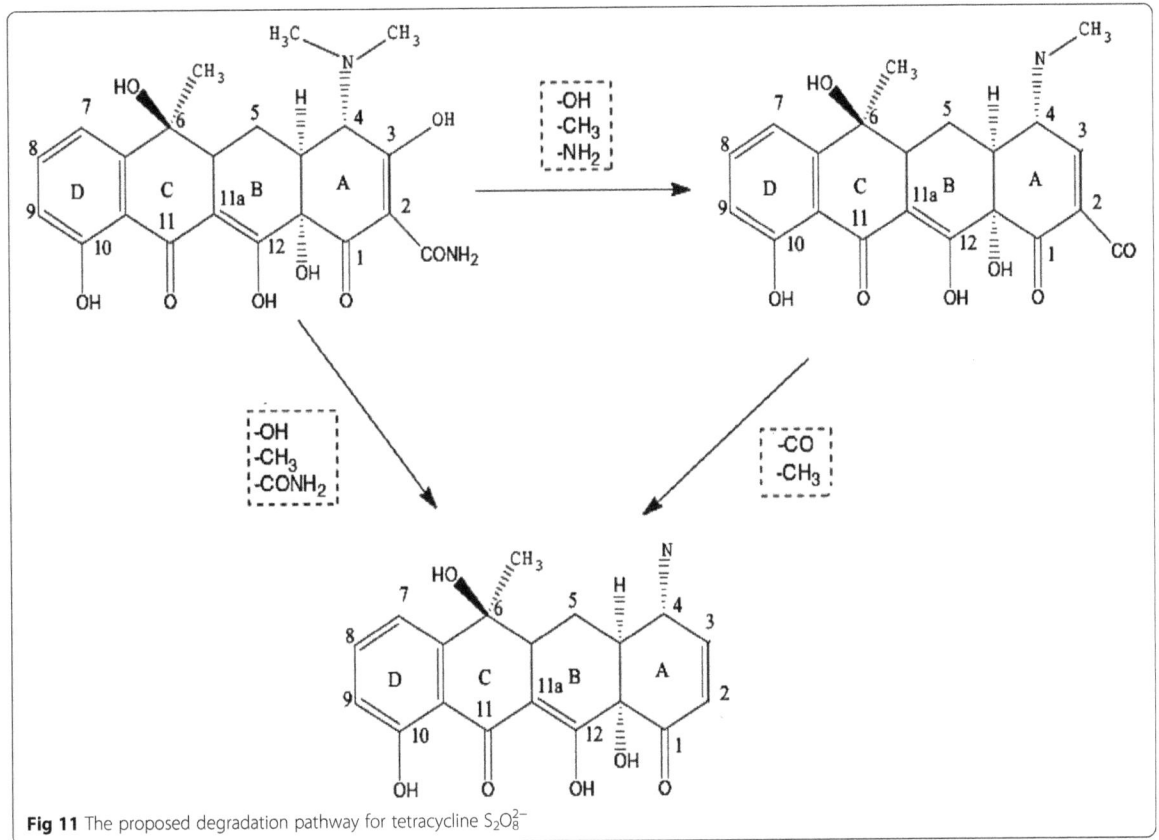

Fig 11 The proposed degradation pathway for tetracycline $S_2O_8^{2-}$

$$LnK = A \; exp \left(-\frac{Ea}{RT} \right) \qquad (16)$$

Arrhenius plot can be used to calculate the Activation Energy at various temperatures by graphing ln k (rate constant) versus 1/T (kelvin). The graph between ln k and 1/T is a straight line with an intercept of ln A and the slop of the graph is equal to $-E_a/R$, where R is a constant equal to 8.314 J/mol-K. According with Arrhenius plot (Fig. 8), the activation energy values of 32.01 (kJ/mol) obtained for degradation of TC by $S_2O_8^{2-}$/US process. It means that for a successful reaction, the colliding molecules must have a

total kinetic energy of 32.01 kJ/mol. The low activation energy indicates that the degradation of TC by $S_2O_8^{2-}$/US process is thermodynamically feasible.

Mineralization, Changes of ultraviolet Visible (UV–Vis) spectra and the proposed degradation pathway

Usually, sonochemical treatment lead to degradation of structure and ultimately mineralization of organic compounds [67]. While perfect mineralization for most antibiotics are difficult because of the structural stability [68]. Therefore, the changes of TOC and COD were evaluated during US/S_2O8^{2-} process and the result are shown

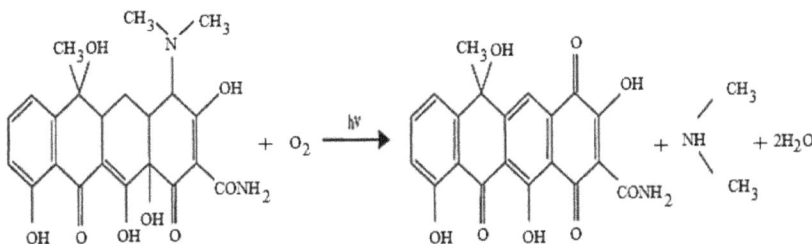

Fig 12 Photodegradation of TC into 4a, 12a-anhydro-4-oxo-4-dedimethylaminotetracycline

Table 8 Degradation of different types of organic pollutants in aqueous solutions using $US/S_2O_8^{2-}$ process

compound	Concentration (mg/L)	operating conditions	Summary of results	reference
Tetracycline	100	$[S_2O_8^{2-}] = 200$ mM; US = 80 W, 20 KHz; pH = 3.7; T = ambient	More than 51 % degradation after 120 min.	Hou et al. [48]
Trichloroethane	50	$[S_2O_8^{2-}] = 0.94$ mM US = 400 kHz, 100 W pH = 7; T = 20 ± 2 °C	100 % degradation after 120 min.	Li et al. [42]
Perfluorooctanoic acid	50	$[S_2O_8^{2-}] = 46$ mM US = 150 W, 40 KHz pH = 4.3; T = 25 °C	More than 98 % degradation after 120 min.	Lin et al. [52]
2,4 Dichlorophenol	50	$[S_2O_8^{2-}] = 4$ Mm US = 40 KHz, pH = 3; T = 30 °C	More than 95 % degradation after 60 min.	Seid Mohammai [74]
Acid Orange 7	30	$[S_2O_8^{2-}] = 1.25$ Mm US = 100 W, 20 kHz pH = 5.8; T = ambient	More than 10 % degradation after 20 min.	Wang et al. [75]
Tetracycline	30	$[S_2O_8^{2-}] = 4$ Mm US = 500 W, 35 KHz pH = 10; T = 25 ± 2 °C	More than 95 % degradation, COD and TOC removal of 72 % and 59 % after 120 min	Present study

in Fig. 9. After 120 min of reaction, TC, COD and TOC were removed approximate 95 %, 73 % and 60 %. The incomplete mineralization implies the potential formation of intermediate products and further identification of the degradation by-products is required.

To evaluate structural changes of TC, the UV–Vis spectra obtained before and after US/ $S_2O_8^{2-}$ process in various time are shown in Fig. 10. The UV–Vis spectra obtained before process indicates two main absorption bands at 275 and 360 nm. The absorption of TC in 360 nm is due to aromatic rings B–D, such as the developed chromophores [68, 69]. With increase of reaction time, the absorption band slightly decreased because of the fragmentation of phenolic groups attached to aromatic ring B [69, 70]. The generation of acylamino and hydroxyl groups led to reduction of absorbance at 270 nm band [70]. The absorption decay at 360 nm band faster than 275 nm. This implies that the ring containing the N-groups (responsible for the absorbance at 276 nm) hardly opened than the other rings, or the created intermediate products absorbed at this wavelength [71]. The proposed degradation pathway for tetracycline based on loses of N-methyl, hydroxyl, and amino groups is shown in Fig. 11. This possible pathway corresponded with conducted studies by other researchers [72]. In addition, TC has a naphthol ring with high stability, which remains unchanged in the reaction and is not easily mineralized. Also, the absorption decay at 360 nm band was found with a relatively small absorption in the visible region. This could be due to the forming of 4a,12a-anhydro-4-oxo-4-dedimethylaminotetracycline according to Fig. 12 [73].

Performance evaluation of $US/S_2O_8^{2-}$ process in the removal of different organics

An overview of performance of $US/S_2O_8^{2-}$ process for removal of different organics along with present study was presented in Table 8. The overview confirm that US/ $S_2O_8^{2-}$ process is an attractive alternative technique for degradation of the wide range of organic compounds in aqueous solutions. This process could effectively decomposed organic pollutants in aqueous solution, and the degradation rate depends heavily on the operating conditions, such as physical and chemical characteristics and initial concentration of pollutant, $S_2O_8^{2-}$ concentration, initial pH, reaction time, ultrasound power, ultrasound frequency, and temperature of the medium. Therefore, the various experimental conditions could lead to the various removal efficiencies of organic compounds using $US/S_2O_8^{2-}$ process.

Conclusion

Sonochemical degradation of TC in the presence of $S_2O_8^{2-}$ was investigated with focusing on the optimizing of the operation parameters such as pH, $S_2O_8^{2-}$ concentration, initial TC concentration and reaction time. This study indicated that RSM was the suitable method to optimizing the best operating conditions to maximizing the TC degradation. The reaction time showed the highest effect on the TC degradation, followed by initial $S_2O_8^{2-}$ concentration, initial TC concentration and pH. Under optimal conditions, the TC degradation rate, COD and TOC removal efficiency were found to be 95.01 %, 72.8 % and 59.7 %, respectively. The degradation process followed the pseudo-first-order kinetics. The activation energy value of 31.71 kJ/mol implies that the degradation of TC by $US/S_2O_8^{2-}$ process is thermodynamically feasible. The ultraviolet visible spectra obtained before and after ultrasound irradiation in the presence of $S_2O_8^{2-}$ indicated that proposed degradation pathway for tetracycline was based on loses of N-methyl, hydroxyl, and amino groups. Overall, $US/S_2O_8^{2-}$ process was found to be a promising technology for TC degradation in aqueous solution.

Competing interests
The authors declare that they no competing interests.

Authors' contributions

GHS was the main investigator, collected the data, performed the statistical analysis, and drafted the manuscript. SN and AHM supervised the study and participated in the design of the study, final revised of manuscript and intellectual helping for analyzing of data. KY, RN, and MA were advisors of the study and carried out statistical and technical analysis of data, participated in design of study and manuscript preparation. All authors read and approved the final manuscript.

Acknowledgments

This research was part of a PhD dissertation of the first author and has been supported by Tehran University of Medical Sciences under grant no. 92-03-46-24084. The authors express their gratitude to all laboratory staff of the Department of Environmental Health Engineering.

Author details

[1]Department of Environmental Health Engineering, School of Public Health, Tehran University of Medical Sciences, Tehran, Iran. [2]Center for Water Quality Research, Institute for Environmental Research, Tehran University of Medical Sciences, Tehran, Iran. [3]Center for Solid Waste Research, Institute for Environmental Research, Tehran University of Medical Sciences, Tehran, Iran. [4]Center for Air Pollution Research, Institute for Environmental Research, Tehran University of Medical Sciences, Tehran, Iran.

References

1. Javid A, Nasseri S, Mesdaghinia A, Mahvi AH, Alimohammadi M, Aghdam RM, et al. Performance of photocatalytic oxidation of tetracycline in aqueous solution by TiO_2 nanofibers. J Environ Health Sci Eng. 2013;11:24.
2. Daghrir R, Drogui P. Tetracycline antibiotics in the environment: a review. Environ Chem Lett. 2013;11:209–27.
3. Jiang WT, Chang PH, Wang YS, Tsai Y, Jean JS, Li Z. Sorption and desorption of tetracycline on layered manganese dioxide birnessite. Int J Environ Sci Technol. 2015;12:1695–704.
4. Lindsey ME, Meyer M, Thurman E. Analysis of trace levels of sulfonamide and tetracycline antimicrobials in groundwater and surface water using solid-phase extraction and liquid chromatography mass spectrometry. Anal Chem. 2001;73:4640–6.
5. Alavi N, Babaei AA, Shirmardi M, Naimabadi A, Goudarzi G. Assessment of oxytetracycline and tetracycline antibiotics in manure samples in different cities of Khuzestan Province. Iran Environ Sci Pollut Res. 2015;1–7.
6. Zhu J, Snow DD, Cassada DA, Monson SJ, Spalding RF. Analysis of oxytetracycline, tetracycline and chlortetracycline in water using solid-phase extraction and liquid chromatography-tandem mass spectrometry. J Chromatogr A. 2001;928:177–86.
7. Guler UA, Sarioglu M. Removal of tetracycline from wastewater using pumice stone: equilibrium, kinetic and thermodynamic studies. J Environ Health Sci Eng. 2014;12:1.
8. Reyes C, Fernandez J, Freer J, Mondaca MA, Zaror C, Malato S, et al. Degradation and inactivation of tetracycline by TiO_2 photocatalysis. J Photochem Photobiol A. 2006;184:141–6.
9. Spongberg AL, Witter JD. Pharmaceutical compounds in the wastewater process stream in Northwest Ohio. Sci total environ. 2008;397:148–57.
10. Karthikeyan KG, Michael TM. Occurrence of antibiotics in wastewater treatment facilities in Wisconsin, USA. Sci total environ. 2006;361:196–207.
11. Deblonde T, Cossu-Leguille C, Hartemann P. Emerging pollutants in wastewater: a review of the literature. Int J Hyg Environ Eealth. 2011;214:442–8.
12. Homem V, Santos L. Degradation and removal methods of antibiotics from aqueous matrices-a review. J Environ Manag. 2011;92:2304–47.
13. Choi KJ, Kim SG, Kim SH. Removal of antibiotics by coagulation and granular activated carbon filtration. J Hazard Mater. 2008;151:38–43.
14. Safari GH, Hoseini M, Seyedsalehi M, Kamani H, Jaafari J, Mahvi AH. Photocatalytic degradation of tetracycline using nanosized titanium dioxide in aqueous solution. Int J Environ Sci Technol. 2014;12:603–16.
15. Kakavandi B, Takdastan A, Jaafarzadeh N, Azizi M, Mirzaei A, Azari A. Application of Fe_3O_4@C catalyzing heterogeneous UV-Fenton system for tetracycline removal with a focus on optimization by a response surface method. J Photochem Photobio A Chem. 2015;2016(314):178–88.
16. Tsitonaki A, Petri B, Crimi M, Mosbaek H, Siegrist RL, Bjerg PL. In situ chemical oxidation of contaminated soil and groundwater using persulfate: a review. Crit Rev Environ Sci Technol. 2010;40:55–91.
17. Li SX, Wei D, Mak NK, Cai ZW, Xu XR, Li HB, et al. Degradation of diphenylamine by persulfate: performance optimization, kinetics and mechanism. J Hazard Mater. 2009;164:26–31.
18. Adewuyi YG, Owusu SO. Ultrasound-induced aqueous removal of nitric oxide from flue gases: effects of sulfur dioxide, chloride, and chemical oxidant. J Phys Chem A. 2006;110:11098–107.
19. Weng CH, Tao H. Highly efficient persulfate oxidation process activated with Fe^0 aggregate for decolorization of reactive azo dye Remazol Golden Yellow. Arabian J Chem. 2015; doi:10.1016/j.arabjc.2015.05.012.
20. Fang JY, Shang C. Bromate formation from bromide oxidation by the UV/persulfate process. Environ Sci Technol. 2012;46:8976–83.
21. Johnson RL, Tratnyek PG, Johnson ROB. Persulfate persistence under thermal activation conditions. Environ Sci Technol. 2008;42:9350–6.
22. Mora VC, Rosso JA, Le C, Roux G, Roux GC, Martire DO, et al. Thermally activated peroxydisulfate in the presence of additives: a clean method for the degradation of pollutants. Chemosphere. 2009;75:1405–9.
23. Asgari G, Seidmohammadi AM, Chavoshani A. Pentachlorophenol removal from aqueous solutions by microwave/persulfate and microwave/H_2O_2: a comparative kinetic study. J Environ Health Sci Eng. 2014;12:94.
24. Chen WS, Su YC. Removal of dinitrotoluenes in wastewater by sono-activated persulfate. Ultrason Sonochem. 2012;19:921–7.
25. Furman OS, Teel AL, Watts RJ. Mechanism of base activation of persulfate. Environ Sci Technol. 2010;44:6423–8.
26. Yang S, Yang X, Shao X, Niu R, Wang L. Activated carbon catalyzed persulfate oxidation of azo dye acid orange 7 at ambient temperature. J Hazard Mater. 2011;86:659–66.
27. Fang G, Gao J, Dionysiou DD, Liu C, Zhou D. Activation of persulfate by quinones: Free radical reactions and implication for the degradation of PCBs. Environ Sci Technol. 2013;47:4605–11.
28. Liang C, Liang CP, Chen CC. pH dependence of persulfate activation by EDTA/Fe (III) for degradation of trichloroethylene. J Contam Hydrol. 2009;106:173–82.
29. Liang C, Guo Y, Chien Y, Wu Y. Oxidative degradation of MTBE by pyrite-activated persulfate: proposed reaction kinetics. J Contam Hydrol. 2010;49:8858–64.
30. Criquet J, Leitmer NKV. Electron beam irradiation of aqueous solution of persulfate ions. Chem Eng J. 2011;169:258–62.
31. Ahmad M, Teel AL, Watts RJ. Mechanism of persulfate activation by phenols. Environ Sci Technol. 2013;47:5864–71.
32. Xu XR, Li XZ. Degradation of azo dye Orange G in aqueous solutions by persulfate with ferrous ion. Sep Purif Technol. 2010;72:105–11.
33. Rivas FJ. Polycyclic aromatic hydrocarbons sorbed on soils: a short review of chemical oxidation based treatments. J Hazard Mater. 2006;138:234–51.
34. Anipsitakis GP, Dionysiou DD. Radical generation by the interaction of transition metals with common oxidants. Environ Sci Technol. 2004;38:3705–12.
35. Zhao J, Zhang Y, Quan X, Chen S. Enhanced oxidation of 4-chlorophenol using sulfate radicals generated from zero-valent iron and peroxydisulfate at ambient temperature. Sep Purif Technol. 2010;71:302–7.
36. Antoniou MG, Cruz AA, Dionysiou DD. Degradation of microcystin-LR using sulfate radicals generated through photolysis, thermolysis and e- transfer mechanisms. Appl Catal B Environ. 2010;96:290–8.
37. Lin YT, Liang CJ, Chen JH. Feasibility study of ultraviolet activated persulfate oxidation of phenol. Chemosphere. 2011;82:1168–72.
38. Liang HY, Zhang YQ, Huang SB, Hussain I. Oxidative degradation of p-chloroaniline by copper oxide activated persulfate. Chem Eng J. 2013;218:384–91.
39. Olmez-Hanci T, Arslan-Alaton I. Comparison of sulfate and hydroxyl radical based advanced oxidation of phenol. Chem Eng J. 2013;224:469–74.
40. Ji Y, Dong C, Kong D, Lu J, Zhou Q. Heat-activated persulfate oxidation of atrazine: Implications for remediation of groundwater contaminated by herbicides. Chem Eng J. 2015;263:45–54.
41. Zhou D, Chen L, Zhang C, Yu Y, Zhang L. novel photochemical system of ferrous sulfite complex: Kinetics and mechanisms of rapid decolorization of Acid Orange 7 in aqueous solutions. Water Res. 2014;57:85–97.
42. Li B, Li L, Lin K, Zhang W, Lu S, Luo Q. Removal of 1,1,1-trichloroethane from aqueous solution by a sono-activated persulfate process. Ultrason Sonochem. 2013;20:855–63.

43. Sivakumar R, Muthukumar K. Sonochemical degradation of pharmaceutical wastewater. Clean Soil Air Water. 2011;39:136–41.

44. Neppolian B, Doronila A, Ashokkumar M. Sonochemical oxidation of arsenic (III) to arsenic (V) using potassium peroxydisulfate as an oxidizing agent. Water Res. 2010;44:3687–95.

45. Neppolian B, Jung H, Choi H, Lee JH, Kang JW. Sonolytic degradation of methyl tert-butyl ether: the role of coupled Fenton process and persulphate ion. Water Res. 2002;36:4699–708.

46. Son HS, Choi SB, Khan E, Zoh KD. Removal of 1, 4-dioxane from water using sonication: effect of adding oxidants on the degradation kinetics. Water Res. 2006;40:692–8.

47. Su S, Guo W, Yi C, Leng Y, Ma Z. Degradation of amoxicillin in aqueous solution using sulphate radicals under ultrasound irridation. Ultrason Sonochem. 2012;19:469–74.

48. Hou L, Zhang H, Xue X. Ultrasound enhanced heterogeneous activation of peroxydisulfate by magnetite catalyst for the degradation of tetracycline in water. Sep Purif Technol. 2012;84:147–52.

49. Gayathri P, Dorathi RPJ, Palanivelu K. Sonochemical degradation of textile dyes in aqueous solution using sulphate radicals activated by immobilized cobalt ions. Ultrason Sonochem. 2010;17:566–71.

50. Kwon M, Kim S, Yoon Y, Jung Y, Hwang TM, Lee J, et al. Comparative evaluation of ibuprofen removal by UV/H_2O_2 and UV/$S_2O_8^{2-}$ processes for wastewater treatment. Chem Eng J. 2015;269:379–90.

51. Chen WS, Huang YL. Removal of dinitrotoluenes and trinitrotoluene from industrial wastewater by ultrasound enhanced with titanium dioxide. Ultrason sonochem. 2011;18:1232–40.

52. Lin JC, Lo SL, Hu CY, Lee YC, Kuo J. Enhanced sonochemical degradation of perfluorooctanoic acid by sulfate ions. Ultrason Sonochem. 2014;22:542–7.

53. Ghafoori S, Mowla A, Jahani R, Mehrvar M, Chan PK. Sonophotolytic degradation of synthetic pharmaceutical wastewater: Statistical experimental design and modeling. J Environ Manage. 2015;150:128–37.

54. Rezaee R, Maleki A, Jafari A, Mazloomi S, Zandsalimi Y, Mahvi AH. Application of response surface methodology for optimization of natural organic matter degradation by UV/H_2O_2 advanced oxidation process. J Environ Health Sci Eng. 2014;12:1.

55. Montgomery DC. Design and Analysis of Experiments. 6th ed. New York: John Wiley & Sons; 2005.

56. Affam AC, Chaudhuri M. Optimization of Fenton treatment of amoxicillin and cloxacillin antibiotic aqueous solution. Desal Water Treat. 2014;52:1878–84.

57. Zuorro A, Fidaleo M, Fidaleo M, Lavecchia R. Degradation and antibiotic activity reduction of chloramphenicol in aqueous solution by UV/H_2O_2 process. J Environ Manage. 2014;133:302–8.

58. Morshedi A, Akbarian M. Application of response surface methodology: design of experiments and optimization: a mini review. In J Fund Appl Life Sci. 2014;54:2434–9.

59. Dominguez JR, Munoz MJ, Palo P, Gonzalez T, Peres JA, Cuerda-Correa EM. Fenton advanced oxidation of emerging pollutants: parabens. Int J Energy Environ Eng. 2014;5:1–10.

60. Xu M, Du H, Gu X, Lu S, Qiu Z, Sui Q. Generation and intensity of active oxygen species in thermally activated persulfate systems for the degradation of trichloroethylene. RSC Advances. 2014;76:40511–7.

61. Hoseini M, Safari GH, Kamani H, Jaafari J, Ghanbarain M, Mahvi AH. Sonocatalytic degradation of tetracycline antibiotic in aqueous solution by sonocatalysis. Toxicol Environ Chem. 2014;95:1680–9.

62. Saeed W. The effectiveness of persulfate in the oxidation of petroleum contaminants in saline environment at elevated groundwater temperature. Ontario, Canada: Waterloo; 2011.

63. Zhao D, Liao X, Xiulan Y, Huling SG, Chai T, Huan T. Effect and mechanism of persulfate activated by different methods for PAHs removal in Soi. J Hazard Mater. 2013;254:228–35.

64. Guan YH, Ma J, Li XC, Fang JY, Chen LW. Effect of pH on the formation of sulfate and hydroxyl radicals in the UV/peroxymonosulfate system. Environ Sci Technol. 2011;45:9308–14.

65. Liang C, Su HW. Identification of sulfate and hydroxyl radicals in thermally activated persulfate. Ind Eng Chem Res. 2009;48:5558–62.

66. Ji YF, Ferronato C, Salvador A, Yang X, Chovelon JM. Degradation of ciprofloxacin and sulfamethoxazole by ferrous-activated persulfate: Implications for remediation of groundwater contaminated by antibiotics. Sci Total Environ. 2014;472:800–8 [38].

67. Jiao S, Zheng S, Yin D, Wang L, Chen L. Aqueous photolysis of tetracycline and toxicity of photolytic products to luminescent bacteria. Chemosphere. 2008;73:377–82.

68. Shaojun JIAO, Zheng S, Daqiang YIN, Lianhong WANG, Liangyan CHEN. Aqueous oxytetracycline degradation and the toxicity change of degradation compounds in photoirradiation process. J Environ Sci. 2008;20:806–13.

69. Wang Y, Zhang H, Zhang J, Lu C, Huang Q, Wu J, et al. Degradation of tetracycline in aqueous media by ozonation in an internal loop-lift reactor. J Hazard Mater. 2011;192:35–43.

70. Zhu XD, Wang YJ, Sun RJ, Zhou DM. Photocatalytic degradation of tetracycline in aqueous solution by nanosized TiO_2. Chemosphere. 2013;92:925–32.

71. Dalmazio I, Almeida MO, Augusti R. Monitoring the degradation of tetracycline by ozone in aqueous medium via atmospheric pressure ionization mass spectrometry. J Am Soc Mass Spectrom. 2007;18:679–87.

72. Ma Y, Gao N, Li C. Degradation and pathway of tetracycline hydrochloride in aqueous solution by potassium ferrate. Environ Eng Sci. 2012;29:357–62.

73. Addamo M, Augugliaro V, Di Paola A, Garcia-Lopez E, Loddo V, Marci G, et al. Removal of drugs in aqueous systems by photoassisted degradation. J Appl Electrochem. 2005;35:765–74.

74. Seid Mohammadi A, Asgari G, Almasi H. Removal of 2,4 Di-Chlorophenol Using Persulfate Activated with Ultrasound from Aqueous Solutions. J Environ Eng. 2014;4:260–68.

75. Wang X, Wang L, Li J, Qiu J, Cai C, Zhang H. Degradation of Acid Orange 7 by persulfate activated with zero valent iron in the presence of ultrasonic irradiation. Sep Pur Technol. 2014;122:41–46.

16

Investigation of dust storms entering Western Iran using remotely sensed data and synoptic analysis

Ali D Boloorani[1], Seyed O Nabavi[2*], Hosain A Bahrami[3], Fardin Mirzapour[4], Musa Kavosi[5], Esmail Abasi[5] and Rasoul Azizi[5]

Abstract

Background: One of the natural phenomena which have had considerable impacts on various regions of the world, including Iran, is "dust storm". In recent years, this phenomenon has taken on new dimensions in Iran and has changed from a local problem to a national issue. This study is an attempt to investigate the formation of the dust storms crossing the Western Iran.

Methodology: To find the sources of the dust storms entering Iran, first we examine three determined dust paths in the region and their temporal activities, using MODIS satellite images. Then, four regions were identified as dust sources through soil, land cover and wind data. Finally, atmospheric analyses are implemented to find synoptic patterns inducing dust storms.

Results and discussion: Source 1 has covered the region between the eastern banks of Euphrates and western banks of Tigris. Source 2 is in desert area of western and south-western Iraq. Finally source 3 is bounded in eastern and south-eastern deserts of Saudi Arabia called Rub-Al-Khali desert, or Empty Quarter. Moreover, south-eastern part of Iraq (source 4) was also determined as a secondary source which thickens the dust masses originating from the above mentioned sources. The study of synoptic circulations suggests that the dust storms originating from source 1 are formed due to the intense pressure gradient between the low-pressure system of Zagros and a high-pressure cell formed on Mediterranean Sea. The dust events in sources 2 and 3 are outcomes of the atmospheric circulations dominant in the cold period of the year in mid-latitudes.

Keywords: Dust source, Dust storm, Dust detection, Remote sensing, Synoptic climatology

Background

Dust storms are the result of the air turbulences which spread a large mass of dust in the atmosphere, and decrease the horizontal visibility to less than 1000 meters [1]. In a general perspective, primary sources of dust plumes lie in the arid and semi-arid regions in East Asia, Middle East, Europe, Latin America, North America, Australia, eastern and southern Africa. Among them, Middle East has many dust sources in Arabian Peninsula, Syria, Egypt, Iraq and Iran [2]. Therefore, in countries with arid climate such as Iran, dust phenomena are not new natural events. However, in recent years there has been a new problem: a sudden increase in the number of dusty days and the level of air dust, especially in western Iran which has seriously affected the people's health and everyday life. As shown in Figure 1, only in the course of 3 years (2005–2008), the dust event spotted in western Iran has considerably become wider in terms of thickness and spatial expansion.

The frequent occurrence of this phenomenon in various regions of the world such as Iran has gained attentions from academia and has triggered many research projects on the issue. Remote sensing is one of the most widely-used methods in dust storm studies. Drawing upon a model based on surface conditions and satellite observations, Ozsoy et al. examined one of the greatest

* Correspondence: s.o.nabavi@gmail.com
[2]Geoinformatics Research Institute (GRI), University of Tehran and Department of Geography and Regional Research, University of Vienna, Vienna, Austria
Full list of author information is available at the end of the article

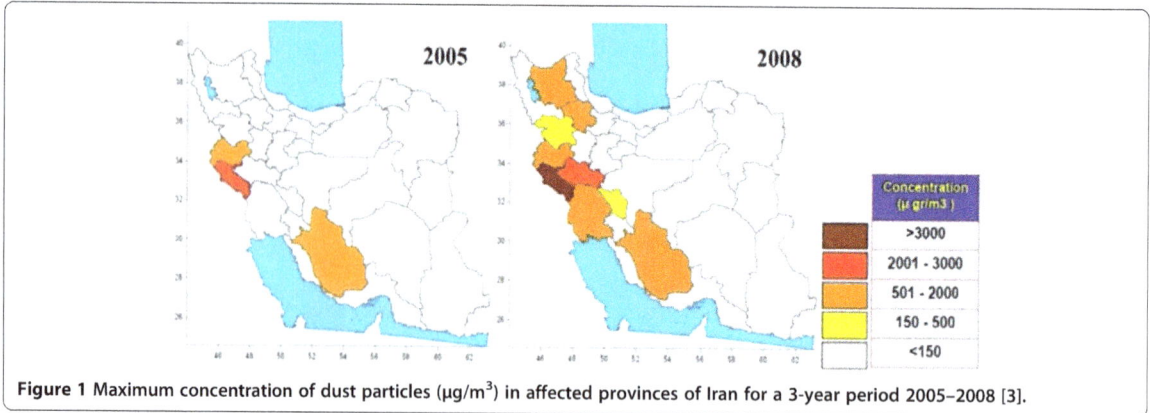

Figure 1 Maximum concentration of dust particles (µg/m³) in affected provinces of Iran for a 3-year period 2005–2008 [3].

dust events in Sahara Desert during the April of 1994 which affected a wide region from Caribbean to Eurasia [4]. El-Askary et al. argued that a synthetic approach is the best way of observing and tracking dust in satellite images [5]. Measuring solar reflectance bands, Qu et al. detected dust masses in satellite images [6]. Alles used air pollution indicatives and MODIS[a] images to determine China's dust sources; the results have shown that the phenomenon has originated in the cold regions and has caused many problems in Gobi desert while crossing and loading enormous dust particles in this region [7]. Li et al. have analysed the greatest dust in the eastern Australia using MODIS images and (BTD[b]) index. He detected the high performance of the method in identification of dust masses [8]. Climatological analysis, especially the synoptic approach - the study of relations between global atmospheric circulations and regional and local climates [9] - has been used in most of the researches performed on the atmospheric conditions during dust events. Investigating the transfer of dust from Sahara to Italy and central Europe, Barkan et al. recognized sub-tropical high pressure system on the south-eastern Mediterranean Sea and a low-pressure centre resulted from Iceland's trough in the western Africa as the primary factors of dust formation and mobilization [10]. Zolfaghari and Abedzadeh adopted a synoptic approach to investigate the atmospheric systems leading to dust in western Iran in a 5-year period. They have concluded that Azur's high-pressure system along with westerly-travelling cyclonic and anti-cyclonic systems are the most effective atmospheric factors on the dust events in the region [11]. Dayan et al. have classified dominant synoptic patterns in the region on dusty days and indicated that 60% of the dust events in the south-eastern Mediterranean occur as a result of Cyprus low-pressure system [12]. Gerivani et al. have adopted a different approach and categorized the sources of the dust storms in Iran based on geological maps and the information on wind erosion of the susceptible lands [13]. Although, all the mentioned

studies have provided considerable results, the fact that a dust event results from multiple factors requires considering synthetic approaches. Utilizing various perspectives such as remote sensing data, synoptic climatology analysis, and ancillary data including soil texture and land cover of the studied regions, as well as the wind speed and direction maps, the present research tries to study the dust events in western Iran more precisely.

On the other hand, the most climatological studies of the dust storms in Iran are limited to a few numbers of case studies. Moreover, in these works the source-based classification of atmospheric circulations has not been considered to account for environmental conditions leading to dust storm. Therefore, in this work, we have increased the number of investigated events to 40 cases and also atmospheric aspects of dust storms are precisely studied. Significant rise in the number of recent dust storms in Iran and also applying new and up-to-date methods to examine this natural event provide another motivation for the present study.

The rest of this paper is organised as follows. Methods and data are presented in the next section. The research findings are discussed in section 3, and section 4 concludes the paper.

Data and methods
Data and methods for the identification of dust sources
To study the dust storms entering western Iran, daily data from 15 synoptic stations (see Figure 2) in a period from 2000 to 2008 are exploited.

In this research, only the dust events satisfying the following criteria were studied: (I) recorded at least in 3 synoptic stations, (II) rising from non-local origins, and (III) leading to a maximum horizontal visibility of 1000 meters.

In cases, the stations in the region under study indicated consecutive days with dust, only those days were specified as independent event when dust particles were injected to atmospheric currents from dust-prone areas.

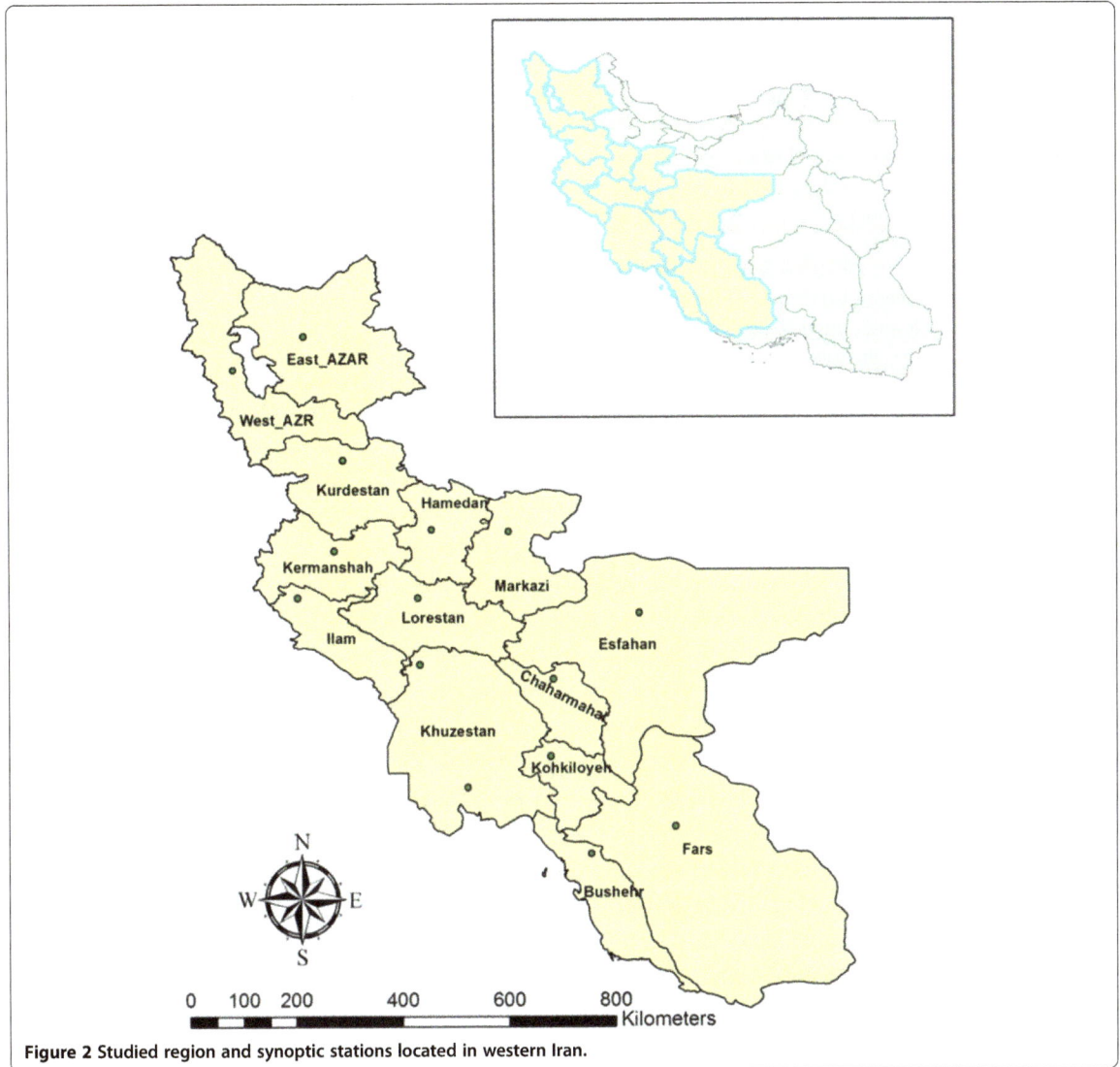

Figure 2 Studied region and synoptic stations located in western Iran.

Whereas, maybe because of air stillness, dust continued for days over a region without any new dust injection, those dusty days were considered as on dust event.

Mobility of dust plumes from the originating sources to the surrounding regions do not necessarily follow density reduction regime, i.e., there may exist some other dust sources throughout the dust paths which inject new dust particles into the primary mass. Therefore, in the present study, the region where dust originates is not considered as the only source of the event. In fact, all the regions located in the dust path are determined using remote sensing imagery and then, supplementary data and information are taken into account to locate the dust sources more precisely. In order to identify dust paths and to estimate primary and secondary dust sources, approximately 100 MODIS images of Middle East were processed within a five-day period before recording time of each event.

In this research, image processing for dust detection is based on Ackerman's method [14], arguing that brightness temperature difference of dust particles in bands 31 and 32 (11 and 12 micrometre wavelengths) is below zero in Kelvin scale. However, it is empirically considered –0.5 in the present work. To distinguish dust particles suspended in the air from land covers, two types of test were used; R-test and BTD-test [15]. After identifying possible dust sources and the relocation paths, more precise identification processes are performed by examining the soil data and land covers of the studied regions, and wind maps.

The following databases have provided the data used in the present study:

– Visibility data from Iranian Meteorology
  Organization database,
– MODIS images from NASA website,[c]
– Soil information from the Harmonized World Soil
  Database [16], and
– Land-cover data from ESA website[d].

Data and methods to identify atmospheric circulations associated with dust storms

The analysis of synoptic circulations associated with dust storms is carried out according to geopotential height at 850 hPa and 500 hPa levels, temperature, and wind data at height of 10 meters above the ground during the beginning days of studied dust storms. All atmospheric data, except

for dust records, were provided by National Centre of Environmental Predicting (NCEP) database [17].

Result and discussion
Identification of western Iran's dust sources

Having examined the recorded dust storm data of the local stations from 2000 to 2008, a number of 40 dust events were selected. The analysis of the satellite images of dust events in the western Iran shows that the dust plumes travel in three main paths (from the source to the region under the study):

– *North western–South eastern path*: The detected dust plumes in Iraq and Syria are carried to the

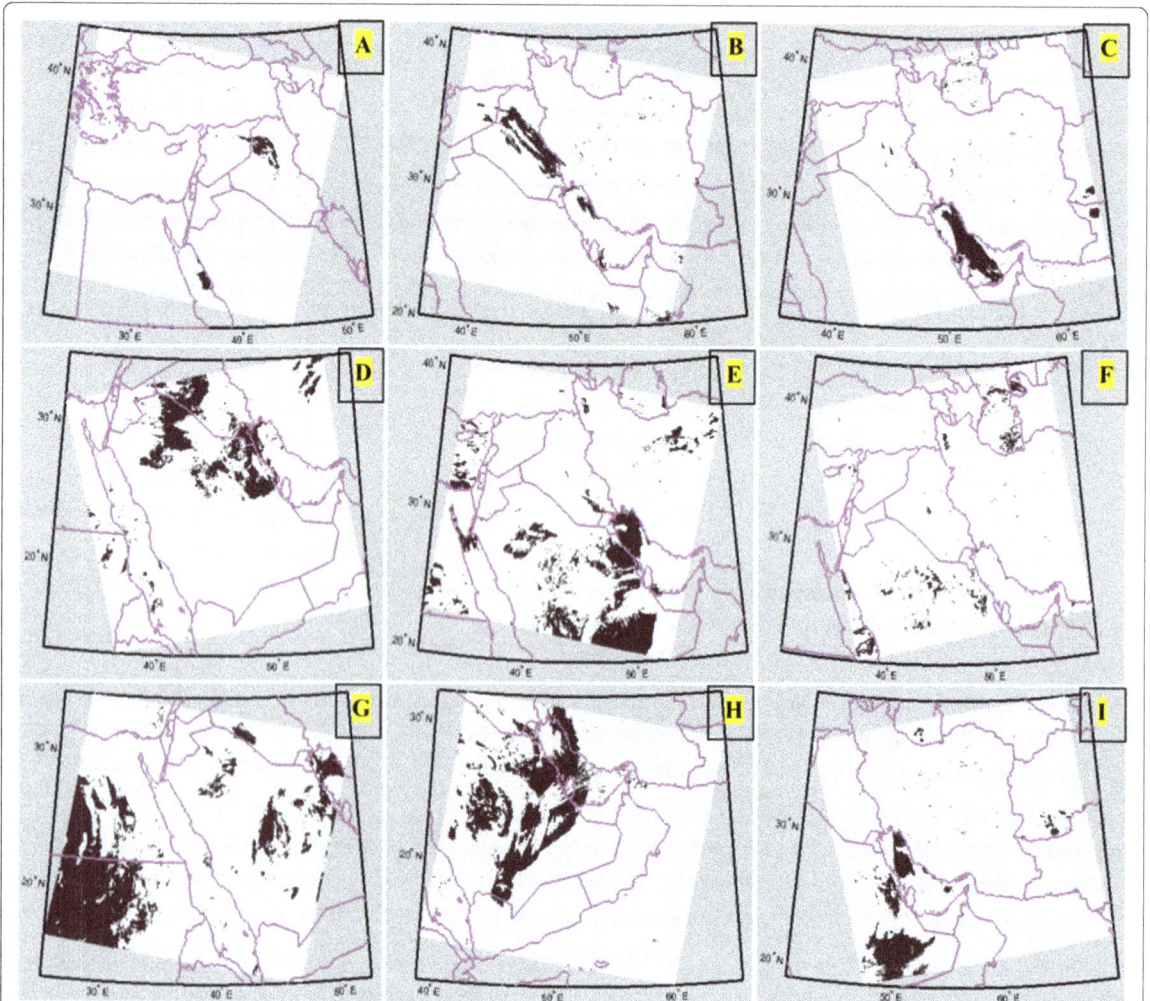

Figure 3 Three specified paths of dust relocation to the western Iran. Images in the first row are the samples of dust plumes moving along the first path (from left to right): the origin of primitive dust mass **(A)**, the dust mass relocating and entering to the study region **(B)**, and its dissipation **(C)** for August 7-8-9,2005. The images shown in the second and third rows show these three steps in the same order for dust events of February 15–16, 2004 **(D, E, and F)** and of March 1–3, 2007 **(G, H, and I)**.

Table 1 The dates of the dust events which have formed and moved through the Northwest-Southeast path

No	Starting date of dust storm	Observation date of dust storm in study area
1	11-Jun-05	12-Jun-05
2	12-Jun-05	13-Jun-05
3	22-Jun-05	23-Jun-05
4	24-Jun-05	25-Jun-05
5	04-Jul-05	05-Jul-05
6	24-Jul-05	25-Jul-05
7	07-Aug-05	08-Aug-05
8	08-Aug-05	09-Aug-05
9	17-Apr-06	18-Apr-06
10	17-Jul-07	18-Jul-07
11	18-Jul-07	19-Jul-07
12	15-Feb-08	15-Feb-08
13	04-Apr-08	05-Apr-08
14	29-Apr-08	01-May-08
15	01-Jun-08	02-Jun-08
16	07-Jun-08	08-Jun-08
17	08-Jun-08	09-Jun-08
18	15-Jun-08	16-Jun-08
19	16-Jun-08	17-Jun-08
20	17-Jun-08	18-Jun-08
21	18-Jun-08	19-Jun-08
22	30-Jun-08	02-Jul-08
23	15-Sep-08	16-Sep-08

Table 2 The dates of the dust events which have formed and moved through the West–east path

No	Starting date of dust storm	Observation date of dust storm in study area
1	21-Dec-02	22-Dec-02
2	22-May-03	22-May-03
3	15-Feb-04	16-Feb-04
4	13-May-04	14-May-04
5	15-May-04	16-May-04
6	06-Jan-05	07-Jan-05
7	24-Jan-05	25-Jan-05
8	03-Apr-05	04-Apr-05
9	05-May-05	05-May-05
10	21-Jan-06	22-Jan-06
11	04-Feb-07	04-Feb-07
12	16-May-07	17-May-07
13	19-Feb-08	20-Feb-08
14	15-Mar-08	15-Mar-08

Table 3 The dates of the dust events which have formed and displaced through the southern-northern path

No	Starting date of dust storm	Observation date of dust storm
1	22-Mar-03	23-Mar-03
2	01-Mar-07	03-Mar-07
3	10-Apr-07	11-Apr-07

western part of Iran by north-westerly winds (Figure 3). Among the 40 dust storms, 23 events have formed along this path (Table 1).

Based on the recorded data, massive dust storms entering Iran began to move in this path since 2005 and hit a peak in 2008. Four consecutive dust events that have affected Iran since June 15 to 19, 2008 are rare instances of this phenomenon resulted from dust source activities in the first dust pathway. Therefore, it may be inferred that dust masses which have entered Iran in the recent years, have mainly moved along this path and are originated from the same sources. Seasonal distribution of dust events (except for an event in February 2008) indicates that this path is mainly active in the warm months of the year, especially June and July. The reasons for these circumstances will be discussed in details in chapter 3.2.

– *West-east path*: dust masses rise from Iraq-Jordan borders and enter Iran (Figure 3). 14 cases of the studied dust events have occurred along this path (Table 2).

Temporal distributions of the west–east dust events indicate a declining trend in their frequency by the end of the study period (in contrast to the dust events corresponding to the first path). According to Table 2, events of this category have mostly occurred in 2004 and 2005. These conditions again reflect the considerable effect of dust origins in recent dust events. The seasonal distribution of the dust storms formed in second path shows a peak in the cold period of year.

– *Southern-northern path*: It begins from the regions located to the south of Persian Gulf and ends at western Iran (Figure 3). Only three dust events travelled on this path (Table 3).

As indicated in Table 3, dust masses moving in the third path do not play a significant role in dust events occurring in Iran and thus, it cannot be considered as an influential region of forming the dust storms occurred in the last decade in the study area. Saudi Arabia dust events occurred in the late cold period of the year that is examined in atmospheric analysis section.

Although it is possible to specify the approximate places of dust formation and its relocation paths by

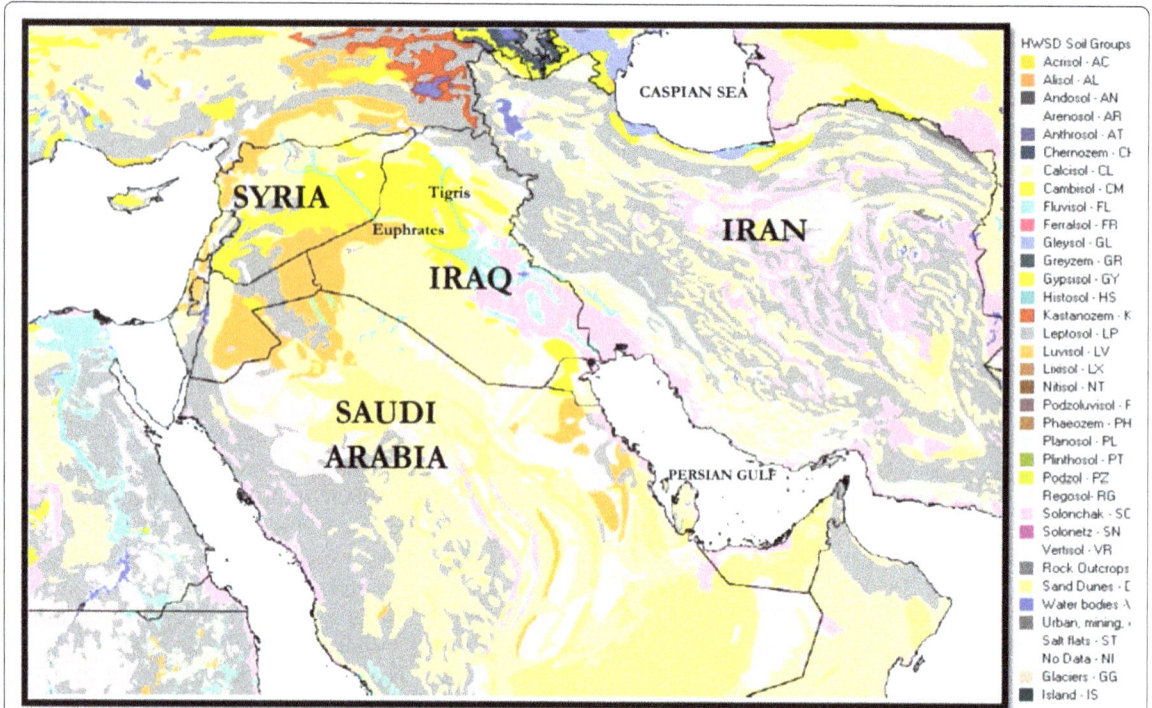

Figure 4 The Middle East Soil Map [16].

processing satellite images, the information on the soil and wind in the region should be analysed to identify the dust sources more accurately. Therefore, we carefully examined the ancillary data corresponding to the mentioned dust paths. These data include soil information of Iraq, Syria, and Saudi Arabia (Figure 4 and Table 4), their land covers data (Figure 5), as well as the maps of wind velocity and direction (Figure 6).

Soil maps of the northwest–southeast dust path in Syria and Iraq indicated a region of gypsum sediments in the north-western Iraq and the eastern Syria. This area has mainly formed in arid regions with poor vegetation called gypsisole (Figure 4). Despite the fact that southernmost region of the path, i.e. the south-eastern Iraq, consists of solonchaks soil, which is also found in arid and semi-arid regions, one of its distinguishing features is that the water table is close to the surface. Since, the soil is almost wet- a condition sometimes leading to water stillness on the soil surface [19]. Although in the recent years, this region has

experienced an increase in depth of groundwater because of drought, and some ponds have changed into limited erodible zones [20], the south eastern Iraq cannot be considered as a main and wide dust source influencing the western Iran. However, in some cases, dust masses originating from other region have been intensified locally in this region and then entered Iran (not shown here). The quantitative characteristics of soil along northwest–southeast path (80% of the soil has a fine texture including clay and silt) verify that the north-western parts of Iraq and the eastern parts of Syria potentially lead to massive dust storms. Furthermore, these conditions are intensified by the regions with low soil moisture (15–50 mm per 1 m soil depth) and spare land covers. It should be noted that despite the fine texture of the soil in south eastern Iraq, higher moisture (150 mm per 1 m soil depth) and more dense land cover lead to a lower chance of dust event in this region (Table 4, Figures 4 and 5). According to the discussion above, the upper zone of

Table 4 Soil characteristics of dust sources [16]

Sources	Dominant soil group	Topsoil USDA texture classification	AWC (mm)	Topsoil sand fraction (%)
The east of Syria and North west of Iraq	GY - Gypsisols	loam	15-50	sand(35)/Silt(45)/clay(20)
South east of Iraq	Sc-Solonchaks	loam	150	sand(36)/Silt(43)/clay(21)
The west and south west of Iraq	CL-Calsisol	loamy sand	15-50	sand(39)/Silt(37)/clay(24)
The east and south east of Saudi Arabia	DS-sand dunes	loamy sand	15-100	sand(84)/Silt(9)/clay(7)

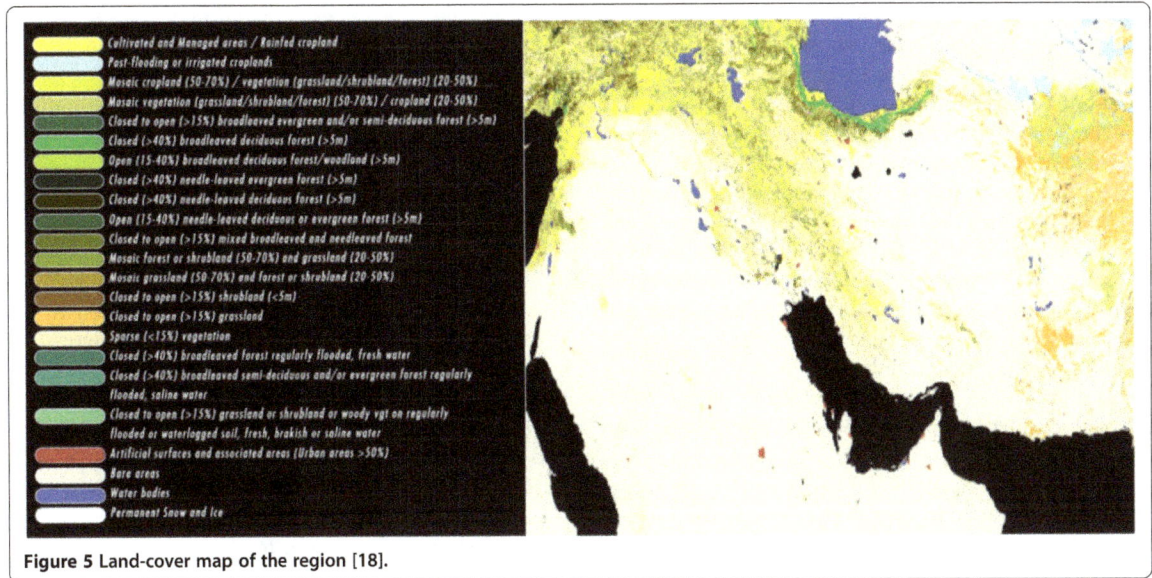

Figure 5 Land-cover map of the region [18].

Mesopotamia (Iraq-Syria border) is the most likely region to provide airborne particles for the dust storms in western Iran along the northwest-southeast path.

The study of the ancillary data on path two shows that the calcisols covers most parts of the west and south western Iraq (Figure 4). Calcisols is mainly formed in arid and semi-arid regions especially in foothills, and is known as an alluvial type of soil with fine texture, which makes land cover potentially erodible [19]. Table 4 gives data for an equal mixture of sand, clay, and silt in the west and south western Iraq. In this region, the amount of soil (dominant soil) moisture is about 50 mm per 1 m soil depth (even as low as 15 mm in some regions), which results in sparse vegetation (Figure 5).

The path three just start from southeast of Arabian Peninsula (the desert region of Rub-Al-Khali or Empty Quarter) to southern coasts of Persian Gulf. This part of Arabia is Regosol soil which mainly is composed of gravel and sand coverings (i.e. 84%) (Figure 4 and Table 4). Since the soil has a coarse texture, the number of dust storms reaching Iran throgh this pathway is much less than that from other dust sources (i.e. 3 out of 40 cases). However, the intense dryness (amount of soil moisture in some regions is around zero) and lack of vegetation (Figure 5) result in a large number of limited daily local sand storms in eastern Arabian Peninsula.

According to Figure 6, in the first day of all the studied dust events, wind speed and direction maps are in accordance with the estimated dust sources. During dusty days of the path one the speedy winds (6 to 12 m/s) skirted on Iraq and south-eastern Syria have the potential to make dust plumes in erosion-prone regions (Figure 6A). The

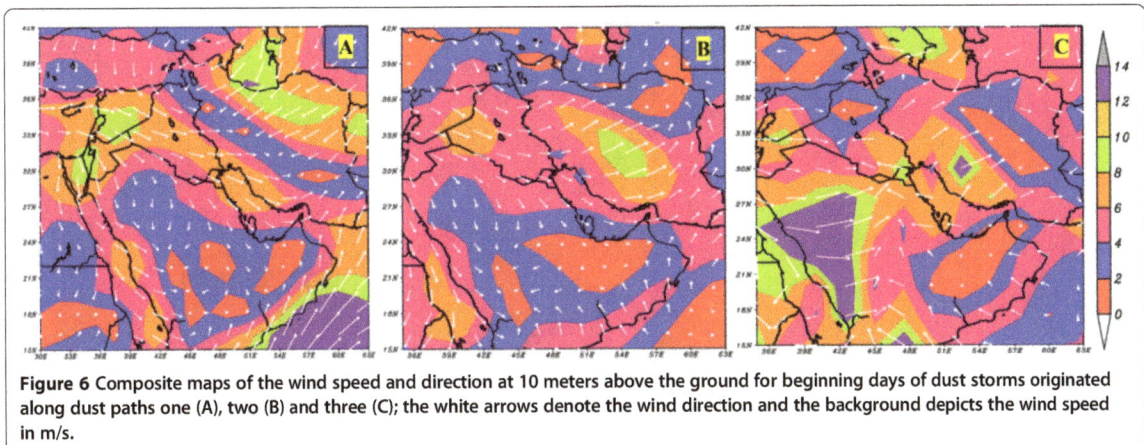

Figure 6 Composite maps of the wind speed and direction at 10 meters above the ground for beginning days of dust storms originated along dust paths one (A), two (B) and three (C); the white arrows denote the wind direction and the background depicts the wind speed in m/s.

composite map of wind speed and direction shows speedy air streams during the beginning days of dust storms in path two. As a result, the western and eastern parts of Iraq are prone to dust if other conditions are available. Along with the formation of dust in path three, two convergent wind currents come together in eastern Saudi Arabia. Surface westerly winds, most likely intensified by temperature differences between the western mountainous regions and the eastern desert area, converge with easterly currents. This circulation transports large amounts of dust masses from the eastern Arabia to upper atmospheric levels and higher latitudes. In fact, because of the coarse soil texture of Saudi Arabia deserts, only such special atmospheric circumstances can make massive dust storms enter regional circulations and subsequently the western Iran (Figure 6C).

Based on our findings the following regions are recognized as the main sources of dust events in the western Iran (Figure 7):

i. *Dust source 1: The region between eastern banks of Euphrates to western banks of Tigris*
The region located in the north western Iraq and the eastern Syria was recognised as the main source of the dust storms entering western Iran. As discussed above, this region was the origin of 23 dust storms out of the 40. Most of dust events originating in this source have formed in Iraq and a few in Syria.

ii. *Dust source 2: The western and south western Iraq*
14 cases out of 40 have formed in this region. It also includes parts of Jordan and Syrian Desert.

iii. *Dust source 3: The eastern and south-eastern Arabian Peninsula called Rub-Al Khali desert*
Some parts of Rub-Al-Khali desert were the source of 3 cases out of the total 40.

iv. *Dust source 4: The south-eastern Iraq*
This source is not considered as a major dust origin, but it intensifies the dust masses originating from other sources, especially from one and two dust formation areas. Therefore, the number of dust events formed in this region is not given separately.

Tracking dust storms and examination of various data and factors gave us most likely dust sources which send dust particles into Iran. However, the study of atmospheric conditions and identification of synoptic patterns during dust events is crucial to clarify predominant environmental conditions leading to dust storms.

Atmospheric conditions and synoptic patterns

Before discussing the atmospheric conditions governing the occurrence of dust storms in the western Iran, it is necessary to note that the activity of each dust source follows certain temporal distribution. All the dust events originating in the source one, have occurred in the warm periods of the year (April to September), while other

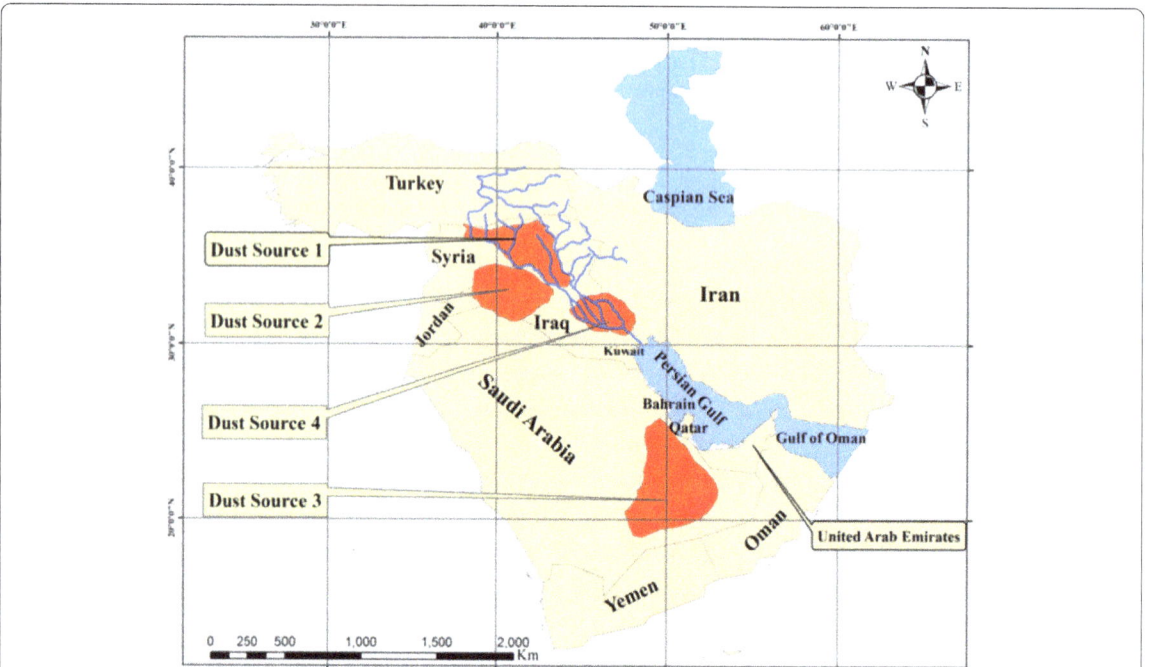

Figure 7 Identified sources of dust storms from 2000 to 2008.

dust cases formed in the western and south western Iraq, and eastern and south eastern Saudi Arabia have been recorded in cold and late cold periods. The activity of the source four has happened in both, the warm and cold seasons. Such a clear distinction in the temporal distribution of the dust activities in the sources can be due to the formation of different atmospheric patterns in specific periods of the year. Synoptic circulations associated with the dust sources are discussed in the following subsections –except for atmospheric conditions affecting the fourth source, because it acts as the secondary origin coincidental to the primary sources.

Synoptic patterns of source one

To identify synoptic circulations leading to dust events as extreme atmospheric patterns, their normal conditions should be examined first. As a dominant atmospheric circulation, a low-pressure centre is usually formed on Iraq during warm times. This low pressure system, formed in the lower troposphere, leads to winds blowing in north-west – southeast direction, known as Shamal Wind [21]. The increase in height causes the low-pressure tongue to expand onto southern regions of Turkey in north of Iran and toward Zagros Mountains in east (Figure 8A). Studies conducted on this circulation system gave the surface heating of Zagros Mountains as the main contributor in formation of this low [22,23]. Zagros low-pressure turns into a ridge and anti-cyclonic circulation in mid-troposphere (Figure 8B) strengthened by surface diabatic processes [23].

The most striking feature of extreme synoptic patterns coincidental to most dust storms originating in source one is the expansion of Iraq low-pressure and its movement toward the western Iran and south of Turkey. This low encounters Azores' high pressure formed on central Europe and northern banks of Mediterranean Sea at 850 hPa (Figure 9A). In 850 hPa composite map, large parts of Iran especially the western part, have

experienced 1430 to 1440 geopotential meters. Whereas, in the long-term 850 hPa the geopotential heights 1475 to 1480 have been recorded for western Iran. These pressure values refer to a very warm and low centre that becomes dominant on the Iranian plateau simultaneous with dust storms in the study area. Another part of this circulation is the progression of Azores' high pressure on Mediterranean Sea. This condition brings a high pressure gradient between these systems and, therefore, speedy winds through Syria and Iraq which causes very strong Shamal winds. These winds have the potential to raise dust plumes from surface to upper levels.

To explain how this circulation pattern is formed, the effects of seasonal thermal processes should be examined. In summertime, apart from the low pressure of the Bay of Bengal, another thermal low forms in Pakistan and sometimes in the north-western India known as Paki-India low. According to Bollasina and Nigam, the intensified surface heating of the Iranian plateau, especially Zagros Mountains, is one of the factors involved in deepening Paki-India low and its movement into Iran [24]. As mentioned before, surface heating of Zagros Mountains in normal conditions also leads to a low-pressure centre at 800–850 hPa levels. But along with Iran dust storms, heating process of Zagros Mountains and eastern deserts of Iran increases dramatically (Figure 9B). Based on the classic definition of thermal atmospheric circulations, this condition also results in the formation of high and low pressure systems in upper and lower-troposphere, respectively. Records of temperature in the composite map at 850 hPa show anomalies[e] over 6 to 9 degrees Celsius in Zagros Mountains. 500 hPa composite map for dusty days occurred in North West Iraq and east Syria origins shows relocation of tropical high air mass, indicated by 5880-gpm contour, to the southern Iran which locates over Saudi Arabia Peninsula in the long term map (Figure 9C). Given that 500 hPa level is not usually affected by low surface thermal conditions, the

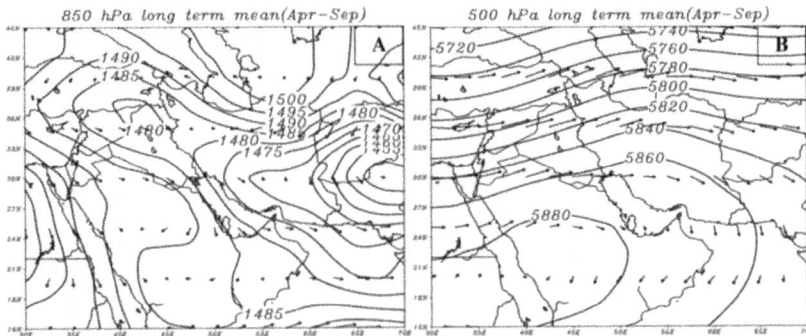

Figure 8 Long-term mean of wind and geopotential height during warm times (Apr-Sep) at 850 hPa (A) and 500 hPa (B) levels. Arrows and solid lines denote wind direction and geopotential height (in meters), respectively.

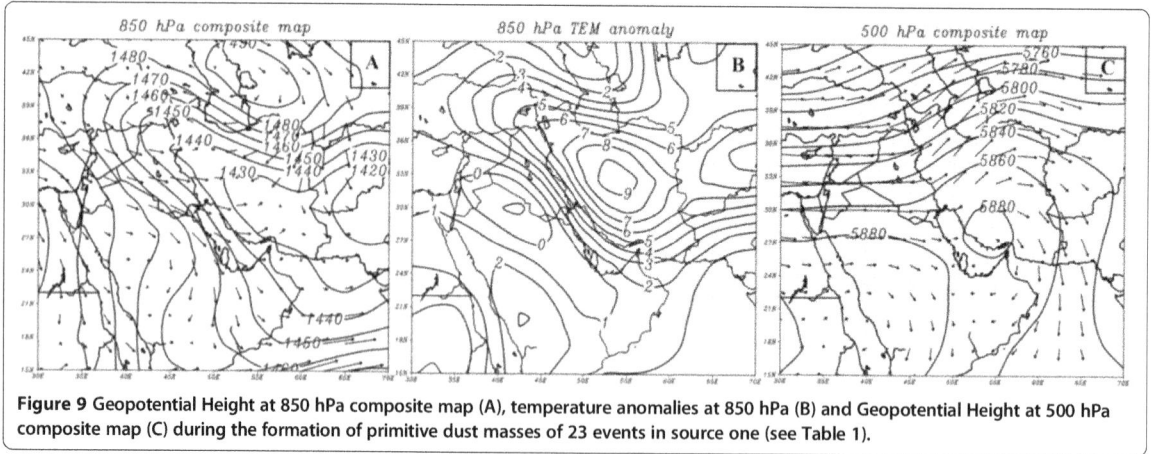

Figure 9 Geopotential Height at 850 hPa composite map (A), temperature anomalies at 850 hPa (B) and Geopotential Height at 500 hPa composite map (C) during the formation of primitive dust masses of 23 events in source one (see Table 1).

locating of this high could be due to the significant rise in surface temperature on Zagrus Mountains. It is necessary to note that the deepening of low-pressure system located on Iraq and western Iran is not only affected by surface diabatic heating but also in some cases it is affected by dynamic processes resulted from travelling troughs [9]. Within the beginning days of dust storms formed in source one, trough patterns, located on Iraq and subsequently Iran, might deepen lower-troposphere thermal cyclones.

Synoptic circulations during dust events originating from sources 2 and 3

Since the temporal distributions of dust activities of the sources happen during cold months of year (October to March), it is expected that synoptic patterns governing these two sources to be similar. The analysis of atmospheric circulations of dust in the western and south western Iraq shows cyclonic and anti-cyclonic circulations as dusty patterns in the region. Almost all dust cases from western and south western Iraq begin when this source is put between travelling cyclonic and anti-cyclonic circulations. In other words, the cold section of cyclones brings high speed and dusty winds. This part of cyclonic circulation can be seen in average temperature for dusty days (Figure 10A). However, a few dust events were formed ahead of cold front or warm section in this region (not given here). Conditions have been partly different during the occurrence of dust events in eastern and south eastern Saudi Arabia. In these cases, dust masses are formed only in warm section of cyclonic circulations (Figure 10B).

The dynamic processes of westerly winds in mid-troposphere have a very significant role in the formation of surface mid-latitude cyclones and anticyclones. As seen in Figure 9C and 9D, cyclonic and anti-cyclonic circulations are formed respectively under the front and back of trough.

During dusty days of the western and south western Iraq, a trough lies at 500 hPa level which leads to the formation of cyclonic (low pressure) and anti-cyclonic (high pressure) circulations in Iran and eastern shores of the Mediterranean sea. In fact, the dust source is strengthened by being located between low and high-pressure centres and subsequently, being affected through pre-cold front (cold sector), west–east, high-speed winds. While trough gets deeper, cyclonic circulation relocates toward lower latitudes on Saudi Arabia. In these cases, speedy winds causing extensive dust storms will form in warm sector of the cyclone with a roughly south–north direction.

Conclusion

The present research uses multiple data and methods to determine the origin of the dust storms affecting the western Iran. By analysing the corresponding satellite images, three main paths of dust paths were recognized:

- *Path 1*: the northwest-southeast path, from the north-western Iraq and the south eastern Syria, to western Iran,
- *Path 2*: the second pathway with eastern-western direction, from Iraq-Jordan borders extended to the western Iran,
- *Path 3*: the last path from southern banks of Persian Gulf to western Iran through a south–north directing.

The role of the first path in the mobility of dust particles to the western Iran is much more significant than other two paths. The temporal distribution of western Iran dust events shows that the first dust path and its dust source have been abruptly activated from 2005 and hit a peak in 2008. Therefore, this path and its dust-prone regions can be considered as the main origin of recent dust storms entering western Iran. Dust events in

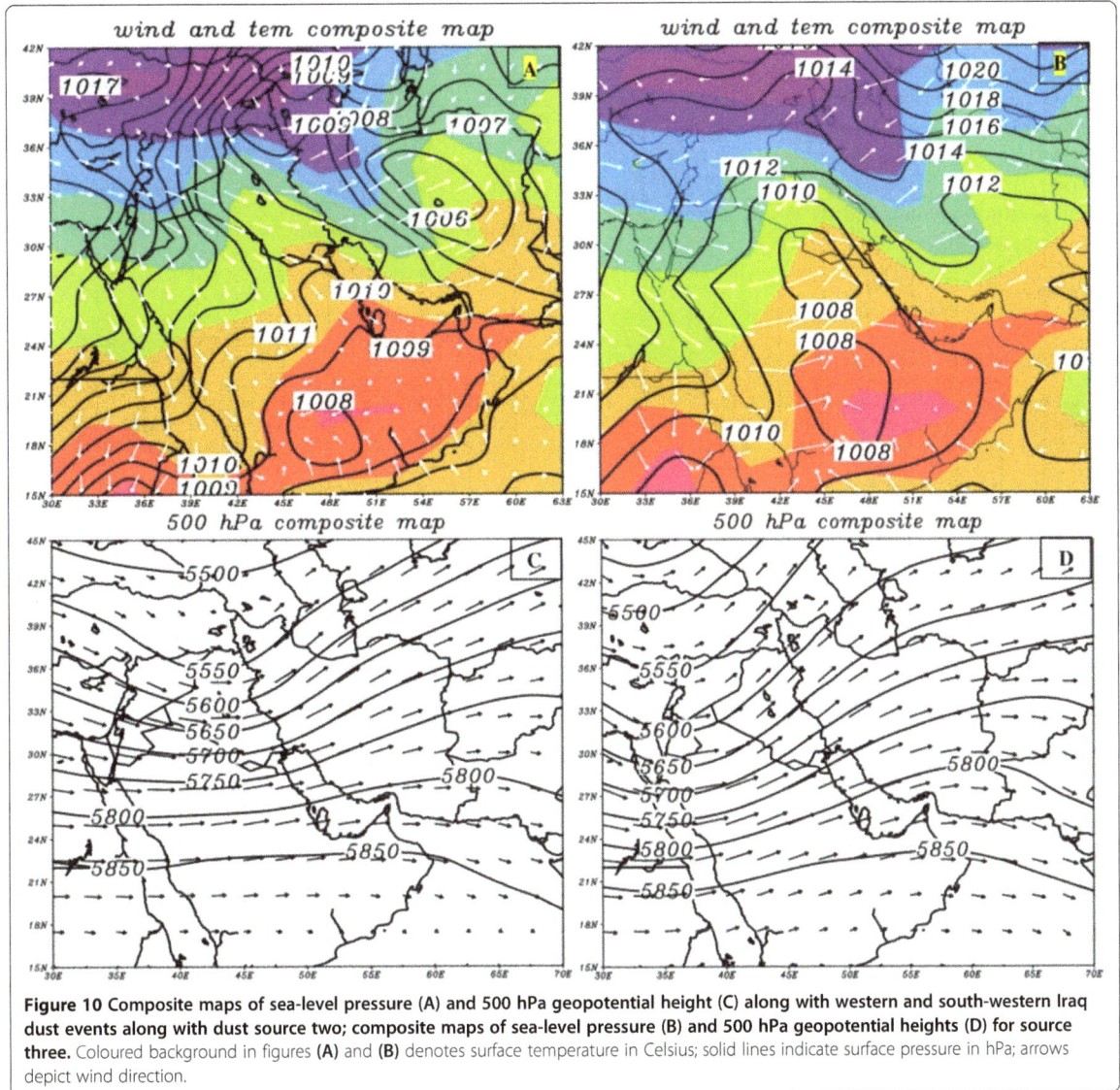

Figure 10 Composite maps of sea-level pressure (A) and 500 hPa geopotential height (C) along with western and south-western Iraq dust events along with dust source two; composite maps of sea-level pressure (B) and 500 hPa geopotential heights (D) for source three. Coloured background in figures **(A)** and **(B)** denotes surface temperature in Celsius; solid lines indicate surface pressure in hPa; arrows depict wind direction.

paths one and two do not show significant rise in recent years. Dust events relocated from northwest-southeast path mainly occur in the warm periods of the year. In contrast, the paths of two and three have maximum seasonal activities in the cold months. The present research has used soil information, the type of land covers, and wind velocity and direction of sea level pressure to locate the dust sources more precisely. Analysing theses information, we found four main dust-prone regions which affect western Iran:

– *Source 1:* the eastern banks of Euphrates and the western banks of Tigris as main western Iran dust sources,

– *Source 2:* the western and south western Iraq, and also small parts of Syria and Jordan,
– *Source 3:* the eastern and south-eastern deserts of Arabia called Rub-Al-Khali Desert,
– *Source 4:* the south-eastern Iraq, as a secondary source.

The examination of the atmospheric conditions governing each dust source indicates that synoptic patterns during dust events originating from source one have been mainly affected by the surface diabatic processes. Whereas, dust storm in sources two and three are caused by cyclonic and anti-cyclonic circulations and are largely outcome of dynamic processes of westerly winds.

Endnotes

[a]Moderate Resolution Imaging Spectroradiometer.
[b]Brightness Temperature Difference.
[c]http://rapidfire.sci.gsfc.nasa.gov/realtime.
[d]http://www.esa.int/Our_Activities/Observing_the_Earth/Space_for_our_climate/ESA_global_land_cover_map_available_online.
[e]Temperature anomaly = (daily temperature − average daily temperature).

Competing interests

The authors declare that they have no competing interests.

Authors' contributions

The most parts of this study including: research scheme, data analysis, and manuscript preparation were implemented by first two authors. Third author has helped to find an effective dust detection algorithm and make it convenient by programing in MATLAB software. Other authors have accompanied in editing, cohesion and translating presented paper. All authors read and approved the final manuscript.

Acknowledgements

This research paper is made possible through the help and financial support from science and technology affairs of Islamic republic of Iran.

Author details

[1]Department of Remote Sensing & GIS & Geoinformatics Research Institute (GRI), University of Tehran, Tehran, Iran. [2]Geoinformatics Research Institute (GRI), University of Tehran and Department of Geography and Regional Research, University of Vienna, Vienna, Austria. [3]Department of Soil Science, Tarbiat Modares University, Tehran, Iran. [4]Faculty of Electrical Engineering, Sadra Institute of Higher Education, Isfahan, Iran. [5]Geoinformatics Research Institute (GRI), University of Tehran, Tehran, Iran.

References

1. Goudie AS, Middleton N: Desert dust in the global system. New York: Springer Berlin Heidelberg; 2006.
2. Shao Y, Wyrwoll KH, Chappell A, Huang J, Lin Z, McTainsh GH, Mikami M, Tanaka TY, Wangh X, Yoon S: Dust cycle: an emerging core theme in earth system science. Aeolian Res 2011, 2:181–204.
3. Ardehjani SS: I.R of Iran national report on regional action plan to combat dust and sand storm. http://icap.atmos.und.edu/AERP/Meeting PDFs/RemoteSensing/Shirkani_Iran_Dust_Report%20(NXPowerLite).pdf.
4. Ozsoy E, Nilgun K: A hemispheric dust storm affecting the Atlantic and Mediterranean in April 1994: Analyses, modeling, ground-based measurements and satellite observations. J Geophys Res 2001, 106:18439–18460.
5. El-Askary HM, Sarkar S, Kafatos M, El-Ghazawi TA: A multisensor approach to dust storm monitoring over the Nile Delta. Geosci Remote Sens 2003, 41:2386–2391.
6. Qu JJ, Hao X, Kafatos M, Wang L: Asian dust storm monitoring combining terra and aqua MODIS SRB measurements. Geosci Remote Sens Lett 2006, 3:484–486.
7. Alles DL: geomorphology and dust storm in the China. http://fire.biol.wwu.edu/trent/alles/ChinaDust.pdf.
8. Li X, Ge L, Dong Y, Chang HC: Estimating the greatest dust storm in eastern Australia with MODIS satellite images. In Proceedings of IGARSS. Honolulu, HI: 2010:1039–1042.
9. Barry RG, Perry AH: Synoptic Climatology: Methods and Applications. London: Methuen; 1973.
10. Barkan J, Alpert P, Kutiel H, Kishcha P: Synoptics of dust transportation days from Africa toward Italy and central Europe. J J Geophys Res 2005, 110:1984–2012.
11. Zolfaghari H, Abedzadeh H: A synoptic analysis of dust systems at the west part of Iran. Geography Dev Q 2006, 6:173–188.
12. Dayan U, Ziv B, Shoob T, Enzel Y: Suspended dust over south-eastern Mediterranean and its relation to atmospheric circulations. Int J Climatol 2008, 28:915–924.
13. Gerivani H, Lashkaripour GR, Ghafoor IM, Jalili N: The source of dust storm in Iran: a case study based on geological information and rainfall data. Carpathian J Earth Environ Sci 2010, 6:297–308.
14. Ackerman SA: Remote sensing aerosols using satellite infrared observations. J Geophys Res 1997, 102:17069–17079.
15. Zhao TXP, Ackerman S, Guo W: Dust and smoke detection for multi-channel imagers. Remote Sens 2010, 2:2347–2368.
16. Harmonized World Soil Database (version 1.1). http://webarchive.iiasa.ac.at/Research/LUC/External-World-soil-database/HTML/.
17. Kalnay E, Kanamitsu M, Kistler R, Collins W, Deaven D, Gandin L, Iredell M, Saha S, White G, Woollen J, Zhu Y, Leetmaa A, Reynolds B, Chelliah M, Ebisuzaki W, Higgins W, Janowiak J, Mo KC, Ropelewski C, Wang J, Jenne R, Bull JD: The NCEP/NCAR 40-year reanalysis project. Am Meteorol Soc 1996, 77:437–472.
18. ENVISAT global land cover map. http://www.esa.int/spaceinimages/Images/2008/12/Envisat_global_land_cover_map.
19. IUSS Working Group WRB: World Reference Base for Soil Resources. http://www.fao.org/fileadmin/templates/nr/images/resources/pdf_documents/wrb2007_red.pdf.
20. UNEP, Partow H: The Mesopotamian Marshlands: Demise of an Ecosystem Early Warning and Assessment Technical Report. http://eoimages.gsfc.nasa.gov/images/imagerecords/1000/1716/meso2.pdf.
21. Aurelius L, Buttgereit V, Cammelli S, Zanina M: The impact of Shamal winds on tall building design in the Gulf Region. https://www.dm.gov.ae.
22. Zaitchik BF, Evans JP, Smith RB: Regional impact of an elevated heat source: the Zagros Plateau of Iran. J Clim 2007, 20:4133–4146.
23. Zarrin A, Ghaemi H, Azadi M, Mofidi A, Mirzaei E: The effect of the Zagros Mountains on the formation and maintenance of the Iran Anticyclone using RegCM4. Meteorol Atmos Phys 2011, 112:91–100.
24. Bollasina M, Nigam S: The summertime "heat" low over Pakistan/north-western India: evolution and origin. Clim Dynam 2010, 37:957–970.

17

Preparation and application of oyster shell supported zero valent nano scale iron for removal of natural organic matter from aqueous solutions

Vali Alipour[1], Simin Nasseri[1,2*], Ramin Nabizadeh Nodehi[1,2], Amir Hossein Mahvi[1,2] and Alimorad Rashidi[3]

Abstract

Background: In this Research, oyster shell supported zero valent iron nanoparticles were prepared and applied for the removal of natural organic matters (NOMs) from aqueous solutions under different experimental conditions.

Methods: The nanoadsorbents prepared by wet impregnation method, then characterized using Scanning Electron Microscopy, Energy Dispersive Spectroscopy, X-Ray Fluorescence and BET analysis. Adsorption test was done in a batch reactor and the effects of different parameters such as initial adsorbate concentration, adsorbent dose, adsorption kinetic, pH, and temperature on removal of NOMs (humic acid as the indicator) were studied.

Results: Results showed that particle size of nanoadsorbent was in the range of 60-83 nm, and surface area and micropore volume as 16.85 m^2/g and 0.021 m^3/g, respectively; the main elements of adsorbent were Ca, O, Fe and Na and lime, as high as about 94.25% was the main structural component of the total weight. Produced nanoadsorbent was not soluble in water. It was also shown that by increasing the nanoadsorbent dose from 0.5 to 5 g/100 ml, the removal of humic acid increased from 62.3% to 97.4%. An inverse relationship was found between initial concentration and adsorption capacity, so that a decreasing rate of 33% for humic acid removal was observed by increasing pH from 5 to 10. Temperature increase from 25°C to 40°C, resulted in an increase in humic acid removal from 76.8% to 91.4% and its adsorption on the adsorbent could be better described by Freundlich isotherm (n = 0.016, K_f = 0.013 and R^2 = 0.74). The most fitted adsorption kinetic model was pseudo-second order model.

Conclusions: The chemical structure of nanoadsorbent was proper and free from harmful substances. Despite the relative good condition of the effective surface, due to the large size of the shell, the overall micropore volume was low. Hence the qualitative characteristics the adsorbent caused the absorption capacity of humic acid to be low (0.96 mg/g).

Keywords: Humic acid, Iron supported oyster shell, Nanoadsorbent, Natural organic matters, Adsorption, Isotherm

* Correspondence: naserise@tums.ac.ir
[1]Department of Environmental Health Engineering, School of Public Health, Tehran University of Medical Sciences, Tehran, Iran
[2]Center for Water Quality Research (CWQR), Institute for Environmental Research (IER), Tehran University of Medical Sciences, Tehran, Iran
Full list of author information is available at the end of the article

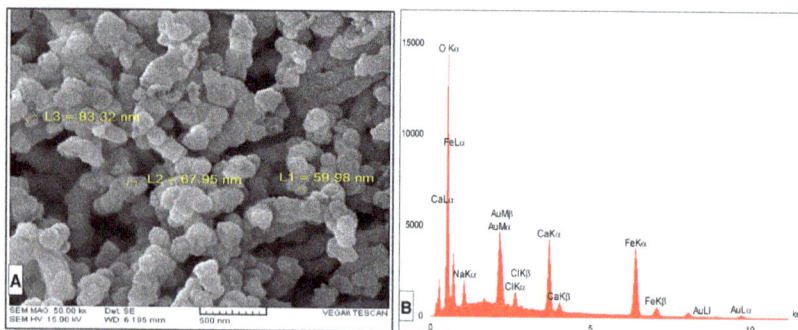

Figure 1 SEM (A) and EDS (B) of prepared nanoadsorbent. SEM **(A)** showing the Nanoscale particles and EDS **(B)** of prepared nanoadsorbent showing the main elements of adsorbent.

Background

Although surface water is an important natural resource used for many purposes, especially drinking, this resource generally contain natural organic matters (NOMs) so many scientific references have pointed out that more than 1500 different organic compounds are suspected to be present in drinking waters [1,2]. Naturally organic materials, a nonhomogeneous mixture of complex organic compounds, including; humic acids (HA), lipids, proteins, hydrophilic acids, carboxylic acids, amino acids, polysaccharides, and hydrocarbons are invariably present in surface and ground water resources, in dissolved or colloidal forms [3]. Various environmental and health problems have been reported that may be related to the presence of NOMs in natural waters, including; (a) the potential of NOMs have to cause undesirable color and taste [4]; (b) NOMs contributing the reactions with heavy metals and biocides to yield high concentrations of these substances and enhance their transportation in water [5]; (c) NOMs role in increasing coagulant and disinfectant dose requirements that lead to more sludge amounts; (d) in treatment plants, NOMs react with chlorine to form harmful organic compounds [4]; (e) NOMs are important factors in fouling which affect various applications of membrane processes [6]; and (f) their tendency to compete with low-molecular weight synthetic organic chemicals and inorganic pollutants, reducing their adsorption rates and equilibrium capacities [7,8].

Considering harmful effects of NOMs on human health, these compounds should be eliminated from water before the chlorination process in drinking water treatment plants and it is important for the public health and drinking water industry to find reliable methods to remove a wide range of organic contaminants from water [9]. During the recent years, a large number of researches have been carried out on the NOMs removal from water in order to minimize its impacts on water quality [3,10-14]. Based on these

researches, the most conventional options for NOM removal include membrane filtration and ion exchange [15,16], ozonation [17], biodegradation [18], ultrasound waves [19], adsorption, and coagulation [20]. Among the methods, membrane separation and adsorption were the most effective and available processes for removing NOMs from water [3]. High removal efficiency and no harmful by-products production are the main advantages for adsorption process and many kinds of adsorbents have been developed for the removal of humic acid from water.

During the past few years, nanoparticles have been proposed as a removal method for a wide range of pollutants from waters [21]. Some important nanoparticles in water treatment include metal-containing nanoparticles, carbonaceous nanomaterials, zeolites, and dendrimers [22]. As nanoscale zero-valent iron (NZVI) particles have unique reactive and sorption characteristics, they have received high attention for the treatment of contaminated waters [23,24].

Several studies have been conducted using various materials such as mesoporous silica beads [25], carbon and titanium dioxide [26], chitosan beads [27], ceramic membranes [12], kaolin [28], and betonite [13] as a bases for nanoparticles. In southern coasts of Iran, there is a considerable amount of oyster shells which may have the potential to be used for a perfect bases for nanoadsorbents, and so far no study and

Table 1 Physical properties of the nanoadsorbent

Parameter	Unit	Value
Density	g/cm^3	1.82
Particle size	nm	60-83
Micropore volume	m^3/g	0.012
BET surface area	(m^2/g)	16.85
Solubility @ (20°C)	%	Insoluble

Table 2 Chemical composition of the nanoadsorbent

Element	Weight %	Atomic %	Element	Weight %	Atomic %
O	43.04	63.68	Si	4.1	2.23
Na	4.06	4.26	Sr	1.25	1.17
Cl	4.61	3.13	Al	2.3	1.64
Ca	17.1	10.28	Mg	0.78	0.43
Fe	9.46	4.08	S	1.5	1.95
Au	0.31	0.04	K	1.18	1.15
C	10.31	5.97	Total	100	100

information are achieved on the use of the shell as adsorbent. Considering the possibility of successful production of air-stable ZVI nanoparticles with a high gravimetric ratio, good stability and no harmful matters, that in this research, it was studied as a bases for NZVI.

This study focused on the oyster shell supported NZVI application for the removal of NOMs under different experimental conditions, in order to remove taste and odor-causing agents from surface water resources.

Methods
Stock solutions
In this study, HA was used as NOMs representative and stock solution of HA was prepared using 1 g of HA powder dissolved in 1 L of deionized water, followed by filtering the solution through a 0.45 μ membrane filter (cellulose acetate). The solution was further diluted to the required concentrations and were stored at 4°C before application.

Preparation of nano-supported shell
The formation of the oyster shell supported iron nanoadsorbent was carried out by wet impregnation technique [26-29]. Shells collected from the coast of the Persian Gulf were transported to research lab and after physically cleaning (removal of mud, rinsing for 10 min and washing by deionized water), were crushed, dried in 100°C oven and then sieved with a mesh of 100. To prepare the NZVI coated shell, the shell beads were placed in the reactor during the synthesis of NZVI, by using the reduction of dissolved

iron method, which comprised of four stages; mixing, separating, washing and drying [30,31].

Initially, 5 g of crushed shell beads were poured in a flask containing 0.1 M FeCl3-6H2O dissolved in absolute ethanol and heated up to 80°C until the solvent was evaporated and dry coated beads were obtained. The beads were dispersed in 150 mL flask containing absolute ethanol and then the flask was placed on an orbital shaker. One hundred mL of 0.16 M NaBH4 aqueous solution was poured into a burette and dropped into stirring flask. During this reaction, ferric iron (Fe^{3+}) was reduced to zero-valent iron (Fe^0) by borohydride and the crushed shell beads started to take black color. This suggested that the ferrous ions attached to the support material were successfully reduced to zero-valent state, according to the following reaction:

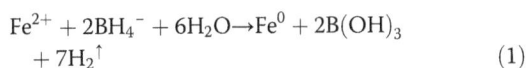

$$Fe^{2+} + 2BH_4^- + 6H_2O \rightarrow Fe^0 + 2B(OH)_3 + 7H_2\uparrow \qquad (1)$$

Afterwards, the contents of the flask were discharged to a funnel containing a millipore filter, under the suction force, leading to dewatering. The shell beads coated with NZVI were washed by ethanol, and then dried at 50°C and then kept in desiccators.

Characterization of nanoadsorbent
Laboratory synthesized nanoadsorbent was analyzed using Scanning Electron Microscopy (SEM) of TESCAN, Energy Dispersive Spectroscopy (EDS) and X-Ray Fluorescence (XRF). The specific surface area of nanoadsorbent was analyzed using BET-N_2 adsorption method.

Batch reactor adsorption system and experiments
Adsorption of HA onto produced adsorbent was carried out by a batch reactor and effect of different parameters such as initial HA concentration, adsorbent dose, pH, and temperature were studied.

Adsorption isotherm
Different doses of nanoadsorbent (0.5gr, 1gr, 3gr and 5gr) were added to the reaction flask containing 100 mL of 5 mg/L HA solution with the initial pH of 7 and temperature of 25°C. The containers were sampled:

Table 3 Chemical composition of the nanoadsorbent

Composition	% Weight	Composition	% Weight	Composition	% Weight	Composition	% Weight
SiO2	1.21	Al2O3	0.37	MgO	0.71	Na2O	1.01
CaO	21.09	K2O	0.65	Cl	0.07	SO3	0.19
FeOOH	4.01	CaCO3	63.38	SrO	0.81	Fe2O3	5.51

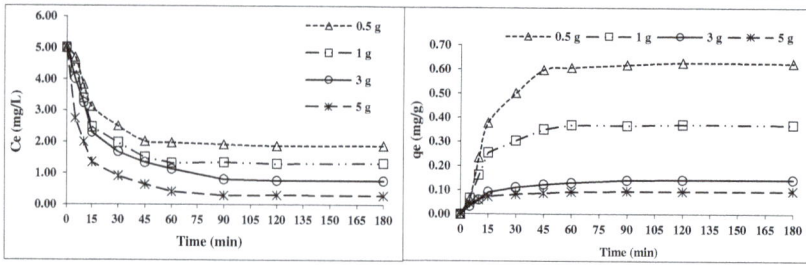

Figure 2 The effect of the adsorbent dose @ (HA = 5 mg/L, Temp = 25°C and pH = 7).

firstly at intervals of 5 and 10 min and then at a frequency of every 15 min to reach equilibrium time. After shaking and settling for 5 min, the samples supernatant were centrifuged followed by membrane filter (0.45 μm, cellulose acetate), absorbance values of solutions remaining without adsorption were measured by using UV–Vis spectrophotometer at wavelength of 254 nm. In the next step, adsorption equilibrium data were correlated with two well-known empirical isotherm models; Freundlich and Langmuir.

The kinetic of HA adsorption
After determining the adsorption isotherm, using collected data and models related to the absorption kinetics, modeling kinetic of HA adsorption study on the nanoadsorbent was determined.

The effect of initial HA concentration
The effect of initial HA concentration on the adsorption rate was studied by contacting 0.5 g of adsorbent at room temperature of 25°C and pH = 7 using four initial concentrations of HA solution (0.5, 2, 5 and 10 mg/L) in 100 mL of samples at time = 0 and at selected time intervals (up to a maximum of 180 min), sample concentration was determined by UV–Vis spectrophotometer.

The effect of pH
In order to survey the pH effect on HA adsorption, pH of the solution was changed in the range of 5.0 and 10.0 at four intervals (5, 7, 8 and 10) using either 0.1 mol/L NaOH or 0.1 mol/L HCl. pH values were determined using a Elmetron CP-501 pH meter, fitted with a combined glass-reference electrode.

The effect of temperature
In order to assess the effect of temperature on the equilibrium adsorption capacity of HA, 0.5gr of the nanoadsorbent was used at different temperatures (25, 30 and 40°C), with 5 mg/L HA solutions, optimum pH = 5 and in equilibrium contact time.

Results and discussion
Characterization of nanoadsorbent
SEM image of the nanoadsorbent is shown in Figure 1, the physical properties of the nanoadsorbent are presented in Table 1, the chemical elements and composition of nanoadsorbent are presented in Tables 2 and 3, respectively.

As shown in the Figure 1, most of the sheet structure has been changed to irregular small particles and depicts the synthesized nanoadsorbent with an approximately 60–85 nm diameter. The main elements of the adsorbent were Ca, O, Fe and Na and as it can be found from Table 3, the highest component of the adsorbents was

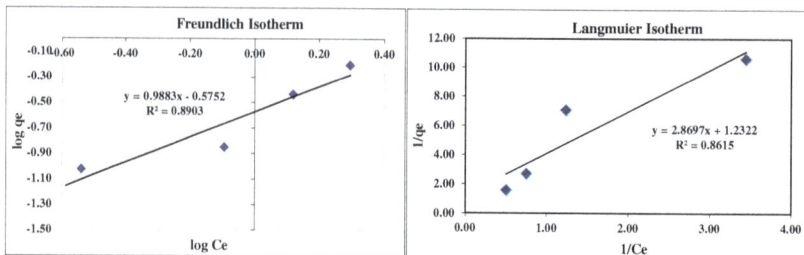

Figure 3 Freundlich and Langmuir adsorption isotherms.

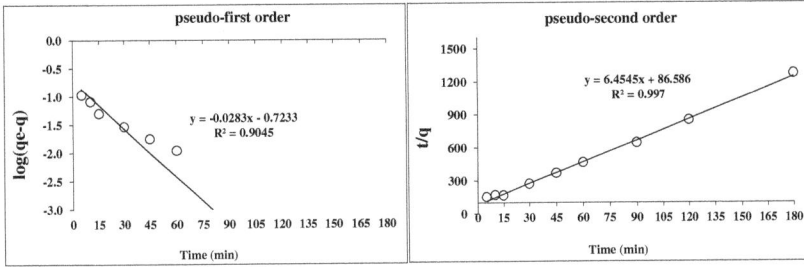

Figure 4 Adsorption kinetics; pseudo-first order and pseudo-second order (nanoadsorbent dose = 3 g/100 mL, C0 = 5 mg/l, Temp = 25°C and pH = 7).

lime ($CaCO_3$ and CaO), whit about 94.25% of the total weight.

The effect of the adsorbent dose

The effect of the initial HA concentration on its adsorption is shown in Figure 2. Although by increasing the nanoadsorbent dose of 0.5 to 10 mg/L, the removal of HA (C = 5 mg/L) was found to increase from 66.4% to 97.4%, the amount of removed HA (mg) per mass of nanoadsorbent showed a decrease in the rate from 0.66 mg/g to 0.09 mg/g. This is because of the fact that increasing adsorbent dosage increases the surface area for adsorption. However, very slow increase in the removal was observed beyond the dose of 3 g/L. This section of our results was in accordance with a study of Doulia etal who studied the adsorption of humic acid on acid-activated Greek bentonite [13].

The data of adsorption equilibrium for HA adsorption by nanoadsorbent are presented in Figures 2 and 3. As it can be seen, Freundlich isotherm had a better correlation with HA removal with R^2 higher than Langmuir. The r-shaped adsorption isotherm indicated that there is a high affinity of contact time with adsorption rate.

Kinetic of HA adsorption

Adsorption kinetics of HA as pseudo-first order and pseudo-second order are depicted in Figure 4.

Based on results of Figure 4, the pseudo second order model generates better fit to the experimental data of the investigated adsorption of HA on nanoadsorbent ($R^2 = 0.997$). The degree of the reaction is determined by sharing or exchange of electrons between adsorbent positive (nzvi) and negative groups (HA).

The effect of initial HA concentration

The effect of initial HA concentration is depicted in Figure 5. It can be observed that the removal rate of HA decreased from 0.64 to 0.84, by increasing the initial HA concentration from 0.5 to 10 mg/L. The equilibration time for the adsorption of HA at different concentrations ranged between 90 and 120 min for 0.5 an10 mg/L of HA, respectively.

As can be seen from Figure 5 when an increase occurs in the initial concentrations, the adsorption capacity of HA would enhance. Driving force is one of the most important factors in the adsorption process so the adsorption capacity of HA is a function of HA initial concentration, as it is important to overcome

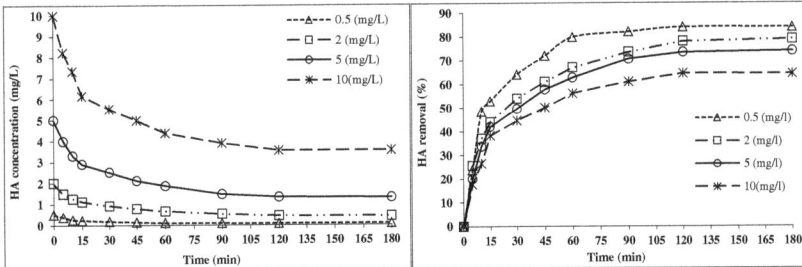

Figure 5 The effect of initial HA concentration and contact time on HA removal by nanoadsorbent @ (nanoadsorbent dose: 0.5 g/100 mL, Temp = 25°C and pH = 7).

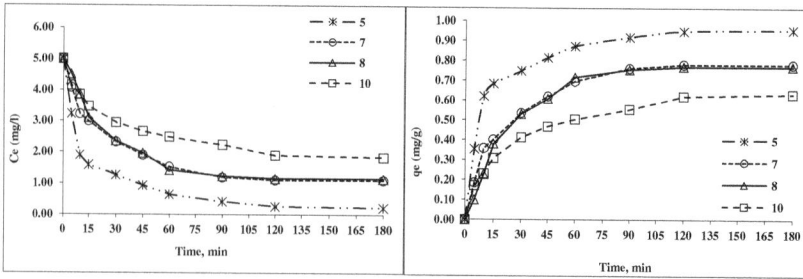

Figure 6 The effect of pH and contact time on HA removal by nanoadsorbent (HA = 5 mg/L, nanoadsorbent dose = 0.5 g/100 mL and Temp = 25°C).

the mass transfer resistance of the HA between the adsorbate and the adsorbent. In other words, the increase in the mass driving force allows more HA molecules to pass from the solution to the adsorbent surface. Therefore there is an inverse relationship between initial concentration and adsorption capacity [32]. This finding were in accordance with Moussavi et al [3] who investigated adsorption characteristics of HA onto single-walled carbon nanotubes and reported that the amount of adsorbed HA was higher at lower initial concentrations of HA.

The effect of pH

The effect of the initial pH on adsorption rate is presented in Figure 6. As it can be seen, a maximum HA adsorption rate of 0.96 mg per gram of adsorbent was observed at pH = 5. This was followed by a decrease in adsorption capacity at the late stage of pH experiments ranging from 5 to 10. Also the minimum adsorption capacity was found to be 0.63 mg/g at initial pH = 10 after reaching the equilibrium time (120 min). Adsorption of HA to nanoadsorbent is strongly influenced by pH. As increasing the pH increases the ionization of HA and hence the concentration of the negatively charged

ions which leads to decreasing the amount of H^+ ions [33].

The effect of temperature

The temperature effect on the HA adsorption on nanoadsorbent is shown in Figure 7, with HA concentration of 5 mg/L, the adsorption rate increased from76.8% to 91.4%, by increasing the temperature from 25°C to 40°C. As depicted in Figure 7, temperature increase led to further the HA adsorption, due to high interaction between HA and ZNVI-shell beads. The temperature increasing creates swelling effect in the internal structure of the adsorbent, causing further penetration HA into the adsorbent [34]. In addition, temperature increase causes more diffusion rate of HA molecules in the external boundary layer and the internal pores of the adsorbent particles. Our finding was in accordance with Zeinali etal results on adsorption of dichloromethane using GAC [35].

Conclusions

The obtained results revealed that surface area and pore volume of nanoadsorbent were 16.85 (m^2/g) and 0.021 m^3/g, respectively; the main elements of the adsorbent were

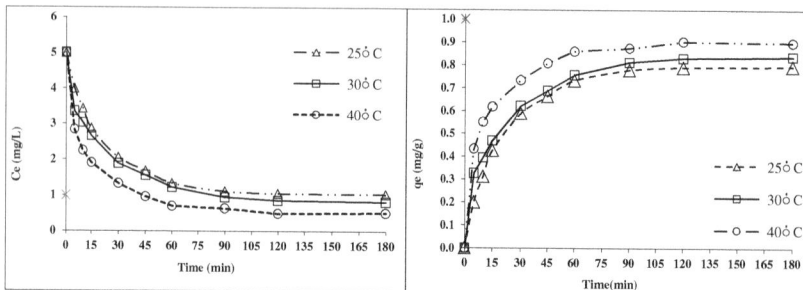

Figure 7 The effect of temperature and contact time on HA removal by nanoadsorbent (HA = 5 mg/L, nanoadsorbent dose = 0.5 g/100 mL).

Ca, O, Fe and Na and most of the adsorbent component was composed of lime ($CaCO_3$ and CaO about 94.25% of the total weight). The produced nanoadsorbent was not soluble in water. HA Adsorption onto the nanoadsorbent increased by increasing of adsorbent dose and temperature and decreased by increasing HA concentration and pH. The r-shaped adsorption isotherm indicated a high affinity of contact time with adsorption rate. In optimal conditions provided in the study (HA = 5 mg/L, nanoadsorbent dose = 0.5 g/100 mL, pH = 5 and temperature = 40°C), maximum HA removal efficiency reached to 94.6%, within 90-120 min at the experimental conditions in all cases is Pseudo-second order and Freundlich isotherm models were fitted to the adsorption kinetic and isotherm, respectively.

Competing interests
The authors declare that they have no competing interests.

Authors' contributions
VA conducted the practical parts of the study and wrote the first draft of the manuscript. AR had shared roles in nanoadsorbent characterization. SN, AHM, and RNN participated in data analysis and provided comments during the writing of the manuscript. All authors read and approved the final manuscript.

Acknowledgements
The authors are grateful to the Dr Elham Boushehri and Mrs. Afsoon Pirouzan, Hormozgan University of Medical Sciences, as well as all laboratory staffs of the Departments of Environmental Health Engineering in both Hormozgan University of Medical Sciences and Tehran University of Medical Sciences.

Author details
[1]Department of Environmental Health Engineering, School of Public Health, Tehran University of Medical Sciences, Tehran, Iran. [2]Center for Water Quality Research (CWQR), Institute for Environmental Research (IER), Tehran University of Medical Sciences, Tehran, Iran. [3]Nanotechnology Research Center, Research Institute of Petroleum Industry (RIPI), Tehran, Iran.

References
1. Yue Z, Economy J: Nanoparticle and nanoporous carbon adsorbents for removal of trace organic contaminants from water. *J Nanoparticle Res* 2005, 7:477–487.
2. Mahvi AH, Alipour V, Rezaei L: Atmospheric moisture condensation to water recovery by home air conditioners. *Am J Appl Sci* 2013, 10(8):917–923.
3. Moussavi G, Talebi S, Farrokhi M, Mojtabaee Sabouti R: The investigation of mechanism, kinetic and isotherm of ammonia and humic acid co-adsorption onto natural zeolite. *Chem Eng J* 2013, 171:1159–1169.
4. Bai R, Zhang X: Polypyrrole-coated granules for humic acid removal. *J Colloid Interface Sci* 2001, 243:52–60.
5. Schmitt D, Saravia F, Frimmel FH, Schuessler W: NOM facilitated transport of metal ions in aquifers: importance of complex-dissociation kinetics and colloid formation. *Water Res* 2003, 37:3541–3550.
6. Yuan W, Zydney AL: Humic acid fouling during ultrafiltration. *Environ Sci Technol* 2000, 34:5043–5050.
7. Klausen J, Vikesland PJ, Kohn T, Burris DR, Ball WP, Roberts AL: Longevity of granular iron in groundwater treatment processes: solution composition effects on reduction of organohalides and nitroaromatic compounds. *Environ Sci Technol* 2003, 37:1208–1218.
8. Tratnyek PG, Scherer MM, Deng B, Hu S: Effects of natural organic matter, anthropogenic surfactants, and model quinones on the reduction of contaminants by zero- valent iron. *Water Res* 2001, 35:4435–4443.

9. Bansal RC, Goyal M: *Activated carbon adsorption*. New York: Taylor & Francis Group; 2005.
10. Jonglertjunya W, Lertchutimakul T: Equilibrium and kinetic studies on the adsorption of humic acid by activated sludge and Bacillus subtilis. *Songklanakarin J Sci Technol* 2012, 34(6):669–677.
11. Giasuddin AM, Kanel S, Choi HC: Adsorption of Humic Acid onto Nanoscale Zerovalent Iron and Its Effect on Arsenic Removal. *Environ Sci Technol* 2007, 41:2022–2027.
12. Harman BI, Koseoglu H, Yigit NO, Beyhan M, Kitis M: The use of iron oxide-coated ceramic membranes in removing natural organic matter and phenol from waters. *Desalination* 2010, 261:27–33.
13. Doulia D, Leodopoulos C, Gimouhopoulos K, Rigas F: Adsorption of humic acid on acid-activated Greek bentonite. *J Colloid Interface Sci* 2009, 340:131–141.
14. Kabsch-Korbutowicz M: Effect of Al coagulant type on natural organic matter removal efficiency in coagulation/ultrafiltration process. *Desalination* 2005, 185(1–3):327–333.
15. Bolto B, Dixon D, Eldridge R, King S, Linge K: Removal of natural organic matter by ion exchange. *Water Res* 2002, 36(20):5057–5065.
16. Zazouli MA, Nasseri S, Mahvi AH, Gholami M, Mesdaghinia AR, Younesian M: Retention of humic acid from water by nanofiltration membrane and influence of solution chemistry on membrane performance. *Iran J Environ Health Sci Eng* 2008, 5(1):11–18.
17. Selcuk H, Rizzo L, Nikolaou AN, Meric S, Belgiorno V, Bekbolet M: DBPs formation and toxicity monitoring in different origin water treated by ozone and alum/PAC coagulation. *Desalination* 2007, 210(1–3):31–43.
18. Leiknes T, Lazarova M, Odegaard H: Development of a hybrid ozonation biofilm-membrane filtration process for the production of drinking water. *Water Sci Technol* 2005, 51(6–7):241–248.
19. Mahvi AH, Maleki A, Rezaee R, Safari M: Reduction of humic substances in water by application of ultrasound waves and ultraviolet irradiation. *Iran J Environ Health Sci Eng* 2009, 6(4):233–240.
20. Ho L, Newcombe G: Effect of NOM, turbidity and floc size on the PAC adsorption of MIB during alum coagulation. *Water Res* 2005, 39(15):3668–3674.
21. Li Q, Mahendra S, Lyon DY, Brunet L, Liga MV, Li D, Alvarez PJ: Antimicrobial nanomaterials for water disinfection and microbial control: Potential applications and implications. *Water Res* 2008, 4(18):4591–4602.
22. Savage N, Diallo MS: Nanomaterials and water purification: Opportunities and challenges. *J Nanoparticle Res* 2005, 7(4–5):331–342.
23. Bezbaruah AN, Shanbhogue SS, Simsek S, Khan E: Encapsulation of iron nanoparticles in alginate biopolymer for trichloroethylene remediation. *J Nanoparticle Res* 2011, 13:6673–6681.
24. Dickinson M, Scott TB: The application of zero valent iron nanoparticles for the remediation of a uranium-contaminated waste effluent. *J Hazard Mater* 2010, 178:171–179.
25. Park H, Yeon KH, Park YM, Lee SJ, Lee SH, Choi YS, Chung Y: Reduction of nitrate by nanoscale zero-valent iron supported on mesoporous silica beads. *J Water Environ Technol* 2008, 6(1):35–42.
26. Petala E, Baikousi M, Dimos K, Filip J, Pechousek J, Tucek J, Safarova K, Douvalis A, Bakas T, Karakassides ZM, Zboril R: Zero valent iron nanoparticles supported on porous matrices: enhanced efficiency and application for water treatment. *J Hazard Mater* 2013, 261:295–306.
27. Liu T, Zhao L, Sun D, Tan X: Entrapment of nanoscale zero-valent iron in chitosan beads for hexavalent chromium removal from wastewater. *J Hazard Mater* 2010, 184(1-3):724–730.
28. Zhang X, Lin S, Lu XQ, Chen ZI: Removal of Pb (II) from water using synthesized kaolin supported nanoscale zero-valent iron. *Chem Eng J* 2010, 163:243–248.
29. WZW Yaacob, Kamaruzaman N, Samsudin AR: Development of nano-zero valent iron for the remediation of contaminated water. *Chemical engineering transactions* 2012, 28:25–30.
30. Pan BC, Chen BJ, Meng XQ, Pan XT, Li MW, Zhang X, Zhang JL, Chen AB, Zhang XQ, Sun Y: Application of an effective method in predicting breakthrough curves of ?xed bed adsorption onto resin adsorbent. *J Hazard Mater* 2005, B,124:74–80.
31. Wan Ngah WS, Kamari A, Koay YJ: Equilibrium and kinetic studies of adsorption of copper (II) on chitosan and chitosan/PVA beads. *Int J Biol Macromol* 2004, 3:155–161.

32. Perez-Aguilar NV, Diaz-Flores PE, Rangel-Mendez JR: **The adsorption kinetics of cadmium by three different types of carbon nanotubes.** *J Colloid Interface Sci* 2011, **364**:279.

33. Bringle CD, Shibi IG, Vinod VD, Anirudhan TS: **Sorption of humic acid from aqueos solutions by lanthana alumina oxide pillared bentonite.** *J Sci Ind Res* 2005, **46**:782–788.

34. Zulfikar MA: **Effect of Temperature on Adsorption of Humic Acid from Peat Water onto Pyrophyllite.** *International Journal of Chemical, Environmental & Biological Sciences,* 2013, **1**(1):88–90.

35. Zeinali F, Ghoreyshi A, Najafpour DD: **Adsorption of dichloromethane from aqueous phase using granular activated carbon; Isotherm and breakthrough curve measurements.** *Middle East J Sci Res* 2010, **5**(4):191–198.

18

Agrometeorological drought in the Romanian plain within the sector delimited by the valleys of the Olt and Buzău Rivers

Ovidiu Murărescu, George Murătoreanu[*] and Mădălina Frînculeasa

Abstract

Background: The last few decades have recorded a high frequency of the meteorological drought phenomenon. Southern and south-eastern Romania make no exception, with such phenomena often occurring from July to November 2011, which brought about an agrometerological drought that lasted from the third decade of July to early December, with a slight improvement in October. This situation led to a decrease in soil water reserves, mainly in the first 20 cm, with a negative impact on agricultural crops and the following agricultural year as well.

Findings: The methodology was based on a correlative analysis between the decadal rainfall quantities and the existing soil water reserve, during the interval between June and November 2011, for eight weather stations.

Conclusion: The statistico-mathematical data analysis showed an intensification of the pedological drought phenomenon in September, with a slight improvement in October and an increase in November.

Keywords: Precipitation, Temperatures, Water climatic, Drought, Soil

Introduction

The increase of the meteorological drought phenomenon leads to the occurrence of the pedological drought and, hence, of aridity, combined with a high soil temperature and evapotranspiration. Such situations, according to statistical history, were recorded in 2000, 2001, 2007, 2011, 2012 and 2013.

The analysed geographical area includes the Romanian Plain, more exactly the sector between the left bank of the Olt river – in the west – as far as its confluence with the Danube; north-eastern boundary is located on the right bank of the Buzău river; the northern is conventionally located north of Slatina, Valea Mare, Potcoava, then the contact between the Cotmeanu Piedmont and the 5th terrace of the Argeş river as far as north of Piteşti, stretching along the contact with the Outer Subcarpathian Curvature Hills; the southern boundary is located at the contact with the Danube valley, while the eastern overlaps the interference between the forest-steppe and the forest – the Argeş river meadow, from the confluence

with the Dâmboviţa river to its flow into the Danube, then Dâmboviţa – Pasărea – Obârşia Mostiştei and Sărata – boundary with the Bărăgan Plain [1] (Figure 1).

There have been concerns regarding the analysis of the phenomena of aridity occurrence ever since early 20th century (Lang – 1920 – the rain factor; De Martonne – 1926 – the aridity index – etc.). In Romania, many climatological geographers have had similar preoccupations [2-19].

Data and working methods

The present study does not aim to calculate the climatic water deficit (CWD) and the De Martonne aridity index (I_{ar}), as these parameters have already been identified, and correlations between them have also been made [11,14,17], therefore the territorial map of the aridity index in Romania – mm/°C has been drawn as well [2].

This approach aims rather to identify and characterise the agrometeorological phenomena of hydric risk by taking into consideration parameters and critical thresholds during the intervals which are specific to phenological processes and phases. Such analyses can be performed decadally, bimonthly, monthly, seasonally, annually or permanently with a view to adopting negative effect prevention and reduction measures [20]. In this regard, we have

---
* Correspondence: muratoreanug@yahoo.com
Department of Geography, Valahia University of Târgovişte, Târgovişte, Romania

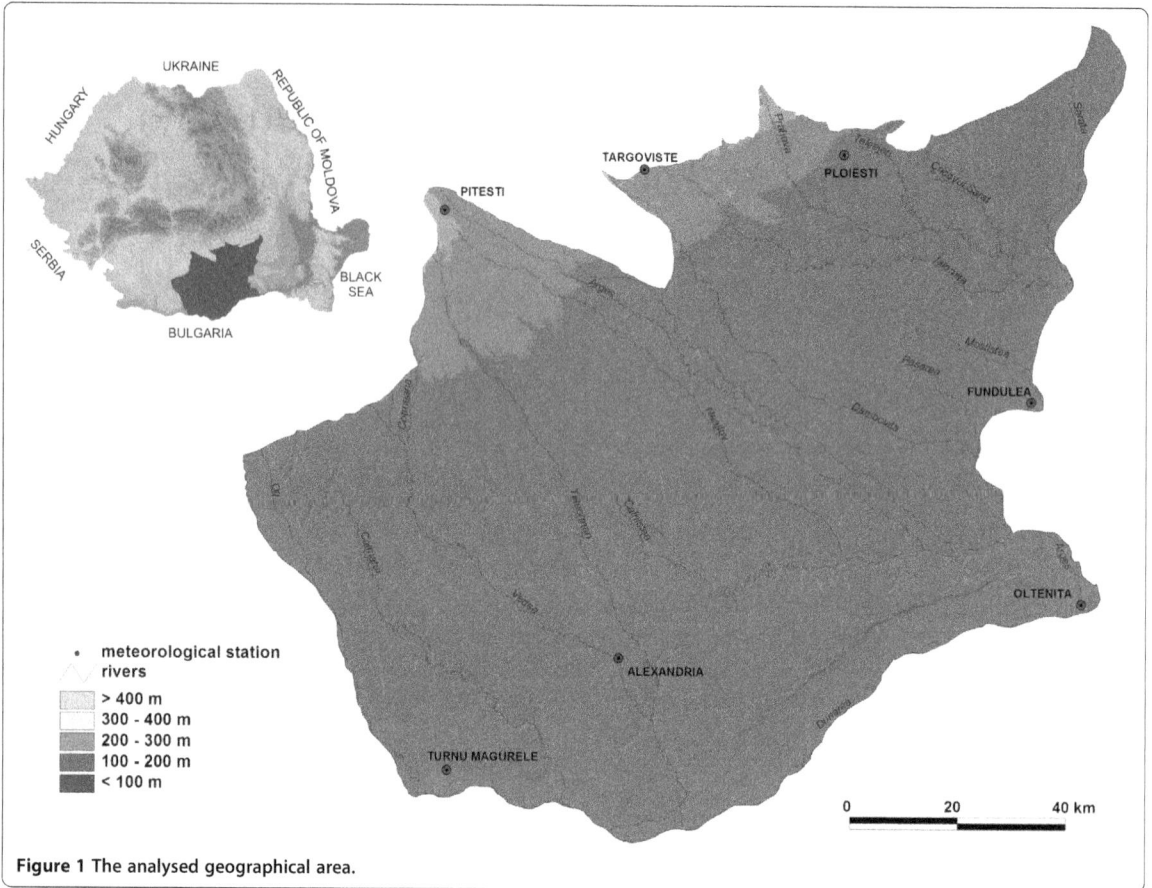

Figure 1 The analysed geographical area.

conducted a correlative analysis between the decadal rainfall quantities and the existing soil water reserve, during the interval between June and November 2011, for eight weather stations (Piteşti, Ploieşti, Târgovişte, Turnu Măgurele, Alexandria, Oltenţia, Fundulea).

The soil moisture considered was determined from the surface to one-metre depth by 20-cm sections. The available moisture is calculated based on the volumetric weight of the soil stratum considered (Vw), existing moisture content (U) and the wilting point (WP), according to the following formula:

$$AWC = 0.1 \times Vw \times (U\% - WP) \times h$$

where AWC = available moisture, and h = depth.

Findings

In the calendar year 2011, the annual mean temperature in Romania was 9.2°C, 0.3°C higher than the climatological normal, with positive deviations between 0.1°C (in March and May) and 2-6°C in September, but also with negative deviations in February, April, October and

November (0.1°C – April up to 2.7°C in November) from the climatological normals.

The country average annual rainfall was 500.4 mm, 22% under the climatological normal, due to the deficits recorded in most months. The excess rainfall was in June-July, while the rest of the months were characterised by deficit, with negative deviations ranging between 10-97%. Months with deficit were March (40%), August (53%), September (72%) and November (97%). As a result, the year 2011 was characterised by low rainfall, with November the driest month, and by pedological drought installed differentially at regional level [21].

The statistico-mathematical data analysis of decadal rainfall quantities recorded at the eight agrometeorological stations located in the field region under study, correlated with the soil water reserve recorded in July-November 2011, showed an intensification of the pedological drought phenomenon in September, with a slight improvement in October and an increase in November.

The average available moisture (AWC) required for plant development, according to norms, is 400 at 20-cm depth, 820 at 40 cm and 1500 at 80 cm.

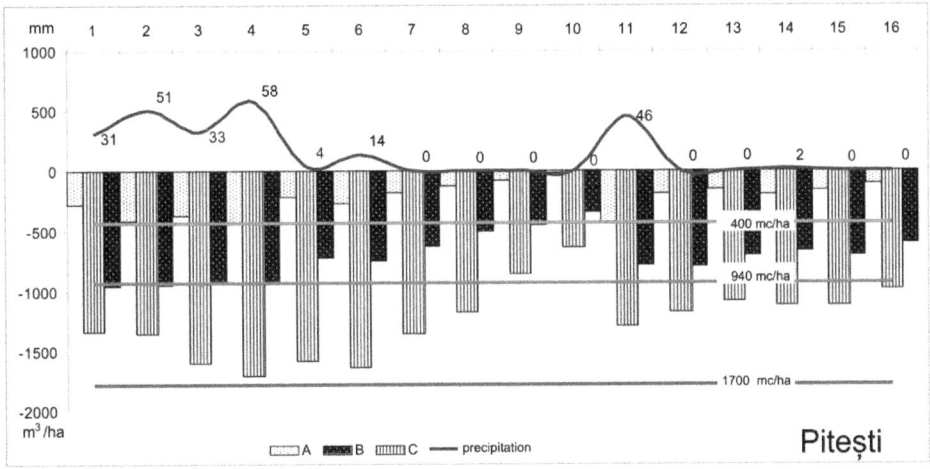

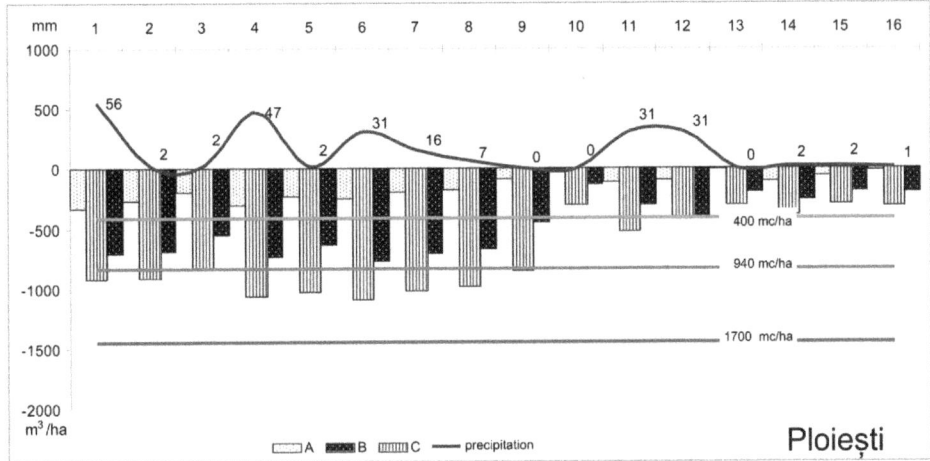

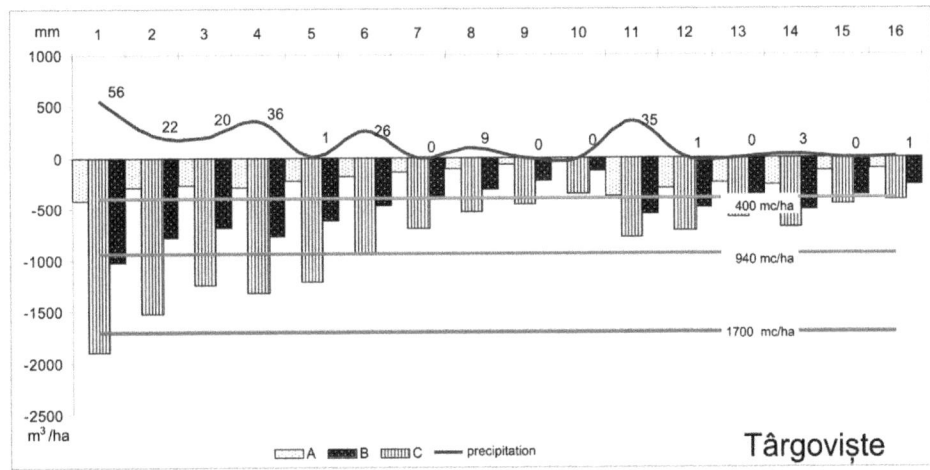

Figure 2 The average available moisture (AWC) and the soil moisture reserve for the agrometeorological station situated in the north of the plain (data source: National Meteorology Administration).

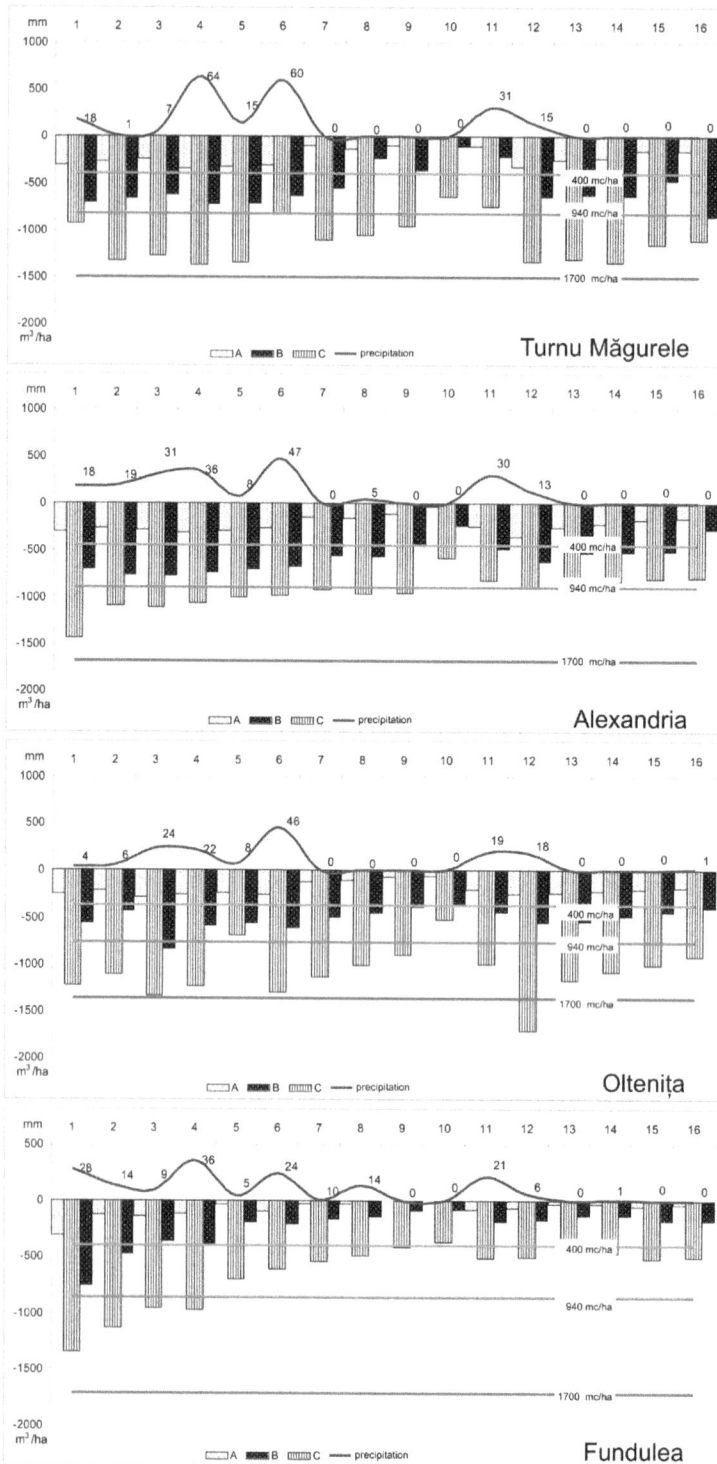

Figure 3 The average available moisture (AWC) and the soil moisture reserve for the agrometeorological station situated in the south of the plain (data source: National Meteorology Administration).

With the agrometeorological stations situated in the north of the plain (Piteşti, Târgovişte, Ploieşti), located on the 30 mm/°C isoline, according to the Romania aridity index map, deficits were observed in August (below 30 mm), September (0 mm) and November (5 mm). Under these conditions, the soil moisture reserve fell drastically, to values much below the normal averages (Figure 2).

As regards the agrometeorological stations located on the 20 mm/°C isoline, according to the same map (Turnu Măgurele, Alexandria, Fundulea and Olteniţa and Buzău), the situation is similar in terms of precipitations, but with a slight extension between the decadal intervals towards the months of July and October (Figure 3).

Conclusions

The results of the study have emphasised the consequences of an agrometeorological phenomenon of risk, with negative effects mainly on summer-autumn agricultural crops, leading to a decrease of the agricultural production, but also with significant implications on the following agricultural year (2012–2013), due to inability to perform specific agro-technical works or, if they had been carried out, they had to be redone because seeding had been compromised.

Abbreviations
AWC: Available water capacity.

Competing interests
The authors declare that they have no competing interests.

Authors' contributions
OM participated in the conception, or acquisition of data, or analysis and interpretation of data, participated in the given final approval of the version to be published; GM participated in the conception of the study and performed the statistical analysis and graphics; MF participated in the analysis and interpretation of data, involved in drafting the manuscript or revising. All authors read and approved the final manuscript.

Acknowledgements
This article was put together with the help of information gathered from the Meteorological and hydrological stations at Târgovişte and from the Basarab I Emergency Inspectorate, Dâmboviţa county.

References
1. Institute of Geography, Romanian Academy: *Geografia României, vol V.* Bucureşti: Ed. Academiei Române; 2006.
2. Cernescu N: **Clasificarea solurilor cu exces de umiditate.** In *Cercetări de Pedologie.* Bucureşti: Academiei RPR; 1961:223–250.
3. Berbecel O, Stancu M, Ciovică N, Jianu V, Apetroaiei Şt, Socor Elena, Rogodjan Iulia, Eftimescu Maria: *Agro-meteorologie.* Bucureşti: Ceres; 1970.
4. Berbecel O, Eftimescu M, Mihoc C, Socor E, Cusurzuz B: **Cercetări privind resursele agro-climatice din RS România.** *Bultinul Informativ ASAS* 1984, **13**:33–43.
5. Apetroaiei Ş: *Evaluarea şi prognoza bilanţului apei în sol.* Bucureşti: Ceres; 1977.
6. Botzan M: *Bilanţul apei în solurile irigate.* Bucureşti: Academiei RSR; 1972.
7. Canarache A: *Fizica solurilor agricole.* Bucureşti: Ceres; 1990.
8. Canarache A: **Indicatori climatici şi regimuri de umiditate şi temperatură a solului.** *Ştiinţa solului, Seria a III-a, SNRSS* 2005, **1–2**:XXXVIII. 66–78.
9. Păltineanu C, Chiţu E, Seceleanu I, Tănăsescu N, Apostol G, Pufu MN, Păltineanu Rodica: *Evapotranspiraţia de referinţă, consumul şi necesarul apei de irigaţie ale principalelor culturi agricole în solurile bazinului hidrografic Argeş – Vedea.* Piteşti: Paralela; 1999:45.
10. Păltineanu C, Chiţu E, Tănăsescu N, Apostol G, Pufu MN: **Irrigation Water Requirements for Some Fruit Trees Specific to the Argeş – Vedea River Basin, Romania.** In *Proceedings of the Third International Symposium on Irrigation of Horticultural Crops.* Lisboa, Portugal: Acta Horticulturae; 2000. 537, **I**, 113–119.
11. Păltineanu C: **Comparison between Experimental and Estimated Crop Evapotranspiration in Romania.** In *International Agrophysics.* Institute of Agrophysics, Polish Academy of Science: Lublin, Poland; 2005:19.
12. Păltineanu C, Mihăilescu IF, Dragotă C, Vasenciuc F, Prefac Z, Popescu M: **Corelaţia dintre indicele de ariditate şi deficitul de apă climatic şi repartiţia geografică a acestora în România.** *Analele Universităţii Spiru Haret Seria Geografie* 2005, **8**:23–28.
13. Păltineanu C, Mihăilescu IF, Seceleanu I, Dragotă C, Vasenciuc F: **Using aridity indices to describe some climate and soil features in Eastern Europe: a Romanian case study.** *Theoretical and Applied Climatology* 2007, **90**(3–4):263–274.
14. Păltineanu C, Mihăilescu IF, Prefac Z, Dragotă C, Vasenciuc F, Nicola C: **Combining the standardized precipitation index and climatic water deficit in characterizing droughts: a case study in Romania.** *Theoretical and Applied Climatology* 2009, **97**(3–4):219–233.
15. Bogdan O, Niculescu E: *Riscurile climatice din România.* Bucureşti: SEGA International; 1999.
16. Bogdan O, Marinică I: *Hazarde meteo-climatice din zona temperată. Geneză şi vulnerabilitate cu aplicaţii în România.* Bucureşti: Universităţii "Lucian Blaga", Sibiu; 2007.
17. Croitoru A-E, Piticar A, Imbroane AM, Burada D-C: **Spatiotemporal distribution of aridity indices based on temperature and precipitations in the extra-Carpathian regions of Romania.** *Theoretical and Applied Climatology* 2013, **112**(3–4):597–607.
18. Peptenatu D, Sîrdoev I, Pravalie R: **Quantification of the aridity process in south-western Romania.** *Iran J Environ Health Sci Eng* 2013 2013, **11**:3.
19. Pravalie R, Sîrdoev I, Peptenatu D: **Changes in the forest ecosystems in areas impacted by aridization in south-western Romania.** *Iran J Environ Health Sci Eng 2014* 2014, **12**:2.
20. Murărescu O, Pehoiu G, Murătoreanu G, Ţurloiu R: **Le déficit d'humidité dans le sol et son impact sur les activités agricoles dans la haute plaine située entre les vallées des rivières argeş et prahova (roumanie) – étude de cas (août-novembre 2011).** *25ème Colloque de l'Association Internationale de Climatologie, Grenoble* 2012, **2012**:565–570.
21. National Meteorological Administration: *Annual Report* 2011, Bucharest; Romania.

19

Determination of aluminum and zinc in infusion tea cultivated in north of Iran

Mahboobeh Ghoochani[1], Sakine Shekoohiyan[2], Masoud Yunesian[1], Shahrokh Nazmara[1] and Amir Hossein Mahvi[1,3,4*]

Abstract

To determine aluminum and zinc levels in black tea cultivated in north of Iran, 105 black tea samples were collected from the tea growing regions of Guilan and Mazandaran provinces and were analyzed for Al and Zn concentration of tea infusion. Contents of all elements were analyzed three times separately by using an Inductively Coupled Plasma Atomic Emission Spectrometry (ICP - AES). The solubility of Al and Zn in infusions at 5, 15 and 60 min with boiling water showed that the mean level of Al in the third infusion was the highest (262.09 mg/kg) and in the first infusion was the lowest (169.40 mg/kg). The mean level of Zn in the third infusion was the highest (51.40 mg/kg) and in the second infusion was the lowest (48.33 mg/kg). The analysis of results also showed that the location factor influences the contents of these metals at different infusions.

Keywords: Black tea, Aluminum, Zinc, Iran

Introduction

Tea is one of the oldest, most popular and non-alcholic drinks in the world and it is provided from the dried leaves of the tea plant [1]. Economic and social benefits in tea are clear from the fact that about 18–20 billion tea cups are consumed daily in the world. Black and green teas are the popular version [2]. Iranians keep one of the highest per capita rates of tea consumption in the world, (about1.6 kg per capita consumption in the period from 2005 to 2007 [3]. Approximately 34 thousand hectares of lands have been cultured for tea in Guilan and Mazandaran provinces, Guilan and Mazandaran have a humid temperate climate with enough annual precipitation. Almost half of the dry tea interior production and the rest come from imports [4]. Many researchers have showed that tea has some useful health effects, including the prevention of diseases such as Parkinson, skin cancer, myocardial infarction, and coronary artery diseases [5]. Tea is exceptionally rich in polyphenolic compounds (flavonoid and phenolic acid), which are powerful antioxidants that protect the body

against disease [6] and contains minerals and trace elements that are essential to human health. The determination of total elements content in different beverages has been a subject of numerous studies. On average, 1 liter of tea is consumed per person per day in Iran and UK which as a percentage of average daily dietary intakes, can provide 58.8 % of Al, 0.44 % of Zn and 58.8 % of Al, 2 % of Cu and 115 % of Mn [7, 8]. Studies by Wang, Su, and (1994) show that of the daily dietary intake of 9–12 mg Al by Chinese population, tea contributes 0.2–1.1 mg Al, assuming that an adult drinks 1–5 g of tea per day. The Provisional Tolerable Weekly Intake (PTWI) for Al of 7 mg/kg body weight (equivalent to 1 mg/kg body weight/day) was considered by the Joint FAO/WHO Expert Committee on Food Additives (JECFA [9]). Al was affirmed as a food contaminant in 1989. Various studies have been conducted, to evaluate the daily dietary intake of Al from a number of food products. Studies on dietary intake of Al show an average intake of 23 mg/day for Indian population [10], 9– 12 mg/day for Chinese population [11], 2–25 mg/day for American population, 2.2–8.1 mg/day for Japanese adult males and 0.6–33.3 mg/day for Dutch adults [12]. The contribution of tea drinking to mineral absorption is not certain, as the bioavailability of many of these metals with tea is not known [13]. Jackson reports that tea which contains a substantial amount of Al may present health

* Correspondence: ahmahvi@yahoo.com
[1]Department of Environmental Health Engineering, School of Public Health, Tehran University of Medical Sciences, Tehran, Iran
[3]Center for Solid Waste Research, Institute for Environmental Research, Tehran University of Medical Sciences, Tehran, Iran
Full list of author information is available at the end of the article

hazards (kidney weakness) for consumers [14]. Moreover, high Al content in the human body has been hypothesized to have possible links with various diseases, such as encephalopathy dementia, oestomalacia, fractures and high levels of bone Al and Alzheimer's disease [15–17]. Although tea leaves and leaf infusions contain high concentrations of Al, only a small proportion of it is available for absorption in the gastro-intestinal tract [13, 18] and the renal excretion of Al is fairly effective [19]. It has been observed that tea, ingested alongside food, inhibits the absorption of inorganic and some forms of organic iron, contributing to iron deficiency, mainly in women on a vegetarian diet of low iron content. Studies show that tea drinking may marginally decrease the availability of divalent metals, such as Cu and Zn [13]. The Zn is known to be essential micronutrients, but can be toxic depending upon the concentration. The low or high amount of dietary intake of Zn was based to create the various physiological and ethological diseases because of these trace element deficiency or toxicity [20].

The origin of Iranian tea is from three varieties of seed from the Northern part of India (from the Assam region of India) [21]. Most of the farms are located the hillsides of Iran like the farms in Darjeeling. These farms produce an orthodox style of black tea. The color of Iranian tea is red with fairly light taste, and it is delicious without adding any milk or sugar. Institute of Standards and Industrial Research of Iran has set the maximum permissible levels of elements in the black tea only for As, Hg, Pb, Cd and Cu, being less than 1, 0.05, 1, 0.1 and 50 μg/g, respectively [22]. In addition, black tea has high amount of fluoride which can be released in tea liquor as well [23]. Also, it should be noted that the heavy metals migration

to plants can be as a matter of fertilizers application [24]. Tea is an indispensable part of everyday life for many people in Iran, so we decided to do this study to ensure that public health is maintained. In the present study, tea samples were collected from different parts of Guilan and Mazandaran provinces to determine their aluminum and zinc contents.

Materials and methods

This research was done between September and November 2010 on tea samples that are cultivated in Guilan and Mazandaran Provinces in north of Iran (Fig. 1). 105 black tea samples were selected randomly from 105 farms in the two different regions of Mazandaran and Guilan provinces.

The weight of each sample was about 100 g. The glassware containers used for analysis were washed with detergent and rinsed several times with tap water to remove absorbance due to detergent; then they were soaked overnight in 6 N HNO$_3$ (Merck) solutions and finally rinsed with deionized water. All aqueous solutions and dilutions were prepared with ultrapure water (c). 5 g of each tea sample was weighted by a digital analytical balance (Mettler Toledo, Switzerland) with ± 0.0001 g precision and then was added to 500 mL of boiling tap water and allowed to infuse for 5, 15 and 60 min. Then samples were filtered under vacuum (using a Whatman No.42 filter paper) to eliminate any turbidity or suspended substance. Liquors from the first, second and third infusions were analyzed for aluminum and zinc contents by Inductively Coupled Plasma Optic Emission Spectrometry End of Plasma (ICP-OES EOP, Spectroacros, Germany).

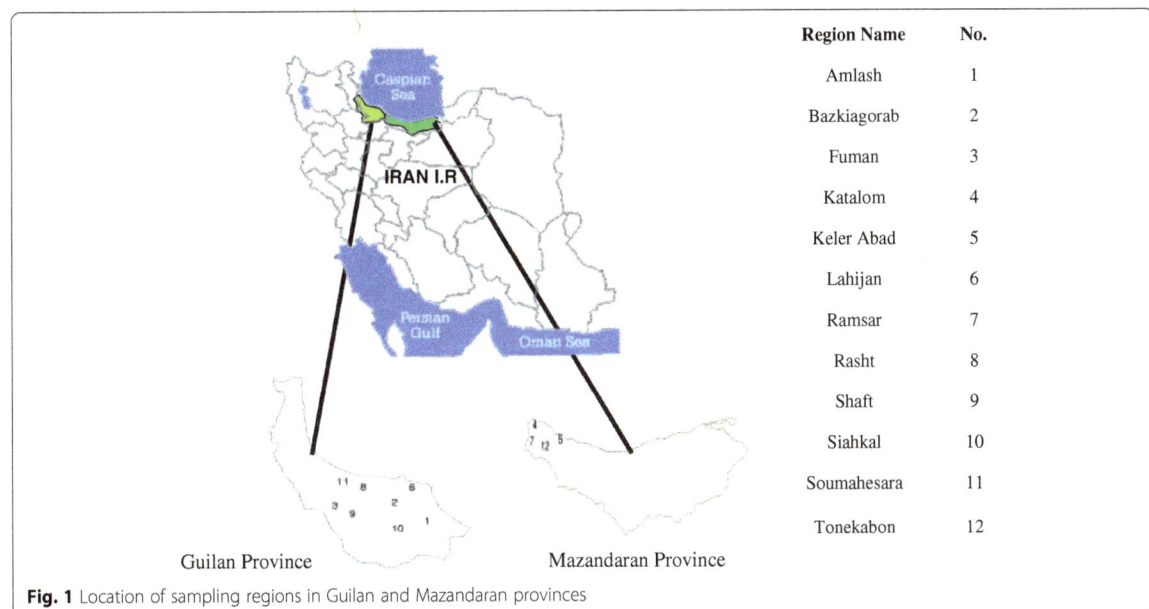

Region Name	No.
Amlash	1
Bazkiagorab	2
Fuman	3
Katalom	4
Keler Abad	5
Lahijan	6
Ramsar	7
Rasht	8
Shaft	9
Siahkal	10
Soumahesara	11
Tonekabon	12

Fig. 1 Location of sampling regions in Guilan and Mazandaran provinces

The blank solution was prepared in similar way without black tea. The purity of argon as carrier gas was 99.999 % (grade 5), with a flow rate of 0.7 L/min for supplementary and Modified Lichte nebulizer and 13 L/min for coolant flow. The speed of 4 channel peristaltic pump was 60 rpm for 45 S in pre-flush condition and 30 rpm for analysis. The power level was adjusted on 1400 KW. Before quantitative analysis of samples, calibration curves of the desired metals were prepared using a series of diluted standard solutions. The recovery percentage, detection limits, and % R.S.D for triplicate measurements of the measured elements were 90 %–95 %, 0.3 ppb, and less than 5 %, respectively. Statistical analysis of the obtained results was performed by SPSS 18 and One Way ANOVA test.

Results

The results of Al and Zn obtained from tea infusion samples in 5, 15 and 60 min are presented in Table 1. Aluminum levels were different from 0.00 to 1516.27 mg/ kg of dry weight (mean = 169.40 mg/ kg) for first infusion, 0.00 to 1851.66 mg/kg (mean = 223.52 mg/kg) for second infusion, 0.00 to 1446.83 mg/ kg (mean = 262.09 mg/kg) for third infusion. Zinc levels were different from 0.00 to 113.66 mg/ kg of dry weight (mean = 49.55 mg/kg) for first infusion, 0.00 to 103.27 mg/kg (mean = 48.33 mg/kg) for second infusion, 0.00 to 141.10 mg/kg (mean = 51.40 mg/kg) for third infusion. One factor analysis of variance (ANOVA) of the data showed that levels of aluminum and zinc in different groups (infusions) were significantly different among the different black tea samples in different locations ($P < 0.001$). Table 2 presents the amount of Al content according to locations and infusion time. The maximum level of Al was determined in Amlash, which the amount of it in infusion time 5, 15 and 60 min were 633.44, 1020 and 982 mg/kg, respectively. The minimum level of Al was determined in Ketalam for first and second infusion and in Kelar Abad for third infusion, the amount in infusion times of 5, 15 and 60 min were 44.89, 54.08 and 49.39 mg/kg, respectively. Also results showed location influenced upon the amount of Al in infusion tea samples ($P < 0.001$). Table 3 showed the amount of Zn content according to locations and infusion time. The maximum level of Zn was determined in Kelar Abad for first and second infusion and in Lahijan for third infusion, the amount of this metal in infusion times of 5, 15

and 60 min were 77.64, 58.50 and 71.28 mg/kg, respectively. The minimum level of Zn was determined in Shaft for first infusion and in Soumahesara for second and third infusion, for which the amount of this metal in infusion times of 5, 15 and 60 min were 32.76, 33.47 and 32.66 mg/kg, respectively. Also results showed location influenced upon the amount of Zn in infusion tea samples ($P < 0.001$).

Discussion

The results of study showed that the infusion time influences the contents of aluminum and zinc but their trend is different. The solubility of Al and Zn in infusions at 5, 15 and 60 min with boiling water showed that the transfer of Al to the brew was positively correlated with the infusion time. The mean level of Al in the third infusion was the highest. The mean level of Zn in first infusion, second and third infusion was 169.40, 223.52 and 262.09 mg/kg, respectively. Relationship between Zn and infusion time is a little different as in first and third infusion the metal levels were positively correlated with the infusion time but in second infusion the metal levels were negatively correlated with the infusion time. This different is very little. The mean level of Zn in first infusion, second and third infusion was 49.55, 48.33 and 51.40 mg/kg, respectively. Moghaddam et al. reported the solubility of Aluminum and Zinc in 31 Iranian consumed tea samples in the first infusion (2 min) was significantly higher than the second infusion (5 min) and the solubility in the second infusion was also significantly higher than the third infusion (10 min) [7]. Mehra et al. [8] reported the solubility of Al, Cu and Mn in infusions at 2, 5 and 10 min with boiling water. Their results showed that in the first infusion, the solubility was the highest and in the third infusion, the solubility was the lowest. The concentration of Al in the first, second and third infusion were 29.7, 10.4 and 3.3 mg/kg [8]. Currently, no national standard for Al and zn in in the black tea was established in Iran, only Institute of Standards and Industrial Research of Iran (ISIRI) has set maximum permissible levels of elements only for As, Hg, Pb, Cd. The study was done by Ebadi et al. [25] in Guilan province (Iran) on green leaf of tea cultivated in Lahijan and Fuman cities. The results showed the amount of Zn was below standard measure (10 ppm) and there was no need to be concerned about the

Table 1 The levels of aluminum and zinc in different infusions (n 630)

Parameters (mg/ kg)		Infusion Times(min)										
		5			15				60			
Min		Max	Mean	SD	Min	Max	Mean	SD	Min	Max	Mean	SD
Al	0.00	1516.27	169.40	220.0	0.00	1851.65	223.52	301.18	0.00	1446.83	262.09	294.86
Zn	0.00	113.66	49.55	21.98	0.00	103.27	48.33	19.49	0.00	141.10	51.40	26.96

Table 2 The levels of aluminum (mg/kg) in different infusions of tea samples (n 315)

| Region | Number of sample | Infusion times (min) | | | | | | | | | | | |
| | | 5 | | | | 15 | | | | 60 | | | |
		Min	Max	Mean	SD	Min	Max	Mean	SD	Min	Max	Mean	SD
Amlash	30	182.65	1516.27	633.44	399.59	465.87	1851.65	1020.47	409.54	539.81	1446.83	982.80	284.57
Bazkiagorab	3	256.97	256.97	256.97	0.00	242.83	242.83	242.83	0.00	133.56	133.56	133.56	0.00
Fuman	24	16.39	223.73	104.30	84.17	16.25	235.08	106.92	86.51	14.64	189.73	82.51	67.25
Katalom	60	4.87	219.67	44.89	46.06	12.78	124.76	54.08	24.31	17.91	482.48	180.66	157.76
Keler Abad	18	3.45	211.60	84.48	86.26	0.00	194.43	69.57	75.21	0.00	138.30	49.39	55.49
Lahijan	45	53.79	256.24	154.79	66.33	84.20	278.33	171.09	57.02	72.17	676.27	308.82	154.08
Ramsar	3	46.37	46.37	46.37	0.00	78.90	78.90	78.90	0.00	288.84	288.84	288.84	0.00
Rasht	21	33.68	384.99	110.19	135.15	17.73	435.71	209.85	168.85	14.77	445.34	203.03	177.50
Shaft	12	62.67	197.84	135.44	55.77	93.63	281.79	209.17	83.37	80.00	282.11	192.37	83.67
Siahkal	48	45.23	541.36	215.92	152.22	0.00	645.84	250.67	212.58	0.00	663.40	216.69	198.12
Soumahesara	6	39.69	61.06	50.38	15.11	38.51	72.18	55.35	23.82	30.95	84.73	57.84	38.03
Tonekabon	45	0.00	354.10	87.07	125.67	0.00	457.37	143.35	161.30	0.00	474.04	152.92	172.29

amount of Zinc in tea samples [25]. In this research the Zn content in all infusion and in all location of Mazandaran and Guilan province is more than 10 ppm. Yemane et al. [1] reported the Zn levels in the clonal tea leaves samples were 67.9 mg/kg (1). Ashraf et al. [2] reported the Zn contents of 17 black tea sample were in the range of 23.7-122.4 mg/kg (mean = 65.7 mg/kg). The international comparison of Zn concentration was in Turkey (Narin et al.) 140.9 [26], Spain (Pedro et al. 2001) 43.2 [27], Japan (Matsuura et al.) 36.6 [28], China (Han and Li) 25.5 [29] and India (Naithani and Kakkar) 39.5 mg/kg [30]. Salahinejad and Aflaki [3] reported the mean of Al and Zn levels in Iranian black tea samples and tea infusion were 1.143, 449.3 and 24.10, 8.86 mg/

kg, respectively [3]. The study of Olivier et al. [6] showed the concentration of Al and Zn in residue of infusion of eight traditional and herbal teas from different geographic regions. The mean concentration of Al and Zn in tea Africa, Ceylon, Oriental and America were 246.3, 344.4, 351.6, 66.9 mg/kg and 12.1, 15.1, 15.7 and 33.4 mg/kg, respectively [6]. The study of Ansari et al. [31] showed the Al and Zn levels in 30 samples of black tea cultivated in Iran were 699.2 and 40.3 mg/kg, respectively. The Al could be accumulated in tea leaves up to 23,000 mg/kg which considered much higher than the other plants accumulation which do not normally exceed 200 mg/kg. As the Al content of tea is dependent to its concentration in soil, applying best agricultural practice (BAP) would be

Table 3 The levels of zinc (mg/kg) in different infusions of tea samples (n 315)

| Region | Number of sample | Infusion times (min) | | | | | | | | | | | |
| | | 5 | | | | 15 | | | | 60 | | | |
		Min	Max	Mean	SD	Min	Max	Mean	SD	Min	Max	Mean	SD
Amlash	30	25.40	85.09	51.00	20.87	30.09	70.31	51.82	14.29	27.29	89.04	46.45	24.99
Bazkiagorab	3	44.31	44.31	44.31	0.00	45.50	45.50	45.50	0.00	36.19	36.19	36.19	0.00
Fuman	24	28.58	60.77	46.49	10.78	30.47	54.38	44.65	8.29	28.97	51.47	41.22	8.68
Katalom	60	17.43	84.49	38.86	14.72	29.67	102.27	46.02	15.35	24.94	141.10	63.82	34.04
Keler Abad	18	27.91	113.66	77.64	31.98	1.82	102.16	58.50	33.81	1.06	65.06	41.80	22.36
Lahijan	45	24.14	78.02	51.88	13.66	29.96	95.45	56.92	17.65	25.82	136.07	71.28	29.66
Ramsar	3	34.19	34.19	34.19	0.00	41.69	41.69	41.69	0.00	56.86	56.86	56.86	0.00
Rasht	21	20.79	44.04	36.81	5.66	26.29	42.87	35.80	6.46	26.49	41.99	35.38	6.57
Shaft	12	27.73	36.03	32.76	3.67	31.78	44.09	36.78	5.49	31.20	44.45	36.24	5.78
Siahkal	48	23.18	99.34	63.35	23.67	0.00	103.27	55.53	28.53	0.00	104.14	50.15	26.05
Soumahesara	6	35.71	36.73	36.22	0.72	29.67	37.27	33.47	5.37	23.10	42.21	32.66	13.52
Tonekabon	45	0.00	88.71	49.67	29.19	0.00	73.57	42.25	17.94	0.00	70.22	39.50	20.53

Table 4 Acceptable Daily Intake (ADI) of toxic metals and
Recommended Daily Intake (RDI) of essential minerals for adults

Element	ADI (mgday-1)	RDI (mgday-1)
Al	<50	-
Zn	-	8-11

considered as an alternative to control the Al contents in tea leaves.

Because of lack of sufficient information about the acceptable contents of Al and Zn in tea, it was decided to interpret our result with their allowable or acceptable daily intakes (ADI).

Table 4 represents the allowable or acceptable daily intake (ADI) of aluminum and zinc and reference or recommended daily intake (RDI) of them [32]. With 1.6 kg per capita consumption of black tea leaves annually, Iranian daily consumption averages 4.3835 g of black tea leaves per person [3]. Based on Table 4 the acceptable daily intake of aluminum is less than 50. According to average daily black tea consumption in Iran (4.3835 g per person) aluminum content in all samples was less than the acceptable daily intake. Salahinejad and Aflaki [3] reported the average daily intake of Al by consumption of black tea infusions was low rather than ADI [3]. However, it must be stated that tea drinking may contribute towards Al toxicity in individuals with impaired absorption or excretion of Al in their systems. Since the tea cultivation lands in north of Iran provide much of black tea being consumed in Iran and the exported amount to other countries, it is recommended all toxic elements and essential mineral elements in black tea samples and their infusions to be determined for safe consumption of black tea.

Since the consumption of tea in Iran is very high, beside regarding the probable release of heavy metals in the infusion tea, considering heavy metal free drinking water or treated with effective methods are suggested [33].

Conclusions

Tea is one of the heavily consumed beverages in the world and on average in Iran one liter of tea is consumed by a person per day. In this regard health aspects related to tea is very important and therefore consumers should be very confident on the absence of any pollutants in black tea. In this study concentrations of aluminum and zinc were measured in black tea samples from different locations in different infusion time. The analysis of results showed that the location factor influences the contents of these metals at different infusions. Due to the lack of standards for all elements (toxic and essential mineral) in tea, it is recommended that the maximum allowable and safe concentrations to be established.

Competing interests
The authors declare that they have no competing interests.

Authors' contributions
This study is a part of a research project. The study was directed by Dr. AHM who is the corresponding author and made the final preparation of article. Engineers MG and SS were engaged in sample preparations and laboratory work. Dr. MY helped on statistical data analysis and engineer SN as ICP expert performed the experiments by this device. All authors have read and approved the final manuscript.

Acknowledgments
The authors would like to thank for financial supports of Institute for Environmental Research, and the staff of Chemistry Laboratory, Department of Environmental Health Engineering, School of Public Health Tehran University of Medical Sciences, and Tea Research Institute of Lahijan because of their help for samples collection.

Author details
[1]Department of Environmental Health Engineering, School of Public Health, Tehran University of Medical Sciences, Tehran, Iran. [2]Department of Environmental Health Engineering, School of Public Health, Hormozgan University of Medical Sciences, Bandar- Abbas, Iran. [3]Center for Solid Waste Research, Institute for Environmental Research, Tehran University of Medical Sciences, Tehran, Iran. [4]National Institute of Health Research, Tehran University of Medical Sciences, Tehran, Iran.

References
1. Yemane M, Chandravanshi B, Wondimu T. Levels of essential and non-essential metals in leaves of the tea plant (Camellia sinensis.) and soil of Wushwush farms, Ethiopia. Food Chem. 2008;107(3):1236–43.
2. Ashraf W, Mian AA. Levels of selected heavy metals in black tea varieties consumed in Saudi Arabia. Bull Environ Contam Toxicol. 2008;81(1):101–4.
3. Salahinejad M, Aflaki F. Toxic and essential mineral elements content of black tea leaves and their tea infusions consumed in Iran. Biol Trace Elem Res. 2010;134(1):109–17.
4. Shekoohiyan S, Ghoochani M, Mohagheghian A, Mahvi AH, Yunesian M, Nazmara S. Determination of lead, cadmium and arsenic in infusion tea cultivated in north of Iran. Iranian J Environ Health Sc Eng. 2012;9(1):1–6.
5. Qin F, Chen W. Lead and copper levels in tea samples marketed in Beijing, China. Bull Environ Contam Toxicol. 2007;79(3):247–50.
6. Olivier J, Symington EA, Jonker CZ, Rampedi IT, Van Eeden TS. Comparison of the mineral composition of leaves and infusions of traditional and herbal teas. S Afr J Sci. 2012;108:1–7.
7. Moghaddam MA, Mahvi A, Asgari A, Yonesian M. Determination of aluminum and zinc in Iranian consumed tea. Environ Monit Assess. 2008;144(1–3):23–30.
8. Mehra A, Baker C. Leaching and bioavailability of aluminium, copper and manganese from tea (Camellia sinensis). Food Chem. 2007;100(4):1456–63.
9. Joint FAO/WHO Expert Committee on Food Additives J. Summary of evaluations performed by the Joint FAO/WHO expertcommittee on food additives, JECFA 1956–2003. Washington, DC, USA: ILSI Press, International Life Sciences Institute; 1999.
10. Rao K. Aluminium content in teas leaves and in differently prepared tea infusions. Food Nahrung. 1994;38(5):533–7.
11. Wang L, Su D, Wang Y. Studies on the aluminium content in Chinese foods and the maximum permitted levels of aluminum in wheat flour products. Biomed Environ Sci. 1994;7(1):91.
12. Greger J. Dietary and other sources of aluminium intake. Aluminium Biol Med. 1992;169:26–9.
13. Powell J, Thompson RH. In vitro mineral availability from digested tea: a rich dietary source of manganese. Analyst. 1998;123(8):1721–4.
14. Jackson ML, Huang PM. Aluminum of acid soils in the food chain and senility. Sci Total Environ. 1983;28(1):269–76.
15. Edwardson J. Aluminum and the pathogenesis of neurodegenerative disorders. Aluminium Food Environ. 1988;2:20–36.
16. Martyn CN. Aluminium and Alzheimer's disease: an epidemiological approach. Environ Geochem Health. 1990;12(1–2):169–71.

17. McLachlan D. Aluminium and the risk for Alzheimer's disease. Environmetrics. 1995;6(3):233–75.
18. Flaten A-K, Lund W. Speciation of aluminium in tea infusions studied by size exclusion chromatography with detection by post-column reaction. Sci Total Environ. 1997;207(1):21–8.
19. Wills M, Savory J. Aluminum toxicity and chronic renal failure. Metal Ions in Biological Systems (Marcel Dekker, Inc New York and Basel) 1988, 24, 315–340.
20. de Romaña DL, Salazar M, Hambidge KM, Penny ME, Peerson JM, Krebs NF, et al. Longitudinal measurements of zinc absorption in Peruvian children consuming wheat products fortified with iron only or iron and 1 of 2 amounts of zinc. Am J Clin Nutr. 2005;81(3):637–47.
21. Willson KC, Clifford MN. Tea: cultivation to consumption. London: Chapman and Hall; 1992.
22. Institute of Standards and Industrial Research of Iran (2008) Black tea - Specifications and test methods, 623, 2nd revision.
23. Mahvi AH, Zazoli MA, Younecian M, Esfandiari Y. Fluoride content of Iranian black tea and tea liquor. Fluoride Res. 2006;39(4):266–8.
24. Atafar Z, Mesdaghinia A, Nouri J, Homaee M, Yunesian M, Ahmadi Moghaddam M, et al. Effect of fertilizer application on soil heavy metal concentration. Environ Monit Assess. 2010;160(1–4):83–9.
25. Ebadi A, Zare S, Mahdavi M, Babaee M. Study and measurement of Pb, Cd, Cr, and Zn in green leaf of tea cultivated in Gillan province of Iran. Pak J Nutr. 2005;4(4):270–2.
26. Narin I, Colak H, Turkoglu O, Soylak M, Dogan M. Heavy metals in black tea samples produced in Turkey. Bull Environ Contam Toxicol. 2004;72(4):844–9.
27. Fernández-Cáceres PL, Martin MJ, Pablos F, González AG. Differentiation of tea (Camellia sinensis) varieties and their geographical origin according to their metal content. J Agric Food Chem. 2001;49(10):4775–9.
28. Matsuura H, Hokura A, Katsuki F, Itoh A, Haraguchi H. Multielement determination and speciation of major-to-trace elements in black tea leaves by ICP-AES and ICP-MS with the aid of size exclusion chromatography. Anal Sci. 2001;17(3):391–8.
29. Han L-x. Determination of minerals and trace elements in various tea by ICP-AES. Spectrosc Spectr Anal. 2002;22(2):304–6.
30. Naithani V, Kakkar P. Evaluation of heavy metals in Indian herbal teas. Bull Environ Contam Toxicol. 2005;75(1):197–203.
31. Ansari F, Norbaksh R, Daneshmandirani K. Determination of heavy metals in Iranian and imported black tea. Iranian J Environ Health Sci Eng. 2007;4(4):243–8.
32. Food and Nutrition Information Center, Dietary Guidance/Dietary Reference Intakes/DRI Tables, Dietary Reference Intakes: Elements, https://www.iom.edu/~/media/Files/Activity%20Files/Nutrition/DRIs/5_Summary%20Table%20Tables%201-4.pdf.
33. Boldaji MR, Mahvi AH, Dobaradaran S, Hosseini SS. Evaluating the effectiveness of a hybrid sorbent resin in removing fluoride from water. Int J Environ Sci Technol. 2009;6:629–32.

20

Highly cadmium tolerant fungi: their tolerance and removal potential

Mehran Mohammadian Fazli[1], Negin Soleimani[1], Mohammadreza Mehrasbi[1], Sima Darabian[2], Jamshid Mohammadi[2] and Ali Ramazani[3*]

Abstract

Background: Soil and effluent of lead and zinc industries contain high concentration of cadmium. The present study was conducted to isolate tolerant fungal strains from cadmium -polluted sites in Zanjan province, Iran.

Methods: Cadmium tolerance and bioremediation capacity of seven isolates including *Aspergilus versicolor, Aspergillus fumigatus, Paecilomyces* sp.9, *Paecilomyces* sp.G, *Terichoderma* sp, *Microsporum* sp, *Cladosporium* sp were determined.

Results: Minimum inhibitory concentration values among 1,000-4,000 mg l^{-1} proved great ability of isolated strains to survive in cadmium polluted environments. The most tolerant fungi, *Aspergilus versicolor*, showed tolerance index of 0.8 in 100 mg l^{-1} cadmium agar media. Fungal resistance against cadmium is depended directly on strain's biological function.
A. versicolor was found to bioaccumulate over 7 mg of cadmium per 1 g of mycelium, followed by 5.878, 5.243, and 5.075, 4.557 by *Paecilomyces* sp, *Aspergilus fumigatus, Microsporum* sp and *Terichoderma* sp, respectively.

Conclusion: It can be noted that tolerance of the strains appears to be independent from bioaccumulation capacity. Finally, the results indicated that *A. versicolor* could be a prospective candidate for bioremediation processes.

Keywords: Cadmium tolerance, Bioremediation, Fungi

Background

Human jump toward industrialization and comfort life is a leap into environmental pollution and consequently deterioration of human health. The environment is polluted by heavy metals from industrial wastewaters during metal processing as well as other pollutant routes. Virtually, any industrial activity using metals has a metal disposal problem [1]. Nature of heavy metals is non-biodegradable and persistent; therefore, environmental compartments (soil and water body) are not able to purify themselves from these toxic pollutants.

Heavy metals can be divided into essential metals such as copper, manganese, zinc, and iron, and nonessential metals such as cadmium, lead, mercury, and nickel [2].

Cadmium and lead are included among the major pollutants because of their high toxicity [3-6]. Cadmium is released to ecosystem by mine tailing, effluents from textile, leather, tannery, electroplating and galvanizing industries, as well as cadmium batteries.

Biomagnification of cadmium in nature and migration through drinking water, food and air to human body cause severe health effects like kidney damage, bronchitis and cancer [3]. As industrial development train is not stoppable, struggle with heavy metal pollution requires novel remediation methods. Conventional treatment systems have failures which include insufficient metal sequestration, high costs, high reagents and/or energy requirements, and generation of toxic sludge or other waste products that require disposal. Restoring metals in an efficient and economical procedure has necessitated the use of different options in metal-separating methods. Research shows that bioaccumulation of metals by organisms has been successful to some extent [3]. Bioremediation of heavy metals from aqueous

* Correspondence: ramazania@zums.ac.ir
[3]Biotechnology Departments, School of Pharmacy, Zanjan University of Medical Sciences, Zanjan, Iran
Full list of author information is available at the end of the article

solutions is a relatively new process that has been confirmed as a promising process in the removal of heavy metal pollution. The major advantages of biosorption are its high effectiveness in reducing the heavy metal ions, and the use of inexpensive biosorbents. Biosorption processes are particularly suitable to treat dilute heavy metal wastewater [7].

Biomass obtained from different sources is used in biotreatment and its key feature and potential controlled for the process. Rather than searching thousands of microbial species for particular metal sequestering features, it is beneficial to look for biomasses that are readily available in large quantities to support potential demand. While choosing of the biomaterial for metal sorption, its origin is a major factor to be taken into account [8].

Among biological sources, fungi possess a number of advantages including producing great biomass, rapid growth, availability, and flexibility to rough circumstances.

The uptake of metals by fungal biomass appears to involve a combination of two processes: bioaccumulation (i.e. active metabolism-dependent processes, which may include both transport into the cell and partitioning into intracellular components) and biosorption (i.e. the binding of metals to the biomass by processes that do not require metabolic energy) [9].

Fungi are ubiquitous members of subaerial and subsoil environments, and often become a dominant grouping metal-rich or metal-polluted habitats [10]. Recent studies have shown that the strains isolated from contaminated areas have remarkable potential to tolerate such toxic conditions. Microorganisms have been shown to possess ability to survive by adapting or mutating at high concentrations of heavy metals [11,12].

This could play a key role to explore newcomer resistant isolates from pool of biomasses. Exhibiting high tolerances to heavy metals, these isolates were selected for bioremediation studies. Generally, fungi have often been proposed as bioagents for metal recovery processes [13].

This study was carried out to isolate fungi from cadmium- contaminated sites from Zanjan province of Iran for the first time and evaluate their resistance level toward cadmium as well as assessing their bioaccumulation capacity in order to expand knowledge about bioremediation science.

Materials and methods
Materials
The aqueous solutions of cadmium were prepared by diluting Cd (II) standard stock solution (concentration 1000 mg L^{-1}) obtained by dissolving Cd $(NO_3)_2.4H_2O$ in deionized water . Fresh dilutions were prepared for each experiment. Cd $(NO_3)_2.4H_2O$ was purchased from Merck, Germany. Potato dextrose agar (PDA), malt extract agar (MEA) and potato dextrose broth (PDB) were used as

solid and liquid medium, respectively. PDA and MEA medium were purchased from Merck, Germany. PDB medium was purchased from Scharlau, European Union. Deionized water was used in all experiments (TKA Smsrt2Pure, Germany). All laboratory glassware and plastic were soaked in 2 M HNO_3 technical grade, rinsed with distilled water and heat dried 2 h, at 180°C before use.

Sampling and experimental sites
The research areas characterized in this study were lead and zinc refinery industries as contaminated sites and municipal wastewater treatment plant as control site located in Zanjan province, Iran. These locations were selected based on their cadmium pollution densities. Soil samples were taken from six different spots to a depth of 20 cm from waste dumping areas. Effluent samples of industries were collected from wastewater discharge of plants. Control sample was obtained from aeration tank of municipal wastewater treatment plant. Samples were placed in sterilized glass bottles, transported on ice (4°C), taken to the laboratory and analyzed within 8 h. The wastewater and soil samples were analyzed for total content of Cd. The soil samples were dried at 105°C manually ground and sieved (500 μm pore size). One g of soil samples were digested with 70% HNO_3 (1 M) and 30% H_2O_2 using microwave digestion (Sineo MDS-10, China). Cadmium concentrations were analyzed by a Varian Atomic Absorption Spectrophotometer (AA 240, Australia) [14].

Isolation of strains
Fungal strains were isolated on PDA and MEA by serial dilution method in order to avoid overlapping colonies. Streptomycin (15 mg l^{-1}) and chloramphenicol (50 mg l^{-1}) were added to mediums after autoclaving at 15 psi for 15 min and 121°C to arrest bacterial growth. The soil samples (1 g) were suspended in 100 ml of sterilized water. The mixture was shaken (200 rpm) for 30 min at room temperature. All samples were diluted up to (10^{-3}). 0.1 ml of different dilutions were spread on Petri plates (diameter 10 cm) containing 20 ml media. Plates were incubated at 28°C in dark condition and monitored every day up to 10 days and each developed colonies were sub-cultured and isolated into fresh PDA plates. Purified isolates were kept on slants at 4°C and recultured every 4 weeks [15].

Screening for Cd- tolerant fungi
In order to select Cd- tolerant strains, 100 mg l^{-1}Cd stress was added to PDA medium. The pH of the solid growth medium was adjusted to 6 with 1 M sodium hydroxide solutions before autoclaving in all experiments in this study. The small agar plugs with young mycelium from the edge of the stock cultures were cut and transferred to surface of solid medium. Plates were incubated

at 28°C for at least 10 days and visually inspected for microbial growth every day. Cadmium- tolerant strains were subjected to resistance studies.

Cadmium tolerance index of fungi

Five mm disks from 10 day old pure cultures of each fungal isolates were inoculated into PDA (three replicates) supplemented with 200 mg l^{-1} Cd [4]. The inoculated plates were incubated at 28°C for 10 days. In parallel, cultures without cadmium were performed as a control. The radial growth was evaluated from four measurements (in millimeters) that passed through the center of the inoculated portion. The initial diameter of the portion was subtracted from the growth diameter [5]. The mean of perpendicular diameter measurements was recorded for each plate on the day10th. The tolerance index (TI), an indication of the organism response to metal stress was calculated from the growth of strain exposed to the metals divided by the growth in the control plate [16]. The higher the TI, the greater the resistance.

Determination of minimum inhibitory concentration

Minimum inhibitory concentration (MIC) was defined as the minimum inhibitory concentration of the heavy metal that inhibited visible growth of test fungi [17]. PDA medium was enriched with increasing concentration of Cd (200, 400, 600, 800, 1,000, 2,000, 3,000, and 4,000 mg l^{-1}Cd). Plates were inoculated with agar plugs from the edge of the 10 day old growing cultures. If no apparent growth of fungi was observed after ten days on the plates, the metal concentration was considered as the highest metal concentration tolerated by the tested fungus.

Identification of selected fungus

All the resistant fungal isolates were initially identified by colony characteristics on PDA characterized to the genus level on the basis of macroscopic characteristics (colonial morphology, color and appearance of colony, and shape), microscopic characteristics (septation of mycelium, shape, diameter and texture of conidia) and the help of the Principles and practice of clinical parasitology [18]. Molecular identification of cultures was carried out with some modifications by extracting chromosomal DNA of potential fungus according to previous study [19]. Briefly, a pure culture of the isolated fungal strain was grown in liquid shaking PDB medium at 28°C and 120 rpm for 72 h. The biomass was then harvested and washed with sterile distilled water. Cells were broken up by liquid nitrogen, re-suspended in Hoffman Winston extraction buffer; then, proteins were removed with phenol–chloroform and DNA was precipitated by adding pure ethanol. DNA concentrations and A260/A280 ratio were determined with biophotometer (Biophotometer plus,

Eppendorf Germany). An A260/A280 ratio of 1.8–2.1 was considered acceptable for PCR (polymerase chain reaction)-based procedures. Part of the 18 s rDNA fragments was amplified using primers 817 F and 1536R [20]. The PCR mixture consisted of 25 µl of 2x PCR Master Mix (Fermentase, USA), 10 pM of each primer, 21 µl water and approximately 2 µg genomic DNA as template in a total volume of 50 µL. The PCR was performed in a Thermo cycler (icycler, USA) using a thermal cyclic condition at 94°C (7 min) followed by 8 cycles at 94°C (1 min), 59.5°C (45 s) and 72°C (1:30 min), then 30 cycles with 94°C (1 min), 56°C (45 s), 72°C (1:30 min) with a final extension at 72°C for 10 min. A sample (5 µL) of the PCR product was analyzed by electrophoresis in 2% agarose gel with 1xTBE buffer. Electrophoresis was performed at 80 V for 40 min. The purified amplicons were sequenced by using an automated sequencer (Bioneer, Korea). The sequences were compared using the BLAST program (http://www.ncbi.nlm.nih.gov/BLAST/) for identification of the isolates [21].

Cadmium bioaccumulation by active fungus

To determine the bioaccumulation ability of the 7 fungal isolates, inoculums (six 5 mm disks of mycelia strain) were prepared from 10-day- old pure fungal culture and inoculated into 250 ml Erlenmeyer flask containing 100 ml potato dextrose broth (PDB) plus 100 mg l^{-1} cadmium. Initial concentrations of Cd (II) in each conical flask were checked by AAS before fungal inoculation. pH was adjusted to 6 [22]. Un-inoculated controls (PDB medium with 100 mg l^{-1} of Cd and without any fungal inoculums) were served to detect any possible abiotic Cd (II) reduction brought about by media components. All flasks were incubated at 28°C on a rotary shaker at 120 rpm in dark conditions.

After 10 days of incubation (logphase ([23], flasks containing fungal biomass were harvested and filtered through Whatman No.42 filter paper. Filtered PDB medium was used for determining total Cd concentration.

Biomass samples were rinsed three times with distilled water and dried in hot air oven at 80°C until a constant weight (24 h) was achieved. The dried fungal biomass was weighed and defined as dry biomass (g) [24]. The amount of heavy metal uptake (q, mg/g) was calculated by using the following equation [25]:

$$q = [C_i - C_f / m] V \qquad (1)$$

In above equation, q (mg/g) is mg of metal ions uptake per gram biomass; C_i (mgL^{-1}) is the initial metal concentration of liquid phase; C_f (mgL^{-1}) is the final metal concentration; m (g) is the amount of dry biomass; and V (L) is the volume of the medium.

Data analysis

All experiments were carried out by triplicate sample. Values reported in this paper are the means ± S.D. The difference in TI and uptake capacity of each isolate was studied by one-way ANOVA followed by post-Hoc multiple comparisons by Duncan's method using SPSS16 (USA, I1, Chicago, SPSS Inc.). The difference was considered as significant when $P < 0.05$.

Results and discussion

Sites characteristic

The range of Cd in earth soils lie between 0.2 and 1.1 mg kg^{-1}. The highest Cd concentrations (in mg kg^{-1}) are reported for soils in the vicinity of metal-processing industries, for example, in Belgium, 1781; in Poland, 270; and in the United States, 1500 [26].

Table 1 demonstrates Cd content of environmental samples. Cd concentrations of soil samples used in this study were 56.90 and 488.25 (mg kg^{-1} soil) for lead refinery industry and zinc refinery industry respectively. Since dumping waste areas were selected for soil samples, it is not surprising that Cd content was too high.

The Cd concentration in effluent of mentioned industries were 50.23 and 70.65 (mg l^{-1}) respectively, whereas Cd in aeration tank of municipal wastewater treatment plant was 0.23 (mg l^{-1}). Clearly these values are 500 times higher than Industry Effluent Guidelines regulated by EPA [27]. These results could be due to the fact that Zanjan province is located close to two major heavy metal processing plants in Iran, as well as neighboring to Mahneshan heavy metal mine.

Number of cadmium- resistant isolates and their origin

As shown in Table 1, sixteen strains could tolerate 100 mg l^{-1} Cd toxicity. These strains were from different sampling sites.

It is known that microorganisms isolated from natural environments contaminated with heavy metals often exhibit tolerance to heavy metal pollutants [28].

Cadmium stress exerted in this study to isolated fungus from municipal wastewater treatment site, made nonfavorable lethal growth medium. Discrepancy in conditions the fungi were adapted to resulted in extinction of isolated fungal population from municipal wastewater. Only one resistant isolate was obtained from low- polluted area. In contrast, the number of resistant isolates from heavy metal industrial sites was significant. It is well known that a long-time exposure of water and sediment to heavy metals can produce considerable modification of their microbial populations, reducing their activity and their number. Generally, pollution of soil and water by heavy metals may lead to a decrease in microbial diversity. This is due to the extinction of species sensitive to the stress imposed, and enhanced growth of other resistant species [29].

Cadmium-resistant assays minimum inhibitory concentration

MIC border line of 1000 (mg l^{-1}) was chosen. Strains with MIC value over 1000 (mg l^{-1}) were subjected to molecular identification. Approximately 700 bp PCR product of 18S rDNA gene was collected by PCR amplification program. The 18S rDNA gene of seven strain deposited in the GenBank database of NCBI under accession numbers KM205077, KM205078, KM205079, and KM205080. Data of top seven MIC values and their relevant isolate identification are shown in Table 2. Results from Table 2 show that soil natural sources trigger microorganisms with higher tolerance ability than aquatic sources. Except *Terichoderma* sp, there was no strain from industrial effluent which could survive in over 1000 (mg l^{-1}) Cd. Evolutionary adaptation to metal-contaminated soils is a well-documented phenomenon, particularly because it is one of the most striking examples of microevolution driven by edaphic factors [30].

Cadmium concentration in sites and the levels of survival against Cd toxicity were not related to each other, as *Terichoderma* sp was able to survive up to 2000 (mg l^{-1}) Cd toxicity which is 10000 orders of magnitude than Cd concentration in urban wastewater. Other examples are about *Paecilomyces* sp.9 and *Paecilomyces* sp.G from lead refinery industry which were able to grow up to 4000 (mg l^{-1}) Cd stress, whereas strains from zinc refinery industry with Cd concentration over eight order of magnitude in soil than lead refinery industry only could tolerate up to 2000 and 1000 (mg l^{-1}) Cd. It implies that some microorganisms

Table 1 Cadmium concentration of samples and number of resistant isolates

Sites		Cadmium concentration[a]	Number of resistant isolates
Soil(mg kg^{-1} of soil)	Industrial dumping area (lead refinery)	56.9 ± 0.55[a]	5
	Industrial dumping area (zinc refinery)	488.25 ± 1.92	3
Wastewater(mg l^{-1})	Industrial effluent(lead refinery)	50.23 ± 1.05	3
	Industrial effluent(zinc refinery)	70.65 ± 1.67	4
	Municipal wastewater treatment plant (aeration tank)	0.23 ± 0.09	1

[a]Mean ± SD.

Table 2 Fungi isolated from different sites and their MIC values

Origin	strain	MIC (mg l $^{-1}$)
Industrial dumping area (lead refinery)	*Paecilomyces* sp.9	4000
	Paecilomyces sp.G	4000
	Microsporum sp	1000
	Aspergillus fumigatus	1000
Industrial dumping area (zinc refinery)	*Cladosporium* sp	1000
	Aspergilus versicolor	2000
municipal waste water treatment plant (aeration tank)	*Terichoderma* sp	2000

without any consideration about their native root could develop ability to keep up living in severe toxic environments. The MIC values suggested that the resistance level against individual metals was dependent on the isolates [31].

The cadmium concentration of 4000 (mg l $^{-1}$) was the highest MIC value in this study that *Paecilomyces* sp.G and *Paecilomyces sp.9* could tolerate. To a lesser degree were *Aspergillus versicolor* and *Terichoderma* sp with MIC value of 2000 (mg l $^{-1}$) and *Microsporum* sp, *Cladosporium* sp, *Aspergillus fumigates* with MIC of 1000 (mg l $^{-1}$).

There are some studies supporting the idea that there is a very little difference in metal tolerance between strains from polluted and unpolluted sites [32]. Indeed, presence of metal may act as a fatal toxicity on microorganism population, but does not have any influence on microbial tolerance ability.

MIC values of 0.328 mM for filamentous fungi and 1 (mg l $^{-1}$) for *Aspergillus*, *Penicillium* and *Fusarium* were reported [29,33]. In another study no determinations were made for cadmium since the majority of the tested fungi were unable to grow in the presence of this metal [34]. In a study conducted in 2007, MIC values of 5,000, 3,000 and 4,000 (mg l $^{-1}$) Cd were reported for Rhizopus sp., *Terichoderma* and *Aspergillus,respectively*. These reported MIC values are relatively similar to the values we observed [31].

Considering previous studies, *Paecilomyces* sp.G *and Paecilomyces* sp.9 are newly introduced fungus with remarkable tolerance potential.

Tolerance index

The reduced tolerance index reflects the inhibitory growth function of heavy metal [35]. To select the most tolerant fungi, the actual resistant potential of fungi must be tested. Tolerance level of fungi can be revealed through both TI and MIC assays. Although it is crucial for scientists to discover a fungus with great ability to survive in extremely high heavy metal concentration, from an environmental engineer view, that fungus is salient and applicable for purification systems which in response to metal toxicity could grow and multiply faster.

It implies that the more rapidly the fungi can adapt to polluted environment and develop its colonies, the more beneficial it is for treatment process. The term adaptation speed is an important armor that prompts one fungus more powerful than other fungi with higher MIC property. TI of each fungus in this study (Figure 1) demonstrated different orders of tolerance, as follows: *Aspergilus versicolor* and *Terichoderma* sp have shifted up to first level of tolerance with TI of 0.85 and 0.69 respectively, followed by *Paecilomyces* sp.G, *Paecilomyces sp.9* and *Aspergillus fumigatus* with TI of 0.56, 0.52 and 0.5 respectively which were fairly tolerant fungi and finally at end of the list were *Cladosporium* sp (0.36) and *Microsporum* sp (0.35).

Aspergilus versicolor from highly polluted zinc industrial site posed minimal reduction in growth (%15). It might be due to the high concentration of cadmium in soil of strain's original environment that induced resistance strategy on fungal metabolism; therefore *Aspergilus versicolor* was able to adjust rapidly to polluted culture media. Based on the result, *Aspergilus versicolor* was a highly adaptable fungus in response to cadmium stress. Excessive Cd in soil could trigger the evolution for higher Cd tolerance in *Suillus luteus* [36]. An exceptional fungus was *Terichoderma* sp again. Despite the fact that this fungus was from least polluted area with % 31 reductions in growth, it adapted better than five remaining fungi. Trying to find rigid general regulation between the microorganism's origin and fungal resistance to heavy metal is cumbersome effort. The trustworthy theory is: fungal resistance to heavy metal is depended directly on biological function of the strain. The Cd-resistance was found to be independent from the pollution level at the site of origin [29,32,37]. TI results in this study illustrated that *Aspergilus versicolor* was the most tolerant and *Microsporum* sp and *Cladosporium* sp with 65% suppression of mycelia growth were the most sensitive fungus.

Results from relative study showed that Cadmium at concentration of 1 mM posed the strongest inhibition toward isolates from the genera *Aspergillus*, *Fusarium*, *Alternaria* and *Geotrichum*. Only *Penicillium* isolates

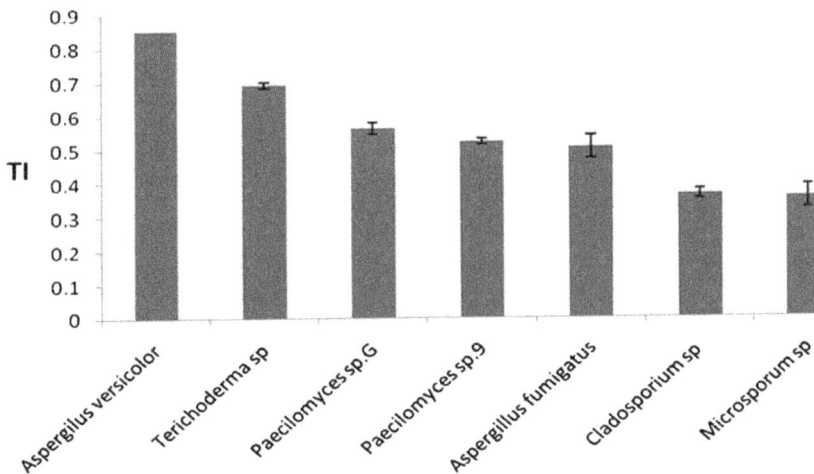

Figure 1 TI of fungus under 200 mg l⁻¹ of cadmium. Cadmium tolerance index of different isolates is significantly different (P < 0.05). The line on each bar is the standard deviation.

expressed tolerance index of 0.8 [34]. In another study, growth of *Aspergillus flavus* was inhibited by 40% at 1 mM Cd concentration [38].

Copious heterogeneity in TI of the isolates, especially in strains of the same genus (*Aspergilus versicolor*, *Aspergillus fumigatus*) and (*Paecilomyces* sp.9, *Paecilomyces* sp.G) proved the theory that various genera and also isolates of the same genus do not necessarily have the same heavy metal tolerance [25,31,34,39]. It comes into mind that tolerance skill is not inherited among microorganisms, in other word it is acquired from ecosystem.

3.4. Various tolerance strategies related to different morphological alteration

On exposure to cadmium, morphological changes were observed in all isolated fungi. Several authors have reported the formation of colorful mycelia in the presence of heavy metals on agar media [28].

Other than *Aspergilus versicolor* and *Cladosporium* sp genera, decoloration of fungus occurred by increasing the cadmium concentration in medial growth. Pink color changed to white in *Paecilomyces* sp. and *Paecilomyces* sp.9 genera (Figure 2). In *Terichoderma* sp and *Aspergillus fumigatus*, green color changed to white. Red color appeared in cracked mycelium of *Aspergilus versicolor* and sides of its colonies in media were polluted with Cd (Figure 3). Red pigmentation in *Aspergilus versicolor* was probably due to the binding of Cd to the protein in the cell wall of mycelium. It has been previously suggested that production of pigments in fungal cell and cell free media is accompanied with precipitation of metal ions on the call wall [23]. Consistent with this study, *Sills hirsutum* produced a yellow-orange pigment both extracellular and in the mycelium, when cultivated in the presence of 0.25 mM or more Cd [40]. Jarosz-Wilkołazka et al. reported that the presence of Cd induced formation of orange–brown pigment which colored the fragile mycelium of *Abortiporus biennis*, as well as the cell-free culture medium [41].

The appearance of condensed, frizzy mycelium was clearly visible in *Aspergillus fumigatus*, *Cladosporium* sp, *Microsporom* sp, *Paecilomyces* sp.G, *and Paecilomyces*

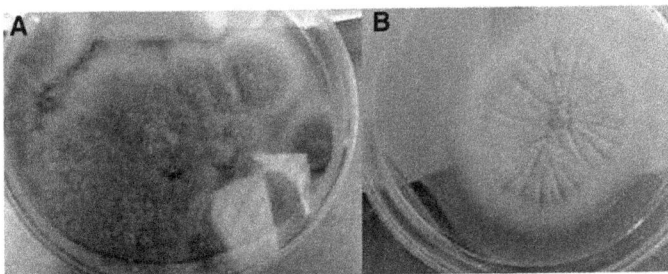

Figure 2 *Paecilomyces* sp.G in control plate (A), and in presence of 200 mg l⁻¹ Cd after 7 days (B).

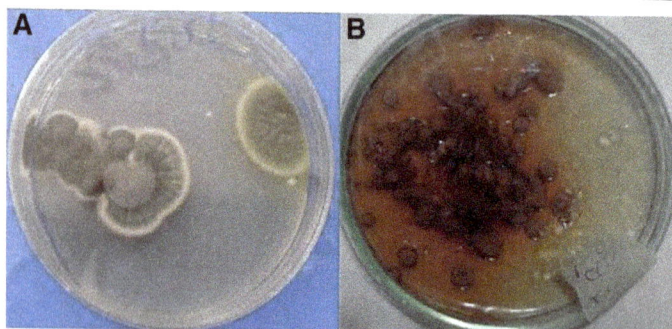

Figure 3 *Aspergilus versicolor* in control plate (A), and in presence of 200 mg l⁻¹ Cd after 7 days (B).

sp.9 generas. In the case of a toxic metal-containing domain, aggregated mycelia could produce high local concentrations of extracellular products such as complexing agents, precipitating agents, polysaccharides and pigments with metal-binding abilities [42]. In *Terichoderma* sp, isolated in this study, aerial mycelium vanished (Figure 4). On Cd-containing agar most of the mycelium of *Paxillus involutes* grew submerged rather than on the surface as occurs on Cd-free agar [40].

Possible explanation for these different morphological changes among isolates may be due to the vast detoxification/tolerance mechanisms that each strain applies. The variation in the metal tolerance might be due to the presence of one or more types of tolerance strategies or resistance mechanisms exhibited by different fungi [31].

Bioremoval of cadmium by fungal isolates

There has been steady progress in studying the biosorption of heavy metals, resulting in the identification of some biomass types that show very promising uptake of metallic ions [43]. The biosorbents used in heavy metal biosorption are usually obtained after screening the heavy metal resistant/tolerant microorganisms from polluted environments [14]. Adaptation of fungal isolates to heavy metal successfully created organisms with greater efficiency in bioaccumulation [35]. Because the putative

specific resistance mechanism(s) could have a potential for biomitigation of contaminated sites, the metal sequestration capacity of the fungus was evaluated [44]. Cadmium bioaccumulation (mg of cadmium uptake per g of dry biomass) of all the tested fungal isolates from liquid media containing 100 mg l⁻¹ of Cd is presented in Figure 5.

The genera, *Aspergilus versicolor, Paecilomyces* sp.G, *Aspergillus fumigatus, Microsporum* sp, *Terichoderma* sp, *Paecilomyces* sp.9 *and Cladosporium* sp *showed* the bioaccumulation capacities of 7.313, 5.878, 5.243, 5.075, 4.557, 2.849 and 2.631 mg g⁻¹ in sequence of decreasing the potential. Surprisingly the best accumulator fungus was the most tolerant strain too. However, for the rest of the fungus this trend was not continued. Except *Paecilomyces* sp.G that appeared semi accumulator and semi tolerant, *Aspergillus fumigatus* and *Microsporum* sp were among the sensitive isolates; however, they were moderately accumulator fungi. These results suggest that removal capacity is not proportional to level of tolerance. Similar observations regarding the lack of correlation between metal tolerance and removal capacity have been reported earlier [25,31,32]. Indeed; uptake capacity was related to the type of tolerance mechanism of fungi. In biotreatment criteria, the resistant mechanism and remediation strategies of microorganism should be

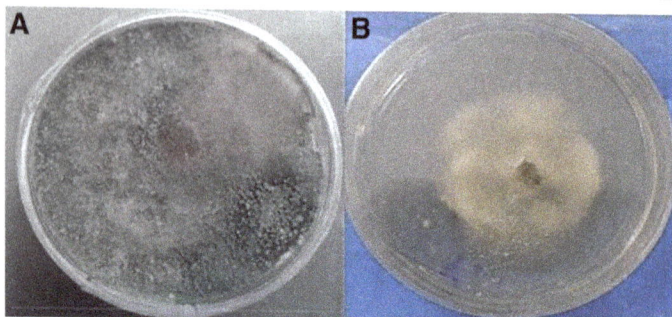

Figure 4 *Terichoderma* sp in control plate (A), and in presence of 200 mg l⁻¹ Cd after 7 days (B).

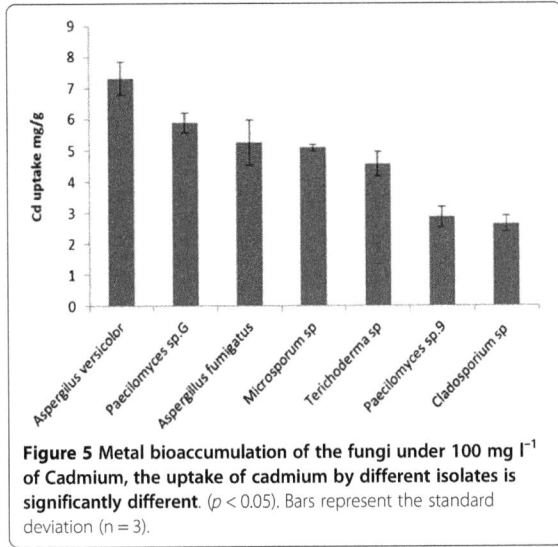

Figure 5 Metal bioaccumulation of the fungi under 100 mg l^{-1} of Cadmium, the uptake of cadmium by different isolates is significantly different. ($p < 0.05$). Bars represent the standard deviation (n = 3).

distinguished and the parts that these topics have in common be selected. Going through mechanisms of tolerance that finally leads to discovering new biouptake activities (bioaccumulation and biosorption) is essential in this field.

A diversity of specific metal accumulation strategies has been known. The physicochemical properties of metals and the physiology of the organism both influence metal uptake [45]. It can be hypothesized that diminished uptake in *Cladosporium* and *Paecilomyces* sp.9 generas contributed to cadmium rejection mechanism of tolerance utilized by this fungus. Fungi are able to restrict entry of toxic metal species into cells by reduced metal uptake and/or increased metal efflux [42]. These microorganisms are known as metal excluders [46]. In contrast, higher Cd removal in *Aspergillus versicolor* may relate to accumulation of cadmium in cell structures. Fungal biomass can act as a metal sink, either by: (1) metal biosorption to biomass (cell walls, pigments and extracellular polysaccharides); or (2) intracellular accumulation and sequestration; or (3) precipitation of metal compounds onto and/or around hyphae [42,47].

Fungi were known to accumulate significant amount of cadmium, for example uptake concentrations of 6.46 mgg^{-1} by *Aspergillus* nijer and 16.25 mg g^{-1} by *Trichoderma viride* have been reported earlier [48]. *Pisolithus tinctorius* presented maximal uptake of 600 mg kg^{-1} dry weight at 10 mg l^{-1} cadmium concentration [5].

Conclusion

The present study declared seven highly tolerant fungi . These fungi exhibited various resistance strategies towards cadmium and they had an ability to sequester cadmium from liquid media.

Aspergilus versicolor remarkably differed in detoxification behavior from other isolated fungi in this study. The fungus showed a remarkable potential to actively grow in presence of Cd and reduce cadmium concentration to less toxic levels. Introducing *Aspergilus versicolor* as scavenger biota is the first step of emerging this fungus in bioremediation science.

Efforts are being made to make bioremediation technically/economically feasible; therefore, we should direct our attention to exploit whole potential of microorganism.

Understanding metal uptake process genetically, manipulation of cell structure such as autoclaving or drying biomass, and using combo strains are innovative technologies in biotreatment studies.

Competing interests
The authors declare that they have no competing interests.

Authors' contributions
NS carried out the experiments and drafted the manuscript. MF and AR designed the study and participated in data analysis. JM helped to draft and edit the manuscript. SD helped to identify the fungi. MM participated in data analysis and study design. All authors read and approved the final manuscript.

Acknowledgment
We thank the research deputy of Zanjan University of Medical Sciences for financial support of this project (MSc thesis, grant number: A-11-343-4).

Author details
[1]Department of Environmental Health Engineering, Zanjan University of Medical Sciences, Zanjan, Iran. [2]Medical Entomology and Mycology Department, School of Medicine, Zanjan University of Medical Sciences, Zanjan, Iran. [3]Biotechnology Departments, School of Pharmacy, Zanjan University of Medical Sciences, Zanjan, Iran.

References
1. Das N, Vimala R, Karthika P. Biosorption of heavy metals-An overview. Indian J Biotechnol. 2008;7:159–69.
2. Grąz M, Pawlikowska-Pawlęga B, Jarosz-Wilkołazka A. Growth inhibition and intracellular distribution of Pb ions by the white-rot fungus Abortiporus biennis. Int Biodeter Biodegr. 2011;65:124–9.
3. Salinas E, Elorza de Orellano M, Rezza I, Martinez L, Marchesvky E, Sanz de Tosetti M. Removal of cadmium and lead from dilute aqueous solutions by Rhodotorula rubra. Bioresource Technol. 2000;72:107–12.
4. Blaudez D, Botton B, Chalot M. Cadmium uptake and subcellular compartmentation in the ectomycorrhizal fungus Paxillus involutus. Microbiology. 2000;146:1109–17.
5. Carrillo-Gonzalez R, Gonzalez-Chavez Mdel C. Tolerance to and accumulation of cadmium by the mycelium of the fungi Scleroderma citrinum and Pisolithus tinctorius. Biol Trace Elem Res. 2012;146:388–95.
6. Jaeckel P, Krauss GJ, Krauss G. Cadmium and zinc response of the fungi Heliscus lugdunensis and Verticillium cf. alboatrum isolated from highly polluted water. Sci Total Environ. 2005;346:274–9.
7. Fu F, Wang Q. Removal of heavy metal ions from wastewaters: a review. J Environ Manage. 2011;92:407–18.
8. Ahluwalia SS, Goyal D. Microbial and plant derived biomass for removal of heavy metals from wastewater. Bioresource Technol. 2007;98:2243–57.
9. Melgar MJ, Alonso J, García MA. Removal of toxic metals from aqueous solutions by fungal biomass of Agaricus macrosporus. Sci Total Environ. 2007;385:12–9.
10. Gadd G. Metal Tolerance, Microbiology of extreme environments. Milton Keynes: Open University Press; 1990. p. 178–210.

11. Anahid S, Yaghmaei S, Ghobadinejad Z. Heavy metal tolerance of fungi. Scientia Iranica. 2011;18:502–8.

12. Yuan H, Li Z, Ying J, Wang E. Cadmium (II) removal by a hyperaccumulator fungus Phoma sp. F2 isolated from blende soil. Curr Microbiol. 2007;55:223–7.

13. Kacprzak M, Malina G. The tolerance and Zn2+, Ba2+ and Fe3+ accumulation by Trichoderma atroviride and Mortierella exigua isolated from contaminated soil. Can J Soil Sci. 2005;85:283–90.

14. Xiao X, Luo S, Zeng G, Wei W, Wan Y, Chen L, et al. Biosorption of cadmium by endophytic fungus (EF) Microsphaeropsis sp. LSE10 isolated from cadmium hyperaccumulator Solanum nigrum L. Bioresour Technol. 2010;101:1668–74.

15. Lopez Errasquin E, Vazquez C. Tolerance and uptake of heavy metals by Trichoderma atroviride isolated from sludge. Chemosphere. 2003;50:137–43.

16. Le L, Tang J, Ryan D, Valix M. Bioleaching nickel laterite ores using multi-metal tolerant Aspergillus foetidus organism. Miner Eng. 2006;19:1259–65.

17. Xu X, Xia L, Huang Q, Gu J-D, Chen W. Biosorption of cadmium by a metal-resistant filamentous fungus isolated from chicken manure compost. Environ Technol. 2012;33:1661–70.

18. Gillespie SH, Pearson RD. Principles and Practice of Clinical Parasitology. West Sussex: Wiley Online Library; 2001.

19. Mirzaei S, Yazdi MT, Sepehrizadeh Z. Secretory expression and purification of a soluble NADH cytochrome b5 reductase enzyme from Mucor racemosus in Pichia pastoris based on codon usage adaptation. Biotechnol Lett. 2010;32:1705–11.

20. Borneman J, Hartin RJ. PCR primers that amplify fungal rRNA genes from environmental samples. Appl Environ Microb. 2000;66:4356–60.

21. Altschul SF, Gish W, Miller W, Myers EW, Lipman DJ. Basic local alignment search tool. J Mol Biol. 1990;215:403–10.

22. Zapotoczny S, Jurkiewicz A, Tylko G, Anielska T, Turnau K. Accumulation of copper by Acremonium pinkertoniae a fungus isolated from industrial wastes. Microbiol Res. 2007;162:219–28.

23. Gruhn C, Miller Jr O. Effect of copper on tyrosinase activity and polyamine content of some ectomycorrhizal fungi. Mycol Res. 1991;95:268–72.

24. Anand P, Isar J, Saran S, Saxena RK. Bioaccumulation of copper by Trichoderma viride. Bioresource Technol. 2006;97:1018–25.

25. Pan R, Cao L, Zhang R. Combined effects of Cu, Cd, Pb, and Zn on the growth and uptake of consortium of Cu-resistant Penicillium sp. A1 and Cd-resistant Fusarium sp. A19. J Hazard Mater. 2009;171:761–6.

26. Kabata-Pendias A. Trace Elements In Soils and Plants. Boka Raton: CRC press; 2010.

27. EPA. Development Document for Effluent Limitations Guidelines and Standards for the Metal Finishing Point Source Category. November 2002 [updated November 2002; cited]; Available from: http://water.epa.gov/scitech/wastetech/guide/taskforce/upload/2000_10_19_guide_taskforce_report5.pdf.

28. Yazdani M, Yap CK, Abdullah F, Tan SG. An in vitro study on the adsorption, absorption and uptake capacity of Zn by the bioremediator Trichoderma atroviride. Environment Asia. 2010;3:53–9.

29. Iram S, Ahmad I, Javed B, Yaqoob S, Akhtar K, Kazmi MR, et al. Fungal tolerance to heavy metals. Pak J Bot. 2009;41:2583–94.

30. Colpaert JV, Vandenkoornhuyse P, Adriaensen K, Vangronsveld J. Genetic variation and heavy metal tolerance in the ectomycorrhizal basidiomycete Suillus luteus. New Phytologist. 2000;147:367–79.

31. Zafar S, Aqil F, Ahmad I. Metal tolerance and biosorption potential of filamentous fungi isolated from metal contaminated agricultural soil. Bioresour Technol. 2007;98:2557–61.

32. Rudawska M, Leski T. Aluminium tolerance of different Paxillus involutus Fr strains originating from polluted and nonpolluted sites. Acta Societatis Botanicorum Poloniae. 1998;67:115–22.

33. Mohamed RM, Abo-Amer AE. Isolation and characterization of heavy-metal resistant microbes from roadside soil and phylloplane. J Basic Microbiol. 2012;52:53–65.

34. Ezzouhri L, Castro E, Moya M, Espinola F, Lairini K. Heavy metal tolerance of filamentous fungi isolated from polluted sites in Tangier, Morocco. Afr J Microbiol Res. 2009;3:35–48.

35. Ge W, Zamri D, Mineyama H, Valix M. Bioaccumulation of heavy metals on adapted Aspergillus foetidus. Adsorption. 2011;17:901–10.

36. Krznaric E, Verbruggen N, Wevers JH, Carleer R, Vangronsveld J, Colpaert JV. Cd-tolerant Suillus luteus: a fungal insurance for pines exposed to Cd. Environ Pollut. 2009;157:1581–8.

37. Baldrian P, Gabriel J. Intraspecific variability in growth response to cadmium of the wood-rotting fungus Piptoporus betulinus. Mycologia. 2002;94:428–36.

38. Akhtar S, Mahmood-ul-Hassan M, Ahmad R, Suthor V, Yasin M. Metal tolerance potential of filamentous fungi isolated from soils irrigated with untreated municipal effluent. Soil Environ. 2013;32:55–62.

39. Muñoz AJ, Ruiz E, Abriouel H, Gálvez A, Ezzouhri L, Lairini K, et al. Heavy metal tolerance of microorganisms isolated from wastewaters: Identification and evaluation of its potential for biosorption. Chem Eng J. 2012;210:325–32.

40. Darlington AB, Rauser WE. Cadmium alters the growth of the ectomycorrhizal fungus Paxillus involutes: a new growth model accounts for changes in branching. Can J Bot. 1988;66:225–9.

41. Jarosz-Wilkołazka A, Grąz M, Braha B, Menge S, Schlosser D, Krauss G-J. Species-specific Cd-stress response in the white rot basidiomycetes Abortiporus biennis and Cerrena unicolor. Biometals. 2006;19:39–49.

42. Gadd GM. Geomycology: biogeochemical transformations of rocks, minerals, metals and radionuclides by fungi, bioweathering and bioremediation. Mycol Res. 2007;111:3–49.

43. Volesky B, Holan Z. Biosorption of heavy metals. Biotechnol Prog. 1995;11:235–50.

44. Cánovas D, Durán C, Rodríguez N, Amils R, De Lorenzo V. Testing the limits of biological tolerance to arsenic in a fungus isolated from the River Tinto. Environ Microbiol. 2003;5:133–8.

45. Vijver MG, van Gestel CA, Lanno RP, van Straalen NM, Peijnenburg WJ. Internal metal sequestration and its ecotoxicological relevance: a review. Environ Sci Technol. 2004;38:4705–12.

46. Alonso J, García M, Pérez-López M, Melgar M. The concentrations and bioconcentration factors of copper and zinc in edible mushrooms. Arch Environ Con Tox. 2003;44:0180–8.

47. Mejáre M, Bülow L. Metal-binding proteins and peptides in bioremediation and phytoremediation of heavy metals. Trends Biotechnol. 2001;19:67–73.

48. Joshi PK, Swarup A, Maheshwari S, Kumar R, Singh N. Bioremediation of heavy metals in liquid media through fungi isolated from contaminated sources. Indian J Microbiol. 2011;51:482–7.

21

Studies on the laccase-mediated decolorization, kinetic, and microtoxicity of some synthetic azo dyes

Hamid Forootanfar[1], Shahla Rezaei[2], Hamed Zeinvand-Lorestani[3], Hamed Tahmasbi[2], Mehdi Mogharabi[2], Alieh Ameri[4] and Mohammad Ali Faramarzi[2*]

Abstract

Background: Enzymatic elimination of synthetic dyes, one of the most environmentally hazardous chemicals, has gained a great interest during the two last decades. The present study was performed to evaluate the decolorization and detoxification potential of the purified laccase of *Paraconiothyrium variabile* in both non-assisted and hydroxybenzotriazole-aided form against six azo dyes.

Results: The obtained results showed that Acid Orange 67, Disperse Yellow 79, Basic Yellow 28, Basic Red 18, Direct Yellow 107, and Direct Black 166 were decolorized up to 65.3, 53.3, 46.7, 40.7, 34, and 26.2 %, respectively, after 1 h treatment with laccase (0.5 U/mL). Addition of HBT up to 5 mM enhanced decolorization percent of all the investigated dyes. The results of kinetic study introduced the monoazo dye of Acid Orange 67 as the most suitable substrate for laccase with K_m of 0.49 mM and V_{max} of 189 mmol/min/mg. Evaluation the toxic effect of laccase-treated dye sample based on the growth inhibition of standard bacterial strains revealed decrease in toxicity of all applied dyes after treatment by laccase.

Conclusions: Application of the *P. variabile* laccase as biocatalyst efficiently decreased the toxicity of all studied synthetic azo dyes.

Keywords: Decolorization, Detoxification, Laccase, Hydroxybenzotriazole, Microtoxicity

Background

Synthetic dyes are being increasingly applied in the textile, paper, cosmetics, pharmaceutical, leather dyeing, color photography, and food industries [1–4]. Azo dyes (contain the N = N group in their structure) identified as one of the most popular synthetic colorant agent because they could be easily and affordably synthesized and found to be stable [5–7]. However, the toxicity, mutagenicity, and carcinogenicity of synthetic azo dyes and/or their metabolites have been well documented [8–10]. Furthermore, the harmful effects of azo dyes on the germination and growth of many environmentally important plants have been well described [1, 10]. So, development of physicochemical [11, 12] and/or biological

techniques for treatment of wastewater rich in such complex aromatic structures received great attention during the two last decades among which enzymatic removal of such pollutants is an economic and environmentally friendly procedure due to the low energy required and the minimal impact on ecosystems [13–15].

Laccases (EC 1.10.3.2), the multicopper-containing oxidases belonging to the superfamily of multicopper oxidase, are mainly produced by plants, fungi especially the white-rot basidiomycetes and some bacterial strains [16–18]. Laccases alone or in assistance with mediators have been found to catalyze the oxidation of a broad range of substrates such as phenol and its derivatives, benzenethiols, aromatic amines, and polycyclic aromatic hydrocarbons (PAHs) [19–21]. The mentioned feature launched these biocatalysts as the main tool in the xenobiotic removal studies [22, 23]. There are many reports on the application of the purified laccases or

* Correspondence: faramarz@tums.ac.ir
[2]Department of Pharmaceutical Biotechnology, Faculty of Pharmacy and Biotechnology Research Center, Tehran University of Medical Sciences, P.O. Box 14155-6451, Tehran 1417614411, Iran
Full list of author information is available at the end of the article

laccase-producing organisms for elimination of synthetic and natural dyes [24, 25]. For example, the ability
of laccase from *Trametes versicolor* for decolorization
of four dyes including Red FN-2BL, Red BWS, Remazol
Blue RR and Blue 4BL was reported by Mendoza et al.
[26]. In the study of Yan et al. [27] two azo dyes including reactive black and congo red decolorized by the
crude laccase of *Trametes trogii* S0301.

In the present study the potential application of the
purified laccase originated from a soil-isolated fungal
strain [25] for decolorization and detoxification of 6 synthetic azo dyes was evaluated. Thereafter, the effects of
factors including pH and temperature as well as enzyme
and mediator concentration on decolorization percent
were also demonstrated. Additionally, the energetic (ΔS,
ΔH, and Ea) and kinetic (K_m and V_{max}) parameters for
laccase-mediated decolorization were determined.

Methods
Chemicals and enzyme
The six applied azo dyes (Fig. 1 and Table 1) were supplied
by Alvan Sabet Co. (Tehran, Iran). 2,2'-Azinobis-(3-ethyl-
benzothiazoline-6-sulfonate) (ABTS) and 1-hydroxyben
zotriazole (HBT) were provided by Sigma-Aldrich (St.
Louis, MO, USA) and Merck (Darmstadt, Germany), respectively. The laccase applied in the present study was
purified as previously described [25].

Laccase assay
The previously reported method of Forootanfar et al.
[25] using ABTS was applied for measuring the oxidative activity of laccase. In brief, into the 0.5 mL freshly
prepared enzyme substrate (ABTS 5 mM in citrate buffer 100 mM pH 5) enzyme sample (0.5 mL) was added
and the reaction mixture was incubated at 37 °C and

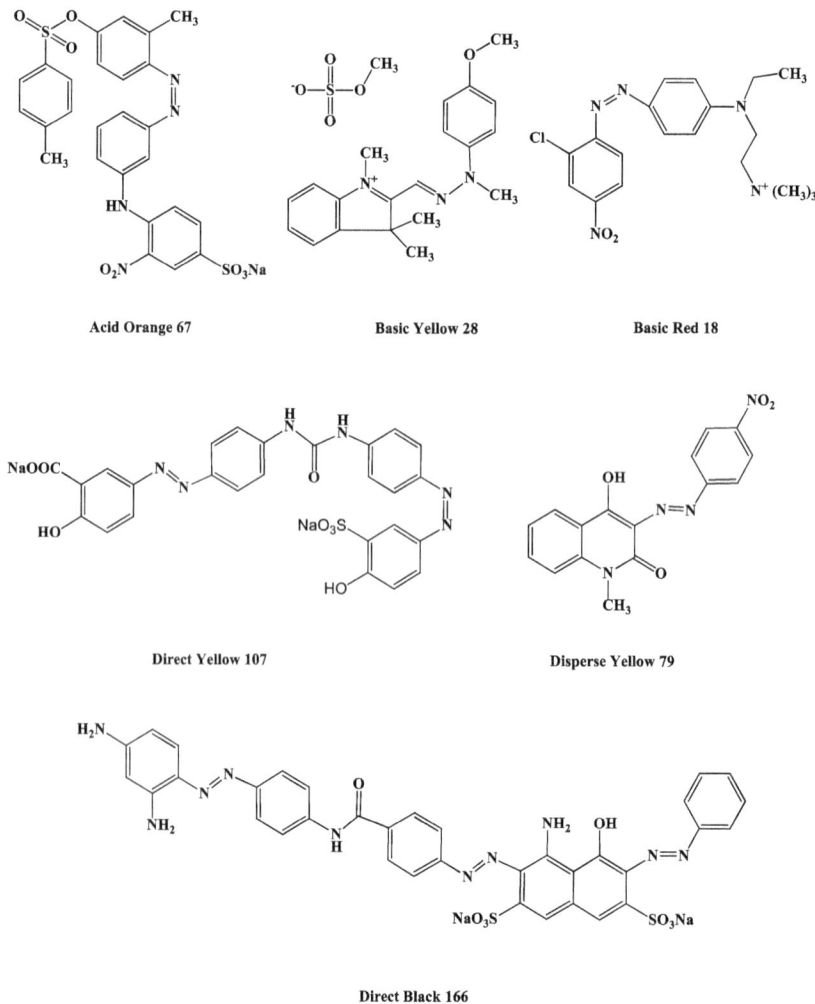

Fig. 1 Chemical structures of azo dyes applied in the present study

Table 1 Kinetic and energetic parameters of the laccase on studied synthetic dyes

Dye name	Type	λ_{max} (nm)	K_m (mM)	V_{max} (mmol/min/mg)	ΔS (Kj/mol/K)	ΔH (kJ/mol)	Ea (kJ/mol)
Acid Orange 67	Monoazo	438	0.49	189	307	93	24
Basic Red 18	Monoazo	489	3.07	64	194	63	28
Basic Yellow 28	Monoazo	450	2.07	100	308	91	29
Disperse Yellow 79	Monoazo	474	2.02	104	189	58	30
Direct Yellow 107	Diazo	368	3.44	47	213	68	26
Direct Black 166	Triazo	602	4.25	33	199	65	31

120 rpm for 10 min. Afterwards, the absorbance was recorded at 420 nm ($\varepsilon_{420} = 36,000$ M^{-1} cm^{-1}) by a Double Beam PC Scanning UV–vis spectrophotometer (UVD 2950, Labomed, Culver City, USA). The amount of enzyme which oxidize 1 μmol of ABTS per minute defined as one unit of laccase activity [28–31].

Dye decolorization experiments
The previously described method of Ashrafi et al. [5] was used in order to assess the decolorization potential of the purified laccase. Firstly, dye solution was prepared by dissolving of each synthetic dye in citrate-phosphate buffer (100 mM, pH 5.0). Thereafter, the purified enzyme (final activity of 0.5 U/mL) was inserted into each dye solution and the reaction mixture was incubated at 35 °C and 50 rpm for 3 h. Subsequently, interval samples were taken each 30 min and the related absorbance (at the λ_{max} of each dye, Table 1) was measured and decolorization percent was calculated using the below equation

$$\text{Decolorization (\%)} = [A_i - A_t / A_i] \times 100$$

where A_i and A_t are the initial absorbance and the absorbance after incubation time of the reaction mixture, respectively [32, 33]. In order to confirm about significant change in decolorization percentage the reaction mixture was further incubated overnight. To examine the abiotic decolorization (negative control) each dye solution was inserted by heat-inactivated laccase and the related decolorization percent was estimated as said above. Three independent experiments were conducted for each dye and means of decolorization percent was calculated.

The influence of pH alteration on enzymatic decolorization
In order to investigate the influence of pH on laccase-mediated decolorization each dye solution was firstly prepared in 0.1 M citrate-phosphate buffer (pH 3–7) or Tris-HCl buffer (0.1 M for pH 8) and the purified laccase (0.5 U/mL) was consequently added to the reaction mixture which was then monitored for decolorization as previously described.

The effect of temperature on laccase-mediated decolorization
After preparation of the reaction mixture including each studied dye and the purified laccase (0.5 U/mL) it was incubated at different temperatures (5–75 °C) and decolorization percent was then determined using the previously described procedure.

Laccase activity influence on dye decolorization
The effect of enzyme concentration on decolorization process was evaluated by insertion of laccase (final activity range of 0.5–5 U/mL) into the freshly prepared dye solution (in 0.1 M citrate-phosphate buffer pH 5) and incubation at 35 °C and 50 rpm for 1 h followed by recording of the related absorbance and calculation of the decolorization percent.

The effect of laccase mediator (HBT) on decolorization
After dissolving of each dye into citrate-phosphate buffer (0.1 M pH 5) HBT was also inserted (final concentrations of 1, 5, and 10 mM) and decolorization was initiated by addition of laccase (0.5 U/mL) into the reaction mixture. The amount of decolorization was then demonstrated as mentioned above.

The kinetics and energetics of dye elimination
Decolorization kinetics
In order to acquire the kinetic parameters of decolorization the related velocity (V) was firstly obtained by performing dye decolorization at different concentrations (C) of each dye. The Michaelis-Menten curve was then drawn by plotting the obtained velocity against concentration. Consequently, Lineweaver-Burk transformation was applied in order to obtain the kinetic constant (K_m and V_{max}) of decolorization for each dye [5, 10].

Decolorization thermodynamics
The related velocity of decolorization was initially determined at various temperature of 10–50 °C and applied for drawing of a curve by plotting the calculated velocity against initial dye concentration. The apparent first-order rate constant (K) was then calculated based on the slope of the straight plot achieved for each dye and employed to drain the linearized Arrhenius curve by

plotting the ln (K) versus $1/T$ ($\times 10^3$ K^{-1}) where T is the absolute temperature (K). The slope of Arrhenius plot indicates $-E_a/R$ in which Ea is the activation energy and R is the gas constant (8.3145 J /mol/K). Thereafter, Van't Hoff curve was drained by plotting the ln (K_{eq}) against $1/T$ ($\times 10^3$ K^{-1}) and applied for determination of the entropy (ΔS) and enthalpy (ΔH) of decolorization reaction. K_{eq} is the apparent equilibrium constant which was determined from difference of initial and remained dye concentration at equilibrium state, when the decolorization percentage become constant and no decolorization was occurred with passing the time [34]. Subsequently, the Gibbs free energy (ΔG) was also calculated using the equation of $\Delta G = \Delta H - T\Delta S$.

Dye toxicity

A modified microtoxicity assay procedure using six standard bacterial strain (three Gram-negative and three Gram-positive species, Table 2) was employed to determine the toxicity of each studied dye before and after treatment by the purified laccase. In brief, a proper dilution of each bacterial suspension (OD$_{600}$ 0.2) in Muller-Hinton broth medium was firstly prepared and successively inserted by each dye or its related laccase-treated solution and incubated at 37 °C for 10 h. Afterwards, interval samples were taken every 2 h and analyzed for changes in the OD$_{600}$ compared to that of the negative control (cultivated bacterial strain in the absence of any dye). The equation of $[(1 - OD_{600S}/ OD_{600C}) \times 100]$ was then utilized for determination of growth inhibition percent (GI%), where OD$_{600S}$ and OD$_{600C}$ are OD$_{600}$ of sample and OD$_{600}$ of control, respectively.

Statistical analysis

Three replicates of each above mentioned experiment was performed and mean ± standard deviation of each value was reported. In order to calculate the statistical significance (probability values less than 0.05) between

mean values the independent sample t-test and one-way analysis of variance (ANOVA) with Dunnett's T3 post hoc test were employed using the SPSS software (version 15.0, SPSS Inc).

Results and discussion
Decolorization studies

As presented in the time course of decolorization of six synthetic azo dyes (Fig. 2), laccase was able to efficiently decolorize all studied dyes within 3 h incubation. Acid Orange 67, Disperse Yellow 79, Basic Yellow 28, Basic Red 18, Direct Yellow 107, and Direct Black 166 were respectively decolorized up to 65.3, 53.3, 46.7, 40.7, 34, and 26.2 % after 1 h treatment with laccase. Elongation of the reaction time to 24 h didn't significantly enhance decolorization percent (data not shown). Furthermore, for abiotic control (the reaction mixture without enzyme) no decolorization was observed indicating the catalytic role of laccase in the decolorization procedure.

Application of laccases and laccase mediated system (LMS) to decolorize mutagenic and carcinogenic synthetic colorants has been received great interest during two last decades [17, 35]. Literature review revealed that laccase assisted decolorization pattern even in the case of one special dye could vary depends on the origin of laccases [1, 10]. There are many reports about decolorization ability of laccase and laccase mediated systems toward a broad range of synthetic dyes including diazo dyes, anthraquinone dyes, triphenylmethane dyes, and textile colorants [2, 5, 24]. The obtained results of the present study revealed the potential application of laccase for decolorization of six synthetic azo dyes (Fig. 2). Literature review introduced azo dyes as recalcitrant dyes using both the enzymatic and physicochemical procedures [1, 13]. In general, resistance of azo colorants to laccase-catalyzed oxidation enhanced by increasing the number of azo group [36, 37]. For example, Ashrafi et al. [5] achieved the lowest and highest decolorization percent for Direct Blue 71 (a triazo dye,

Table 2 Growth inhibition percent (GI%) of untreated and laccase-treated dyes against six bacterial strains. Values are averages of three replicates ± standard deviation

Bacterial strains												
Dye names	E. coli		P. aeroginosa		S. typhi		B. subtilis		S. aureus		M. luteus	
	U^a	T^b	U	T	U	T	U	T	U	T	U	T
Acid Orange 67	28 ± 1.1	19 ± 0.5*	21 ± 0.4	13 ± 0.6*	24 ± 0.8	16 ± 0.4*	37 ± 1.1	28 ± 1.0*	33 ± 1.8	20 ± 0.8*	42 ± 2.3	31 ± 0.6*
Basic Red 18	25 ± 1.1	14 ± 0.4*	18 ± 1.2	8 ± 0.3*	23 ± 1.2	11 ± 0.7*	34 ± 1.5	21 ± 0.5*	30 ± 1.1	19 ± 0.6*	37 ± 0.8	23 ± 0.7*
Basic Yellow 28	32 ± 1.8	20 ± 0.4*	26 ± 1.4	16 ± 0.3*	29 ± 1.3	17 ± 0.6*	41 ± 1.9	30 ± 0.7*	36 ± 0.7	28 ± 0.6*	49 ± 1.2	35 ± 1.1*
Disperse Yellow 79	21 ± 0.9	9 ± 0.4*	14 ± 0.7	5 ± 0.3*	18 ± 0.3	7 ± 0.4*	27 ± 0.4	12 ± 0.2*	23 ± 0.9	9 ± 0.3*	30 ± 1.1	16 ± 0.5*
Direct Yellow 107	46 ± 1.4	34 ± 0.4*	38 ± 1.0	29 ± 0.3*	41 ± 0.8	31 ± 0.8*	57 ± 0.6	44 ± 0.6*	49 ± 0.9	41 ± 0.9*	62 ± 0.7	48 ± 0.7*
Direct Black 166	51 ± 1.8	37 ± 0.7*	43 ± 1.2	30 ± 0.6*	47 ± 1.9	31 ± 0.8*	61 ± 2.2	50 ± 0.9*	56 ± 2.1	45 ± 0.8*	68 ± 2.9	54 ± 1.0*

^{a}GI% in presence of untreated dye solution; ^{b}GI% in presence of laccase-treated dye solution; *Significance was determined using independent sample t-test (p-value < 0.05)

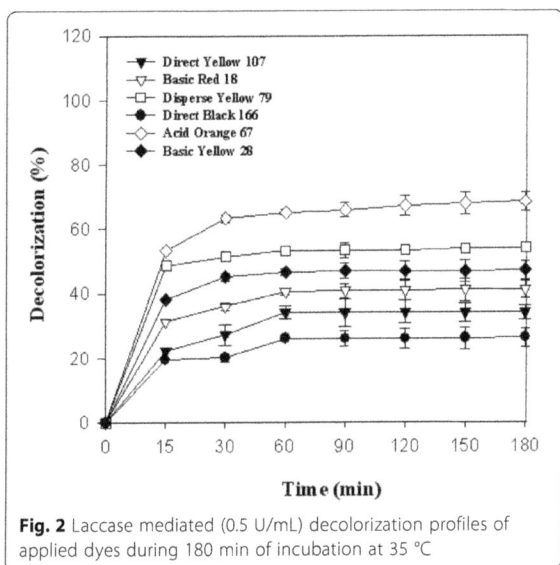

Fig. 2 Laccase mediated (0.5 U/mL) decolorization profiles of applied dyes during 180 min of incubation at 35 °C

30 %) and Acid Red 18 (a monoazo dye, 97 %), respectively, after 15 min treatment using the secreted laccase of *P. varibile*. In the study of Zeng et al. [38] it was found that the laccase originated by *T. trogii* SYBC-LZ didn't decolorize Acid Red 1 (a monoazo dye) and Reactive Black 5 (a diazo dye) while the highest decolorization percent (80.4 %) achieved for the anthraquinonic dye of Remazol Brilliant Blue R (RBBR). Grassi et al. [8] demonstrated that only 7 % of Fast Blue RR (an azo dye) was removed after 1 h incubation in the presence of the purified laccase of *T. trogii* while more than 80 % of anthraquinonic dye of RBBR and Indigo Carmine was eliminated at the same time.

The influence of pH and temperature on laccase-catalyzed decolorization

The results of pH alteration on decolorization of investigated dyes were illustrated in Fig. 3a. Maximum decolorization of all dyes occurred at the optimum pH of the applied laccase (5) which previously reported by Forootanfar et al. [25]. Alteration from the optimum pH of laccase negatively affected decolorization percent of all synthetic dyes. In the study of Ashrafi et al. [5] and Mirzadeh et al. [10] laccase mediated decolorization of all applied dyes was maximally achieved at acidic pH of 5. It was demonstrated that most of laccases of fungal origins maximally work at acidic pH and the enzyme activity at higher pH is decreased due to binding of hydroxide anion to the T2/T3 coppers of laccase and as a result interrupting with the internal electron transfer from T1 to T2/T3 centers [39].

Decolorization percent of all studied dyes was gradually increased by enhancing the reaction temperature

and maximum decolorization of Disperse Yellow 79 (60 %), Acid Orange 67 (71.3 %), and Basic Yellow 28 (58.4 %) was observed at the temperature of 45 °C (Fig. 3b). Basic Red 18 (48 %), Direct Yellow 107 (50.2 %), and Direct Black 166 (46 %) were maximally decolorized at 55 °C (Fig. 3b). The amount of decolorization of all studied dyes was dropped below 20 % (p-value < 0.05) by elevating of temperature to 70 °C which was in agreement with the acquired results of Ashrafi et al. [5] who observed that maximum dye decolorization occurred between the temperatures of 40–60 °C. In general, most of the fungal derived laccases optimally work at the temperature range of 50–70 °C [39].

The effect of laccase activity on decolorization

As represented in Fig. 4a, the decolorization percent of all studied dyes (except for Acid Orange 67) was significantly enhanced (p-value < 0.05) after increasing of laccase activity from 0.5 U/mL to 1 U/mL. Maalej-Kammoun et al. [6] determined decolorization percent of 80 % for laccase-catalyzed (1 U/mL) elimination of Malachite green within 2 h of initiation of the reaction. In the study of Ashrafi et al. [5] it was showed that increasing of laccase activity from 0.025 to 0.1 U/mL significantly increased decolorization percent of all thirteen studied synthetic dyes.

The influence of HBT on laccase mediated decolorization

As represented in Fig. 4b, decolorization percent of all synthetic dyes was positively affected by increasing of HBT concentration up to 5 mM and decreased by enhancing the HBT concentration more than 5 mM.

Treatment of hazardous organic pollutants and synthetic dyes using laccase-mediated systems, especially in the case of laccases with low redox potential and recalcitrant pollutants, has been well documented [22]. For example, Ostadhadi-Dehkordi et al. [40] investigated on the mediating activity of the phenol derivatives of vanillic acid (VA) and 2,6-dimethoxyphenol (DMP) and the non-phenolic mediators (ABTS and HBT), among which HBT was the most efficient one increased the removal percent of four investigated benzodiazepines (one of the most prescribed pharmaceuticals). Grassi et al. [8] evaluated the effect of synthetic (HBT) and several natural laccase mediators including ρ-hydroxybenzoic acid (HBA), tyrosine, vanillin, vanillic acid, anisaldehyde, and ferulic acid on decolorization of Azure B, Xylidine, and Gentian Violet in the presence of *T. trogii*'s laccase. They introduced HBT as the best mediator for maximum decolorization [8]. Laccase-HBT system is generally more effective compared to unaided laccase due to stronger oxidative activity of free-radical HBT species. Furthermore, high redox

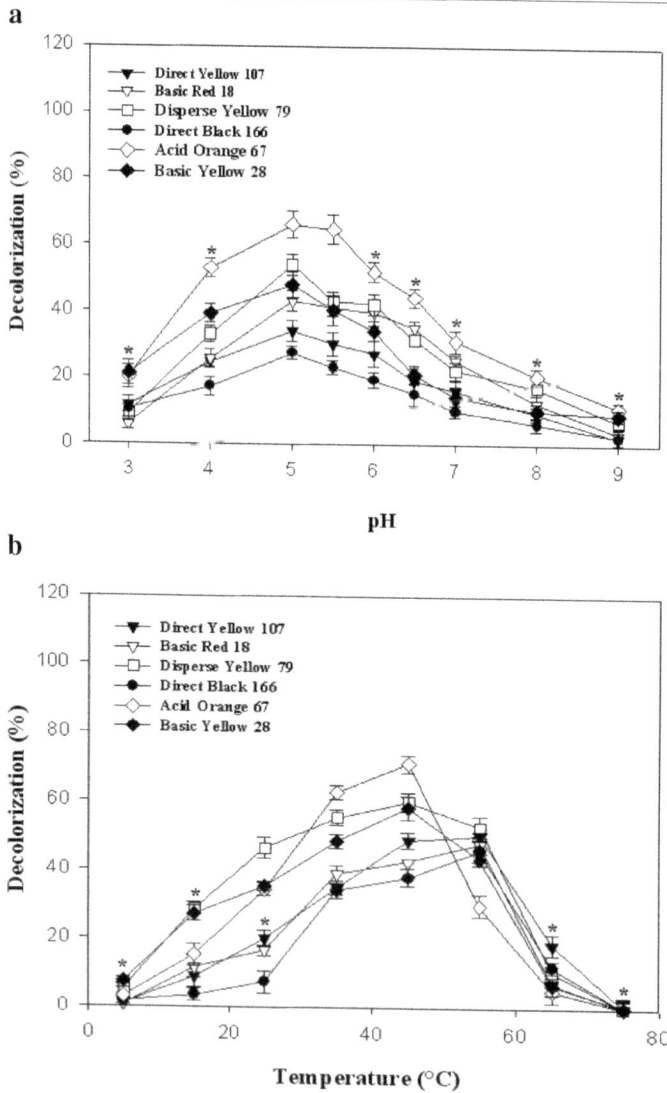

Fig. 3 Effect of various (**a**) pH, and (**b**) temperatures on decolorization of azo dyes using the purified laccase (0.5 U/mL) after 60 min of incubation. Obtained means were analyzed by ANOVA with Dunnett's T3 post hoc test (*, p-value <0.05)

potential of HBT (1084 mV) together with the stabilizing activity of HBT on laccases makes it suitable for elimination of recalcitrant colorants using oxidative activity of laccases [22, 40]. However, depend on the source of laccases applied in the elimination studies, there is a critical concentration above which the assisting activity of this non-phenolic laccase mediator is negatively affected by the destructive effect of produced HBT radical on enzyme structure [22]. So, literature review revealed many reports on determination of optimum concentration of HBT to get maximum decolorization using laccase-HBT system [5, 40].

Kinetics and energetic studies

Based on the kinetic parameters of the laccase toward studied dyes (Tabale 1) the lowest and the highest K_m value belonged to the monoazo dye of Acid Orange 67 (0.49 mM) and triazo dye of Direct Black 166 (4.25 mM), respectively. Unsuitability of Direct Black 166 as a laccase substrate is evident from the high value of the calculated K_m of this colorant (4.25 mM) and its related maximum velocity (33 mmol/min/mg). The resistance of azo dyes to laccase-catalyzed oxidation compared to other classes of colorants has been well documented. Grassi et al. [8] reported the lowest

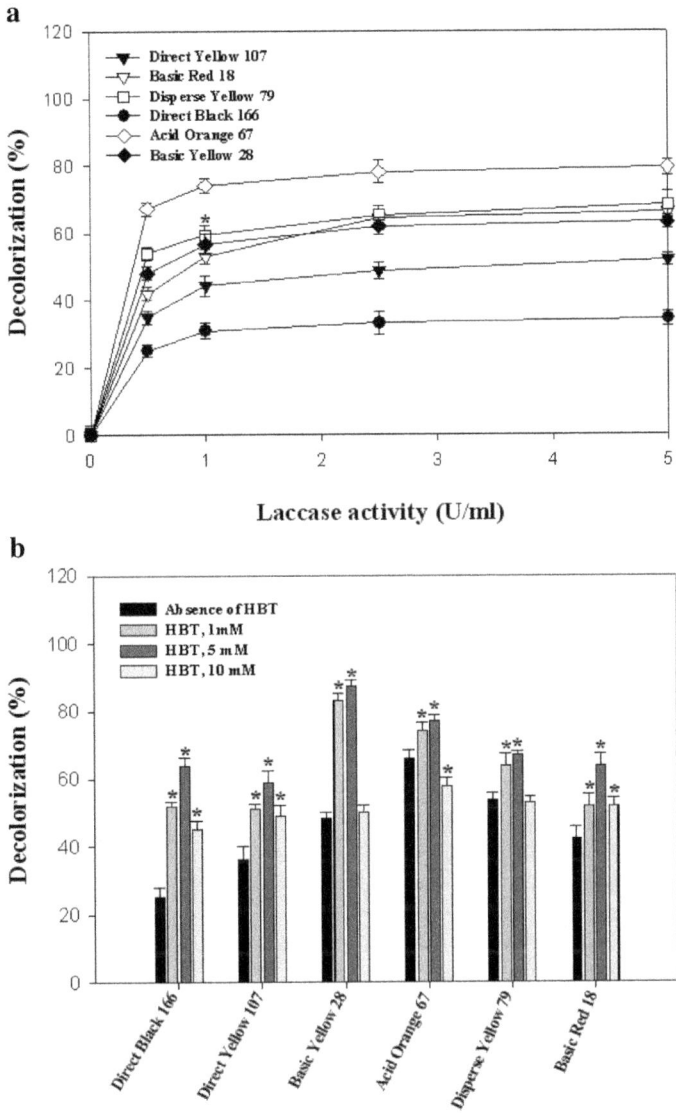

Fig. 4 Influence of (**a**) laccase activity and (**b**) HBT concentration on decolorization of investigated dyes after 60 min of incubation at 35 °C. Significant values (*, p-value <0.05) were achieved after ANOVA analysis with Dunnett's T3 post hoc test

decolorization percent for the azo dye of Fast Blue RR using the purified laccase of *T. trogii* B6J among tested synthetic dyes. Generally, presence of electron withdrawing groups such as $-SO_3H$, $-SO_2NH_2$ together with the high molecular weight decreases the ability of laccases for elimination of azo dyes [1]. Application of laccases harboring higher redox potential such as bacterial laccases and addition of synthetic or natural redox mediators are two main solutions to overcome such problem [1]. As an example, the recombinant bacterial CotA-laccase from *B. subtilis* was able to decolorize Direct Blue 1 (a diazo dye), Reactive Black

5 (a diazo dye), and Sudan Orange G (a monoazo dye) by 88, 85 and 98 %, respectively at alkaline pH and in the absence of redox mediators [9].

The obtained results of energetic assessment for laccase-mediated decolorization (Table 1) indicated the significant influence of temperature on decolorization of all investigated synthetic dyes as decolorization rate increased from 10 °C to 50 °C. The activation energies (calculated from the slope of Arrhenius plot) together with the estimated values for ΔH and ΔS were also represented in Table 1. Considering the negative amount of slope obtained for van't Hoff plot or the positive signs of ΔH (Table 1) it was demonstrated that the decolorization reactions are endothermic.

Microtoxicity study

The results of microtoxicity study were summarized in Table 2. The highest GI% belonged to the Direct Black 166 (a triazo dye) suppressing *M. luteus*, *S. aureus*, *B. subtilis*, *E. coli*, *P. aeruginosa*, and *S. typhi* by 68, 56, 61, 51, 43, and 47 %, respectively, when exposed to each bacterial strain for 10 h. On the other hand, the monoazoic dye of Disperse Yellow 79 represented GI% of 30, 23, 27, 21, 14, and 18 % for *M. luteus*, *S. aureus*, *B. subtilis*, *E. coli*, *P. aeruginosa*, and *S. typhi*, respectively (Table 2) which was the lowest observed value. The GI% of all synthetic azo dyes was significantly reduced following laccase treatment as presented in Table 2.

In order to evaluate the probable toxicity of colorants and their metabolite(s) after physicochemical or biological treatment many biological assays such as inhibition the growth of bacterial, yeast, and mammalian cell lines as well as phytotoxicity evaluations on plant seeds were developed [5, 9, 41, 42]. Like the acquired results reported by Ashrafi et al. [5] the attained results of the present investigation showed that the toxicity of all applied dyes was significantly decreased after addition of laccase to dye solution. Younes et al. [13] found that the GI% of *Bacillus megaterium* was enhanced from 2 to 99 % after incubation of malachite green in the presence of purified laccase originated from *S. thermophilum*. Maalej-Kammoun et al. [6] reported that decolorization of malachite green resulted in the removal of its toxicity against *Phanerochaete chrysosporium*. However, Mendes et al. [14] found lower toxicity for intact azo dyes of Direct Black 38, Direct Red 80, Reactive Black 5, Reactive Yellow 145, Acid Black 194, and Acid Red 266 compared to that of laccase-treated sample in the presence of the yeast *S. cerevisiae*.

Conclusion

The potential activity of the purified laccase of *P. variabile* for decolorization of 6 azo dyes in the presence and absence of HBT as the laccase mediator was evaluated. All the synthetic dyes were decolorized using non-assisted laccase. In addition, decolorization percent of all applied colorants improved in the presence of HBT (up to 5 mM) as laccase mediator. Evaluation the toxicity of laccase-treated dye samples assisted by growth inhibition percentage (GI%) of standard bacterial strains revealed significant decrease in the toxicity of all studied dyes after treatment by laccase.

Competing interests

The authors declare that they have no competing interests.

Authors' contributions

Decolorization studies were performed by HT and HZ-L. Laccase of *P. variabile* was purified by HF. SR involved in the studies of decolorization kinetic. MM performed decolorization studies and writing of the manuscript. Microtoxicity investigations were carried out by AA. MAF involved in purchasing of required materials and instruments, designing of decolorization experiments, analyzing of data and reviewing of the manuscript. All authors read and approved the final manuscript.

Acknowledgement

This work was financially supported by a grant from Tehran University of Medical Sciences, Tehran, Iran to M.A.F.

Author details

[1]Department of Pharmaceutical Biotechnology, Faculty of Pharmacy, Kerman University of Medical Sciences, Kerman, Iran. [2]Department of Pharmaceutical Biotechnology, Faculty of Pharmacy and Biotechnology Research Center, Tehran University of Medical Sciences, P.O. Box 14155-6451, Tehran 1417614411, Iran. [3]Department of Pharmacology and Toxicology, Faculty of Pharmacy, Tehran University of Medical Sciences, P.O. Box 14155-6451, Tehran 1417614411, Iran. [4]Department of Medicinal Chemistry, Faculty of Pharmacy, Kerman University of Medical Sciences, Kerman, Iran.

References

1. Saratale RG, Saratale GD, Chang JS, Govindwar SP. Bacterial decolorization and degradation of azo dyes: a review. J Taiwan Inst Chem Eng. 2011;42:138–57.
2. Forootanfar H, Moezzi A, Aghaie-Khouzani M, Mahmoudjanlou Y, Ameri A, Niknejad F, Faramarzi MA. Synthetic dye decolorization by three sources of fungal laccase. J Environ Health Sci Eng. 2012;9:27.
3. Rezaie S, Tahmasbi H, Mogharabi M, Firuzyar S, Ameri A, Khoshayand MR, Faramarzi MA. Efficient decolorization and detoxification of reactive orange 7 using laccase isolated from paraconiothyrium variabile, kinetics and energetics.
   J Taiwan Inst Chem Eng. 2015;56:1–9.
4. Rezaie S, Tahmasbi H, Mogharabi M, Ameri A, Forootanfar H, Khoshayand MR, Faramarzi MA. Laccase-catalyzed decolorization and detoxification of Acid Blue 92: statistical optimization, microtoxicity, kinetics, and energetics. J Environ Health Sci Eng. 2015;13:31.
5. Ashrafi SD, Rezaei S, Forootanfar H, Mahvi AH, Faramarzi MA. The enzymatic decolorization and detoxification of synthetic dyes by the laccase from a soil-isolated ascomycete, *Paraconiothyrium variabile*. Int Biodeter Biodegr. 2013;85:173–81.
6. Maalej-Kammoun M, Zouari-Mechichi H, Belbahri L, Woodward S, Mechichi T. Malachite green decolourization and detoxification by the laccase from a newly isolated strain of *Trametes* sp. Int Biodeter Biodegr. 2009;63:600–6.
7. Shirmardi M, Mesdaghinia A, Mahvi AH, Nasseri S, Nabizadeh R. Kinetics and equilibrium studies on adsorption of acid red 18 (Azo-Dye) using multiwall carbon nanotubes (MWCNTs) from aqueous solution. E-J Chem. 2012;9:2371–83.
8. Grassi E, Scodeller P, Filiel N, Carballo R, Levin L. Potential of *Trametes trogii* culture fluids and its purified laccase for the decolorization of different types of recalcitrant dyes without the addition of redox mediators. Int Biodeter Biodegr. 2011;65:635–43.
9. Pereira L, Coelho AV, Viegas CA, Santos MM C d, Robalo MP, Martins LO. Enzymatic biotransformation of the azo dye Sudan orange G with bacterial CotA-laccase. J Biotechnol. 2009;139:68–77.
10. Mirzadeh SS, Khezri SM, Rezaei S, Forootanfar H, Mahvi AH, Faramarzi MA. Decolorization of two synthetic dyes using the purified laccase of *Paraconiothyrium variabile* immobilized on porous silica beads. J Environ Health Sci Eng. 2014;12:6.
11. Maleki A, Mahvi AH, Ebrahimi R, Zandsalimi Y. Study of photochemical and sonochemical processes efficiency for degradation of dyes in aqueous solution. Korean J Chem Eng. 2010;27:1805–10.
12. Mahvi AH, Ghanbarian M, Nasseri S, Khairi A. Mineralization and discoloration of textile wastewater by TiO2 nanoparticles. Desalination. 2009;239:309–16.
13. Younes SB, Bouallagui Z, Sayadi S. Catalytic behavior and detoxifying ability of an atypical homotrimeric laccase from the thermophilic strain *Scytalidium thermophilum* on selected azo and triarylmethane dyes. J Mol Catal B Enzym. 2012;79:41–8.

14. Mendes S, Farinha A, Ramos CG, Leitao JH, Viegas CA, Martins LO. Synergistic action of azoreductase and laccase leads to maximal decolourization and detoxification of model dye-containing wastewaters. Bioresour Technol. 2011;102:9852–9.

15. Asadgol Z, Forootanfar H, Rezaie S, Mahvi AH, Faramarzi MA. Removal of phenol and bisphenol-A catalyzed by laccase in aqueous solution. J Environ Health Sci Eng. 2014;12:93.

16. Faramarzi MA, Forootanfar H. Biosynthesis and characterization of gold nanoparticles produced by laccase from *Paraconiothyrium variabile*. Colloid Surf B. 2011;87:23–7.

17. Mogharabi M, Faramarzi MA. Laccase and laccase-mediated systems in the synthesis of organic compounds. Adv Synth Catal. 2014;356:897–927.

18. Forootanfar H, Faramarzi MA. Insights into laccase producing organisms, fermentation states, purification strategies, and biotechnological applications. Biotechnol Prog. 2015;31:1443–63.

19. Forootanfar H, Movahednia MM, Yaghmaei S, Tabatabaei-Sameni M, Rastegar H, Sadighi A, Faramarzi MA. Removal of chlorophenolic derivatives by soil isolated ascomycete of *Paraconiothyrium variabile* and studying the role of its extracellular laccase. J Hazard Mater. 2012;209–210:199–203.

20. Dehghanifard E, Jafari AJ, Kalantary RR, Mahvi AH, Faramarzi MA, Esrafili A. Biodegradation of 2,4-dinitrophenol with laccase immobilized on nano-porous silica beads. J Environ Health Sci Eng. 2013;10:25.

21. Rahmani K, Faramarzi MA, Mahvi AH, Gholami M, Esrafili A, Forootanfar H, Farzadkia M. Elimination and detoxification of sulfathiazole and sulfamethoxazole assisted by laccase immobilized on porous silica beads. Int Biodeter Biodegr. 2015;97:107–14.

22. Canas AI, Camarero S. Laccases and their natural mediators: biotechnological tools for sustainable eco-friendly processes. Biotechnol Adv. 2010;28:694–705.

23. Ashrafi SD, Nasseri S, Alimohammadi M, Mahvi AH, Faramarzi MA. Optimization of the enzymatic elimination of flumequine by laccase-mediated system using response surface methodology. *Desalin Water Treat* doi:10.1080/19443994.2015.1063462.

24. Aghaie-Khouzani M, Forootanfar H, Moshfegh M, Khoshayand MR, Faramarzi MA. Decolorization of some synthetic dyes using optimized culture broth of laccase producing ascomycete *Paraconiothyrium variabile*. Biochem Eng J. 2012;60:9–15.

25. Forootanfar H, Faramarzi MA, Shahverdi AR, Tabatabaei Yazdi M. Purification and biochemical characterization of extracellular laccase from the ascomycete *Paraconiothyrium variabile*. Bioresour Technol. 2011;102:1808–14.

26. Mendoza L, Jonstrup M, Hatti-Kaul R, Mattiasson B. Azo dye decolorization by a laccase/mediator system in a membrane reactor: enzyme and mediator reusability. Enzyme Microb Technol. 2011;49:478–84.

27. Yan J, Chen D, Yang E, Niu J, Chen Y, Chagan I. Purification and characterization of a thermotolerant laccase isoform in *Trametes trogii* strain and its potential in dye decolorization. Int Biodeter Biodegr. 2014;93:186–94.

28. Sadighi A, Faramarzi MA. Congo red decolorization by immobilized laccase through chitosan nanoparticles on the glass beads. J Taiwan Inst Chem Eng. 2013;44:156–62.

29. Zeinvand-Lorestani H, Sabzevari O, Setayesh N, Amini M, Nili-Ahmadabadi A, Faramarzi MA. Comparative study of in vitro prooxidative properties and genotoxicity induced by aflatoxin B1 and its laccase-mediated detoxification products. Chemosphere. 2015;135:1–6.

30. Alberts JF, Gelderblom WCA, Botha A, Vanzyl WH. Degradation of flatoxin B1 by fungal laccase enzymes. Int J Food Microbiol. 2009;135:47–52.

31. Tahmasbi H, Khoshayand MR, Bozorgi-Koushalshahi M, Heidary M, Ghazi-Khansari M, Faramarzi MA. Biocatalytic conversion and detoxification of imipramine by the laccase-mediated system. Int Biodeter Biodegr. 2016;108:1–8.

32. Couto SR. Decolouration of industrial azo dyes by crude laccase from *Trametes hirsuta*. J Hazard Mater. 2007;148:768–70.

33. Khlifi R, Belbahri L, Woodward S, Ellouz M, Dhouib A, Sayadi S, Mechichi T. Decolourization and detoxification of textile industry wastewater by the laccase-mediator system. J Hazard Mater. 2010;175:802–8.

34. Annuar MSM, Adnan S, Vikineswary S, Chisti Y. Kinetics and energetics of azo dye decolorization by *Pycnoporus sanguineus*. Water Air & Soil Poll. 2009; 202:179–88.

35. Roriz MS, Osma JF, Teixeira JA, Couto SR. Application of response surface methodological approach to optimize reactive black 5 decolouration by crude laccase from *trametes pubescens*. J Hazard Mater. 2009;169:691–6.

36. Gholami-Borujeni F, Mahvi AH, Nasseri S, Faramarzi MA, Nabizadeh R, Alimohammadi M. Enzymatic treatment and detoxification of acid orange 7 from textile wastewater. Appl Biochem Biotechnol. 2011;165:1274–84.

37. Gholami-Borujeni F, Faramarzi MA, Nabizadeh-Barandozi R, Mahvi AH. Oxidative degradation and detoxification of textile azo dye by horseradish peroxidase enzyme. Fresen Environ Bull. 2013;22:739–44.

38. Zeng X, Cai Y, Liao X, Zeng X, Li W, Zhang D. Decolorization of synthetic dyes by crude laccase from a newly isolated *Trametes trogii* strain cultivated on solid agro-industrial residue. J Hazard Mater. 2011;187:517–25.

39. Baldrian P. Fungal laccases –occurrenceand properties. FEMS Microbiol Rev. 2006;30:215–42.

40. Ostadhadi-Dehkordi S, Tabatabaei-Sameni M, Forootanfar H, Kolahdouz S, Ghazi-Khansari M, Faramarzi MA. Degradation of some benzodiazepines by a laccase-mediated system in aqueous solution. Bioresour Technol. 2012; 125:344–7.

41. Zhuo R, Ma L, Fan F, Gong Y, Wan X, Jiang M, Zhang X, Yang Y. Decolorization of different dyes by a newly isolated white-rot fungi strain *Ganoderma* sp. En3 and cloning and functional analysis of its laccase gene. J Hazard Mater. 2011;192:855–73.

42. Gholami-Borujeni F, Mahvi AH, Nasseri S, Faramarzi MA, Nabizadeh R, Alimohammadi M. Application of immobilized horseradish peroxidase for removal and detoxification of azo dye from aqueous solution. Res J Chem Environ. 2011;15:217–22.

22

Accumulation of intermediate denitrifying compounds inhibiting biological denitrification on cathode in Microbial Fuel Cell

Abdullah Al-Mamun[1*] and Mahad Said Baawain[2]

Abstract

Background: Bio-cathode denitrifying microbial fuel cell (MFC) is a promising bio-electrochemical system (BES) where both the reactions of anodic oxidation and cathodic reduction are catalyzed by microorganisms. In this nitrogen removal process, a complete biological denitrification from nitrate (NO_3^-) to molecular nitrogen (N_2) was achieved by four reduction steps, forming nitrite (NO_2^-), nitric oxide (NO) and nitrous oxide (N_2O) as intermediate compounds. These enzymatic catalysis reductions are often slowed down on cathode electrode at the higher cathodic nitrate loading. This study investigated the cause for inhibition of the biological denitrification in a three-chambered MFC where the middle chamber acted as denitrifying bio-cathode and the two chambers at the side acted as bio-anode. Carbon fiber brushes were used as electrodes and nafion membranes were used as separator between the chambers.

Results: The maximum power obtained was 14.63 W m^{-3} net cathodic compartment (NCC) (R_{ext} =11.5Ω) at an optimum nitrate loading of 0.15 kg NO_3^--N m^{-3} NCC d^{-1}. The accumulation of one of the intermediate denitrifying compound, e.g., NO_2^- adversely affected biological denitrification rate on cathode. According to chemical kinetics, the accumulated NO_2^- will form free nitrous acid (FNA, HNO_2) in aqueous chemical system spontaneously. The study showed that approximately 45 % of the current production and 20 % of the total denitrification was decreased at a FNA concentration of 0.0014 ± 0.0001 mg $HNO_2 - N$ L^{-1} with an equivalent nitrite concentration of 6.2 ± 0.9 mg $NO_2^- - N$ L^{-1}.

Conclusions: The novel biological process indicates the potential of using denitrifying bio-cathode MFC for green energy production.

Keywords: Microbial fuel cell, Bio-cathode, Biological denitrification, Bioremediation, Process inhibition

Background

Microbial fuel cell (MFC) could be a sustainable hybrid technology for removal of bio-degradable organics and generation of electricity from various types of pure organic substrates and wastewater [1, 2]. In an MFC, the biochemical reactions are catalyzed by electrogenic bacteria on the anaerobic anode surface and the electrons and protons are produced from the degradation of organics.

Concurrently, another set of bio-electrochemical reactions are taken place in the aerobic cathode surface, whereby the anode produced electrons and protons are consumed for reduction [3–6]. Fig. 1 shows the electron harvesting procedure in an MFC. A typical MFC comprises of an anaerobic anode and an aerobic cathode separated by a proton exchange membrane (PEM) or by an electrolytic solution (wastewater) by which the protons diffuse from the anode toward the cathode. The flow of electrons and the potential difference between the respiratory enzymes of anodic microbes and the oxygen reduction reaction on cathode generates current and voltage, respectively [7, 8].

* Correspondence: aalmamun@squ.edu.om
[1]Department of Civil & Architectural engineering, Sultan Qaboos University, Al-Khodh, P.C. 123, P.O. Box 33, Muscat, Sultanate of Oman
Full list of author information is available at the end of the article

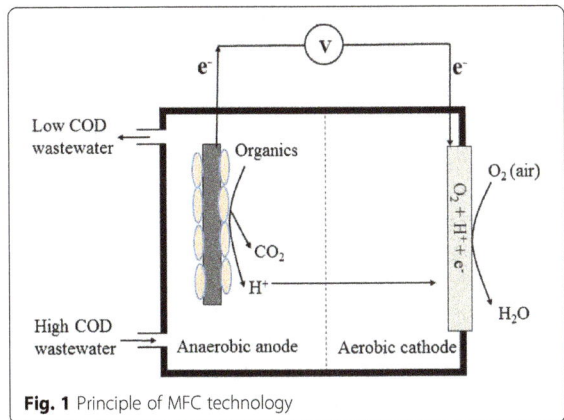

Fig. 1 Principle of MFC technology

In MFC, the most commonly used electron acceptor for cathodic reaction is oxygen due to its abundant availability, higher redox potential and eco-friendly character [9]. Rather than oxygen, some other chemical electron acceptors such as Fe (III), Mn (IV), and ferricyanide or bio-reducing substance such as NO_3^-, SO_4^{2-} have been used by different researchers [10–14].

The added benefits of treating wastewater in MFC compared to that of the conventional activated sludge process are – (a) no requirement of aeration for COD removal, which reduces the power consumption; (b) direct conversion of substrate into electricity without any intermediate steps; (c) low generation rate of sludge; (c) generated electricity can be used to operate other units of the treatment plant [10, 15, 16]. Despite the advantages, the MFC technology still faces serious limitations in terms of large-scale application due to the use of costly catalyst (Pt) coating on cathode electrode and the lack of cost-effective proton exchange membrane or the knowledge of optimum operating conditions [17]. Due to these limitations, a number of issues have to be resolved through research and development before having this technology as a cost-effective alternative for green energy production. Among them, the major concerns are the selection of cheaper electrode materials [18] and the optimization of the reaction processes in order to maximize electrical power output and reduce installation

costs. An MFC operating with a cost-effective bio-cathode would hence be of major interest for practical application of this technology for wastewater treatment. This research investigates the potential application of denitrifying bio-catalyst on cathode electrodes as an eco-friendly alternative of the costly Pt catalyst in MFC.

Denitrifying bio-cathode MFC is a complete bio-electrochemical system (BES) where both the anodic oxidation and the cathodic reduction reactions are catalyzed by microorganisms [19]. The complete biological denitrification from nitrate (NO_3^-) to molecular nitrogen gas (N_2) is a step down reduction reaction forming nitrite (NO_2^-), nitric oxide (NO) and nitrous oxide (N_2O) as intermediate compounds. The biochemistry details of the biological denitrification process are shown in Fig. 2 with the redox potential values and the name of the enzymes at each step [20]. Several studies on biological denitrification with environmental media, such as soil and sediments have demonstrated that the N_2O reductase is inhibited at high oxygen and NO_2^- concentrations or at low pH [21–23]. The inhibition of the N_2O reductase will lead to the accumulation of intermediate denitrifying compounds and decrease the total rate of biological denitrification.

There are several mechanisms involved to inhibit the biological denitrification process. That is why, the dominating mechanism of inhibition to biological denitrification is still ambiguous. One study on the biological nitrogen removal from wastewater have reported that the accumulation of nitrite (10 mg NO_2^-- N L^{-1}) is a possible cause of N_2O release from denitrifying sludge [24]. Some other studies have speculated that the enzymatic competition between the terminal electron acceptors is a cause of inhibition to biological denitrification [25]. Kinetic studies on N_2O reductase using pure bacterial cultures have demonstrated that the activity of nitrous oxide reductase enzyme is not only dependent on accumulated NO_2^- concentration, but also on the pH of the media [26]. According to chemical kinetics, the accumulated NO_2^- will spontaneously form free nitrous acid (FNA) in the aqueous chemical system depending on pH of the aqueous media [27]. Incorporating the concept of FNA generation inside the NO_2^- containing aqueous solution, some recent studies on biological denitrification

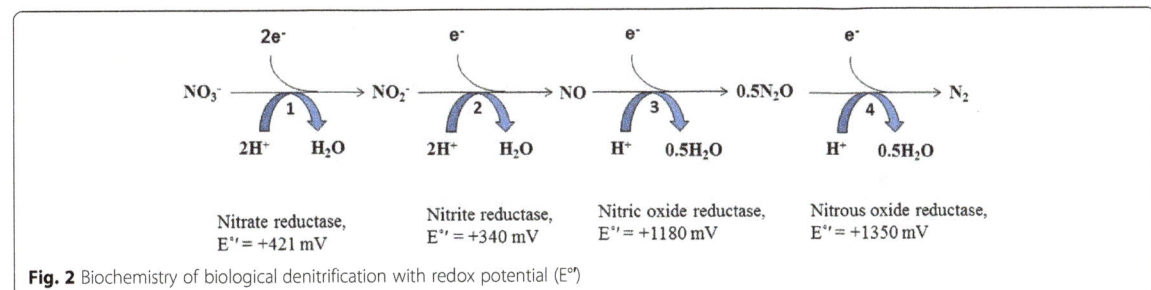

Fig. 2 Biochemistry of biological denitrification with redox potential ($E^{o'}$)

have claimed that FNA inhibits the growth and energy generation into a variety of phylogenetic organisms, including denitrifiers, denitrifying poly-phosphate accumulating organisms [28, 29], ammonia and nitrite oxidizing bacteria [30]. However, all of those studies investigated the inhibitory effects of NO_2^- as well as FNA only on heterotrophic denitrifying microorganisms. None of the previous study has investigated the inhibitory effects of NO_2^- and FNA on autotrophic denitrifying microbes. The main goal of this study is to have an improved understanding of any possible inhibition of the accumulated intermediate denitrifying compounds that takes place on the autotrophic denitrifiers, which act as biocatalysts on the cathode surface of MFC.

The study was carried out to (i) *understand the behavior of autotrophic denitrifying microbes under various cathodic nitrate loading for electricity production in MFC*, (ii) optimize cathodic nitrate loading rate using synthetic acetate solution as electron donor for denitrification, (iii)

determine the attainable electricity and power production based on this cathodic denitrification, (iv) determine the rate of nitrogen removal in cathode chamber, and (v) *determine the optimal conditions for the bio-cathode denitrification without inhibition of the intermediately formed denitrifying compounds.*

Materials and methods
Design of MFC

The autotrophic denitrifying bio-cathode MFC system used in this study was illustrated in Fig.3a showing the flow directions of anodic and cathodic liquid streams, and bio-chemical species. The MFC contained three upflow baffle-channeled chambers. In the assembled MFC, the middle chamber acted as denitrifying bio-cathode and two chambers at the sides acted as bio-anode. The bacteria grown on the anode and cathode electrode surfaces as attached biofilm catalyzed the anodic and cathodic reactions. Acetate containing synthetic wastewater

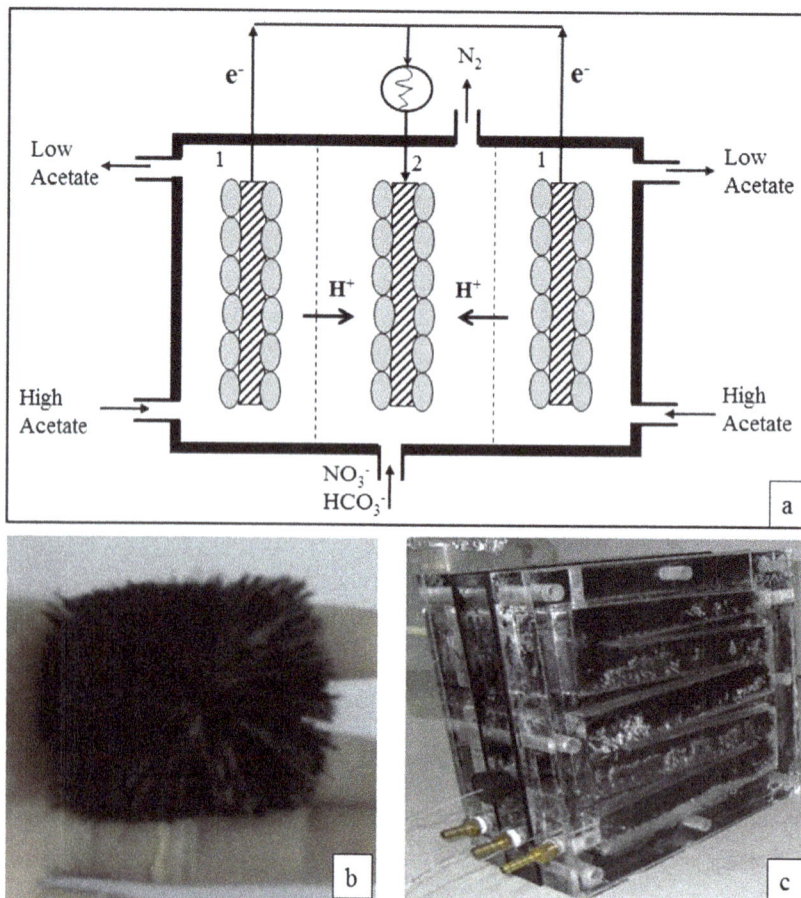

Fig. 3 a Bio-cathode denitrifying MFC system: (1) anode chamber, (2) cathode chamber; **b** The cross-sectional view of the carbon brush electrode; and **c** The whole assembly of the MFC

continuously pumped through the anodic chambers, where bacteria utilized the organic substrate in its metabolic pathways to produce CO_2, protons and electrons. The electrons produced at the metabolic process were collected at the anode electrode and diverted towards the cathode electrode, where autotrophic denitrifying bacteria retrieved them to reduce NO_2^- to N_2 gas. A synthetic wastewater containing nitrate was continuously pumped into the cathode chamber by a peristaltic pump (master flex, USA). Due to a proton motive force, the produced protons in the anode chamber migrated to the cathode surface by means of diffusion within the electrolytes and throughout the cation exchange membrane.

The actual internal design of the baffled channel in each chamber was shown in Fig. 3c. Each channel had a 3-cm square uniform cross-sectional area along the entire length. The midline length of each channel was 0.9 m. The channel was made by thorough cutting of a 3-cm thick acrylic plate with CNC machine. The horizontal portions of the channel were inclined at 4.1° upward direction to facilitate the up-ward movement of produced gas (biogas in anode chamber and N_2 gas in cathode chamber) inside the channel. The internal volume of channel in each chamber was 0.75×10^{-3} cubic meter.

Both the anode and the cathode chambers were filled with a long and square cross-sectional fiber brush electrode. The brush electrode was made of carbon fibers (PANEX35 CONTINUOUS TOW, 50 K, ZOLTEK) with the following properties: tensile strength of 3.8×10^6 KPa, Tensile Modulus of 241×10^6 KPa, and Density of 1.81 g cm^{-3}. The carbon fiber bunches were cut to a designated length and wound using an industrial brush manufacturing system into a twisted core consisting of two stainless steel wires (wire diameter of 0.5 mm). Then the circular cross sectional brush was trimmed to a 3 cm square cross-sectional brush (Fig. 3b) using a hair dressing electric razor so that the brush was completely filled in the 3 cm square cross-sectional up-flow channel. Prior to use, the carbon fibers were treated with ammonia gas as described by Logan et al. [31, 32]. The mass of the carbon fibers used in each channel was 26 ± 1.5 g.

According to the mass of fibers used in each channel of the MFC and an average fiber diameter of 7.2 µm, the estimated brush surface area was 7.98 ± 0.46 m^2 or $10,650 \pm 600$ m^2 m^{-3} of the reactor volume at an attained brush packing density of 7 % in the reactor. The volume of the total cathodic chamber (TCC) is 0.75 L and the liquid volume after packing of brush in the channel, the net cathodic chamber (NCC), was 0.69 L. The total volume of the two anodic chambers (TAC, total anode chambers) was 1.5 L and the volume of the net anodic chamber (NAC) was 1.38 L. Both the anode and cathode chamber was separated by a proton exchange membrane (nafion-112, 2

mill thickness, Gashub Technology). The pretreatment of the PEM was done by sequentially boiling it at 80 °C in H_2O_2 (3 %), deionized water, 0.5 M H_2SO_4 and then deionized water, each for 1 h and then stored in deionized water prior to being used [33].

Bacterial inoculation

Both the anodic and the cathodic liquid streams were circulated (1.25×10^{-3} L min^{-1} or 0.075 L h^{-1}) in an upflow mode for inoculation. In the anode chambers, sieved sewage (355-µm diameter sieve, effluents from primary settling tank, Sultan Qaboos University wastewater treatment plant, Oman) was circulated continuously as a substrate and inoculums during the start-up period by a peristaltic pump (Master flex, USA). The characteristics of domestic wastewater averaged 312 ± 32 mg L^{-1} of total COD, 92 ± 13 mg L^{-1} of soluble COD, 217 ± 64 mg L^{-1} of SS, 195 ± 56 mg L^{-1} of VSS, 405 ± 5 mg L^{-1} of TDS, 38 ± 4 mg L^{-1} of TN and 39 ± 34 mg L^{-1} of PO_4^{3-}. The pH was maintained at 7.2 ± 0.1 (these data shows averages and standard deviations based on a series of 4 samples).

At inoculation stage, the cathodic liquid stream was recirculated in up-flow mode by an external recirculation vessel. The cathodic liquid streams were consisted of a modified buffer medium. The detail composition of the buffer medium [34] and the trace minerals and vitamins solution [35–37] were collected from previous studies on bacterial culturing procedure. Different types of aerobic and anaerobic sludge (activated sludge, digester sludge) and sediments (sediments from primary and secondary settling basins) were mixed in order to obtain cathodic inoculums (2.5 mL of dewatered sludge in 1 L of cathodic liquid stream) with sufficient microbial diversity. A concentrated KNO_3 solution was added into the recirculation vessel twice a day to achieve a desired volumetric loading rate of 0.1 kg NO_3^-- N m^{-3} net cathodic compartment (NCC) d^{-1} (33.33 mg NO_3^--N L^{-1} of buffer solution pumped at a flow rate of 1.25×10^{-3} L min^{-1} or 0.075 L h^{-1}).

Operational conditions

After inoculating the anode electrode with domestic wastewater, the anodic streams were switched to a synthetic acetate solution containing the same modified buffer medium (the buffer medium used in cathode chamber) with sodium acetate. During this operational period, the sole electron donor for the anodic microbes was acetate and the sole electron acceptor for the cathodic microbes was nitrate. The concentration of nitrate and acetate for different cathodic nitrate loading are mentioned in Table 1. Each specific nitrate loading was fed approximately 30 days, keeping the constant feeding

Table 1 Operating conditions when synthetic nitrate and acetate solution were used

[a]Cathodic nitrate loading (kg NO_3^-- N m^{-3} NCC d^{-1})	[a]Acetate concentration in anodic liquid (mg NaAc L^{-1})	Anodic and cathodic liquid flow rate (L min^{-1})	Operating resistances (Ω)	pH of the cathodic liquid stream
0.05	96	2×10^{-3}	10.5	7.0 ± 0.10
0.1	192	2×10^{-3}	10.5	7.0 ± 0.10
0.125	240	2×10^{-3}	10.5	6.7 ± 0.1
0.15	286	2×10^{-3}	10.5	6.3 ± 0.2
0.2	381	2×10^{-3}	10.5	6.0 ± 0.1
0.25	450	2×10^{-3}	10.5	5.8 ± 0.1

[a]Acetate loading to anode chamber was 2.5 times higher than that needed as stoichiometric requirement for nitrate reduction in cathode chamber (stoichiometric ratio of Acetate carbon/ Nitrate nitrogen = 2.5)

of acetate solution at the anodic chamber at an stoichiometric ratio of 2.5 (Acetate carbon/Nitrate nitrogen).

Calculations

Cell potential (E_{cell}, V) was measured with a multimeter connected to a computer by a data acquisition system (PC1604, TTi, RS) at every 30 min interval. Power (P, W) was calculated as: $P = I \cdot E_{cell}$, where current (I, A) was determined according to the ohm's law: $I = E_{cell} / R_{ext}$, and R_{ext} (Ω) is the fixed external resistance. Volumetric power (P_v, W m^{-3} NCC) was determined by $P_v = E_{cell}^2 / (V \cdot R_{ext})$, where V (m^3) is the net volume of cathodic compartment.

The open circuit voltage (OCV) of an MFC was the maximum cell potential generated by the system under infinite resistance (no current). Polarization and power densities were obtained by varying the external circuit resistance from infinity to 1 Ω using a resistor box (RS-201 precision resistance substitute, IET LABS, INC). The cell potential values were recorded only after the pseudo-steady-state conditions had been established. The establishment of this pseudo-steady-state might took several minutes or more, depending on the cathodic nitrate concentration and the external resistance. By changing the external resistance, we obtained a new cell potential, and hence a new current density and power density.

According to Anthonisen et al. [27], the spontaneous generation of FNA was calculated by the concentration of accumulated nitrate (NO_2^-) formed as intermediate compounds in the biological denitrification process, pH, and temperature as follows:

$$\text{FNA as } HNO_2 \left(mg \ L^{-1} \right) = \frac{46}{14} X \frac{NO_2^- - N \left(mg \ L^{-1} \right)}{K_a \ X \ 10^{-pH}} \tag{1}$$

In which, K_a is the ionization constant of the nitrous acid equilibrium equation and it is also varied with temperature [27]. The value of K_a is related to temperature (°C) by

$$K_a = e^{-\frac{2300}{(273 + °C)}} \tag{2}$$

Analytical methods

The concentration of NO_3^-, NO_2^-, PO_4^{3-}, NH_4^+ and SO_4^{2-} in the anodic and cathodic liquid stream were determined by an Ion Chromatograph (DIONEX-500 fitted with GP50 Gradient pump and CD20 conductivity detector) with IonPac CS12A cation and IonPac AS9-HC anion column. In those measurements, samples were first filtered through a 0.2-μm pore sized membrane before analysis. Acetate was analyzed using a gas chromatograph (Shimadzu, AOC-20i) equipped with a FIT detector and a 25 m × 0.32 mm × 0.5 μm HP-FFAP column. Samples were also filtered through a 0.2 μm pore sized membrane and acidified using formic acid before analysis. Produced N_2 gas analyses were performed using a gas chromatograph (GC-17A, Shimadzu) with charlston 80/100 porapak column using Helium gas as carrier. Total nitrogen was measured using a Shimadzu TNM-1 unit coupled with a TOC-V analyzer. In both case, samples were pre-filtered through a 0.2 μm pore sized membrane.

Results and discussion

Effect of different cathodic nitrate loading on current and power generation of MFC

Two similar set of MFCs were operated simultaneously for better understanding of the results. During the operational period, both the MFCs were operated with the synthetic acetate solution as the anodic influent and the synthetic nitrate solution as the cathodic influent. The MFCs were operated at different cathodic nitrate loadings using 10.5 Ω as external resistance. Each combination of substrate loading (i.e., nitrate in cathode chamber and acetate in anode chamber) was operated for 30 days. During each combination of substrate loading, the current production and the denitrification rate was increased gradually with the increase of microbial population inside the MFC reactor and subsequently, reached to a saturated value. The saturation conditions were achieved by approximately 18 to 20-day continuous feeding of substrate at every specific substrate loading. Here the recorded current production and denitrification rates were the saturated values for each specific substrate loading. Fig. 4 showed the saturated current generation profiles with the gradual increase of specific cathodic nitrate (substrate for the cathodic denitrifying bacteria) loading from 0.05-0.25 kg NO_3^-- N m^{-3} NCC d $^{-1}$. The maximum saturated current production obtained in this bio-cathode MFC system at an external resistance of 10.5 Ω was 53.8 ± 1.6 A m^{-3} NCC with 0.15 kg NO_3^--N m^{-3} NCC d^{-1} of cathodic nitrate loading. The current production rate was slightly higher than that found by Virdis et al., 2008, 2009 [19, 38].

The current generation profiles showed that the rate of current production was decreased instead of increasing at a cathodic nitrate loading of more than 0.15 kg NO_3^-- N m^{-3} NCC d^{-1}. A detailed study on why current production was decreased at higher cathodic nitrate loading was investigated by monitoring the accumulation of possible intermediate denitrifying compounds. Fig. 4 showed the inhibitory effect of the intermediate denitrifying product - nitrite (NO_2^-), which was formed during the denitrification and current generation process of the denitrifying bio-cathode MFC. The formation of nitrite is associated with the formation of free nitrous acid (FNA, the protonated form of nitrite ion, HNO_2) in the aqueous chemical system spontaneously. A recent study revealed that the formation of FNA was enhanced

at the higher acidic conditions of the aqueous system [39]. The similar findings were also observed in this study. Table 1 showed that the pH values of the cathodic liquid stream were neutral (pH = 7.0 ± 0.10) at lower nitrate concentrations and it was dropped gradually for the nitrate concentration of more than 0.1 kg NO_3^-- N m $^{-3}$ NCC d^{-1}. These data confirmed that the acidity of the cathodic liquid stream was increased with the increase of the cathodic nitrate-loading rate for more than 0.1 kg NO_3^-- N m^{-3} NCC d^{-1}. Fig. 4 showed that the formation of FNA increased exponentially at nitrate concentration of more than 0.1 kg NO_3^-- N m^{-3} NCC d^{-1}. Therefore, the total understanding of these two data confirmed that the increased acidity level of the cathodic liquid stream at higher cathodic nitrate loading rate was enhanced the exponential formation of FNA. The formation of nitrite (NO_2^-, denitrifying product) at the lower cathodic nitrate-loading rate did not show any inhibition to the microbial process, but the formation of FNA at higher cathodic nitrate loading inhibited the microbial process on cathode surface. These findings demonstrated that the observed degree of inhibition correlated much more strongly with the FNA, rather than nitrite concentration, indicating FNA as the true inhibitor on the activity of denitrifying microbes. Similar inhibitory effects of FNA were found to impart a wide range of phylogenetic types of bacteria, including both ammonia and nitrite oxidizing bacteria [27, 30, 40], denitrifiers and also denitrifying enhanced phosphorus removal organisms [28, 41]. The results showed that both the current generation and the

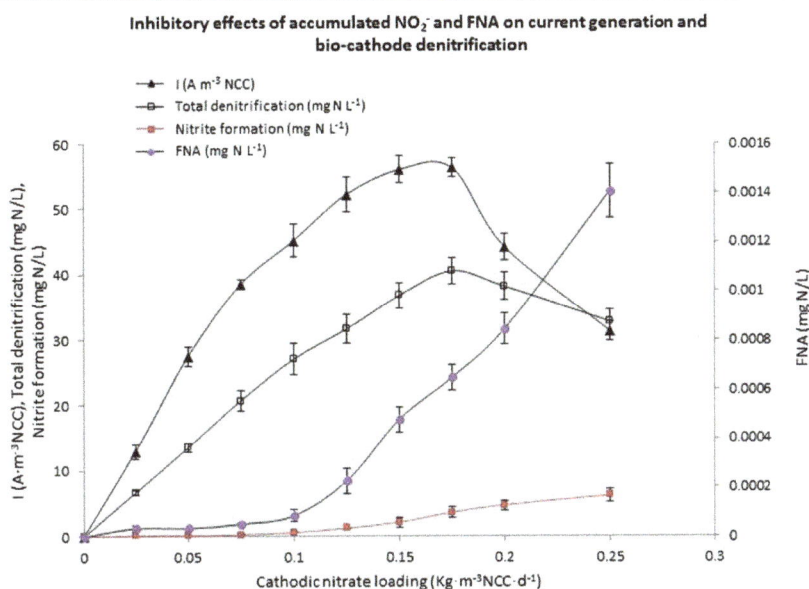

Fig. 4 The inhibitory effect of intermediate denitrifying products, nitrite as well as FNA on the current generation and bio-cathode denitrification using 10.5 Ω fixed external resistance (Results showing as averages and standard deviations of 3 samples)

denitrification activity were decreased at a cathodic nitrate loading rate of more than 0.175 kg NO_3^--N m^{-3} NCC d^{-1}. Approximately 45 % of the current production and 20 % of the denitrification activity in this bio-cathode MFC was decreased at a FNA concentration of 0.0014 ± 0.0001 mg HNO_2-N L^{-1} (equivalent to the nitrite concentration of 6.2 ± 0.9 mg NO_2^--N L^{-1} at a pH of 7 ± 0.1). The results showed that the autotrophic denitrifying bacteria is more tolerant than that of the heterotrophic denitrifying bacteria as the FNA concentration of 0.0014 ± 0.0001 mg HNO_2-N L^{-1} caused only 20 % decrease of the autotrophic denitrification in the present study, whereas the FNA concentration of 0.0007-0.001 mg HNO_2-N L^{-1} caused 50 % decrease of the heterotrophic dentrification [41].

Polarization and Power generation

Table 2 shows that the maximum power production increased gradually with a gradual increase of initial cathodic nitrate loading. But the initial cathodic nitrate loading of more than 0.15 kg NO_3^--N m^{-3} NCC d^{-1} could not produce higher power. This is due to the inhibitory effect by the intermediately formed denitrifying compounds (nitrite and FNA) to the denitrifying organisms on bio-cathode.

The maximum volumetric power obtained was 14.63 W · m^{-3} NCC ($R_{ext} = 11.5 \ \Omega$) at a cathodic nitrate loading of 0.15 kg NO_3^--N m^{-3} NCC d^{-1}, indicating the potential of using denitrifying bio-cathode MFC for energy production. The obtained maximum power was approximately 83 % higher than that obtained by Clauwaert et al. [34]. The higher maximum power obtained was due to the special internal design of the MFC reactor that achieves a plug flow condition. The cathodic nitrate loading beyond 0.15 kg NO_3^--N m^{-3} NCC d^{-1} could not produce more power indicated the limitations of acetate as electron donor for biological denitrification.

Conclusions

In this study, a bio-cathode MFC was investigated to achieve complete biological denitrification on cathode surfaces by the help of electrons supplied from the organics oxidation during synthetic wastewater treatment

at anode surfaces. The prime goal of this study was to explore the feasibilities of a continuous up-flow baffled channel MFC using carbon fiber brush electrode for treating synthetic wastewater in anode and biological nitrogen removal (denitrification) in cathode with attainable electricity and power production. The results of the experiments performed over 6 months of operation with acetate demonstrated the feasibilities of combined carbon and nitrogen removal process with electrical energy recovery. The optimum cathodic nitrate loading corresponds to the electrons supplied from the COD removal from synthetic acetate solution was 0.15 kg NO_3^--N m^{-3} NCC d^{-1} indicating that the bio-cathode denitrification is relatively dependent on the electron supplying capacity of the anodic substrate. Hence, the anodic substrate has a limit to act as electron donor for this type of autotrophic denitrification. At the same time, the MFC reactor design and the selection of materials should be optimized to achieve sufficient microbial density at the vicinity of the MFC electrode to harvest maximum electrons. The highest volumetric power obtained was 14.63 W m^{-3} NCC with concurrent denitrification of 148.3 ± 1.4 g N m^{-3} NCC d^{-1}. This power generation rate was significantly higher than that found by the similar studies using granular graphite electrode, indicating the good internal design of this MFC reactor.

The initial nitrate loading to the cathodic bio-film has an effect on the denitrification and current generation process. The inhibitory effect of the intermediate denitrifying product- nitrite (NO_2^-) as well as FNA was observed. The current production and the denitrification activity of this bio-cathode MFC system was reduced approximately 45 % and 20 % respectively, at a FNA concentration of 0.0014 ± 0.0001 mg HNO_2-N L^{-1} (equivalent to the nitrite concentration of 6.2 ± 0.9 mg NO_2^--N L^{-1} at a pH of 7 ± 0.1). However, this study shows the practical relevance of a novel denitrification process using MFC even though it has not been fully optimized at this stage. Future investigation might be required to understand the reductase enzyme activity for the step down biological denitrification and the possible electron competitions for the reductase enzyme under respective redox potential.

Table 2 Maximum OCV, current and power obtained at different cathodic nitrate loadings

Cathodic nitrate loading (kg NO_3^--N m^{-3} NCC d^{-1})	Maximum power (W m^{-3} NCC)	Maximum OCV (V)	External R at max^m power (Ω)
0.050 (20.83 mg NO_3-N L^{-1})	6.92	0.492	18.5
0.100 (20.83 mg NO_3-N L^{-1})	9.56	0.523	18.5
0.125 (26.05 mg NO_3-N L^{-1})	13.52	0.599	11.5
0.150 (31.25 mg NO_3-N L^{-1})	14.63	0.624	11.5
0.200 (41.67 mg NO_3-N L^{-1})	11.79	0.603	14.5

Abbreviations

MFC: Microbial fuel cell; BES: Bio-electrochemical system; NCC: Net cathodic compartment; FNA: Free nitrous acid; TCC: Total cathodic compartment; TAC: Total anodic compartment; NAC: Net anodic compartment; OCV: Open circuit voltage.

Competing interests

I confirm that I have read BioMed Central's guidance on competing interests. None of the authors have any competing interests in the manuscript.

Authors' contributions

Dr. Abdullah Al-Mamun is the principle investigator of the project. His contributes were to design and fabrication of the MFC reactor, operating the MFC, collecting data, analyzing data, drafting the article.
Dr. Mahad S. Baawain is the co-principle investigator of the project. His contributes were analyzing the data and reviewing the draft for final submission. Both authors received and approved the final manuscript.

Authors' information

Dr. Abdullah Al-Mamun
Ph. D. (Civil and Environmental Engineering, National University of Singapore, Singapore)
Assistant Professor, Dept. of Civil and Architectural Engineering, Sultan Qaboos University, Muscat, Oman
Ex-faculty, Singapore Republic Polytechnic
Ex-Research fellow, National University of Singapore
Member: American Chemical Society
Member: Singapore Water Association
Dr. Mahad Said Baawain
Ph. D. (Civil and Environmental Engineering, University of Alberta, Canada)
Associate Professor, Dept. of Civil and Architectural Engineering, Sultan Qaboos University, Muscat, Oman
Director, Centre for Environmental Studies and Research, Sultan Qaboos University, Muscat, Oman

Acknowledgements

We express our gratitude to Mr. Nafeer Alhuda, Mr. Induka Nilubul and Mr. Hamood for their technical assistance during fabrication of Microbial fuel cells used in this study. Thanks to Miss Raya Al-Sharji and Mr. Fauzul Marikar to assist us in operating the equipments used for different analysis. Sultan Qaboos University Internal Grant (CR/ENG/CIVL/11/01), Oman, supported the work.

Author details

¹Department of Civil & Architectural engineering, Sultan Qaboos University, Al-Khodh, P.C. 123, P.O. Box 33, Muscat, Sultanate of Oman. ²Centre for Environmental Studies and Research, Sultan Qaboos University, Al-khodh, P.C. 123, Muscat P.O. Box 17, Muscat, Sultanate of Oman.

References

1. Kim BH, Kim HJ, Hyun MS, Park DH. Direct electrode reaction of Fe (III))-reducingbacterium Shewanella putrefaciens. J Microbiol Biotechnol. 1999; 9(9):127–31.
2. Rabaey K, Clauwaert P, Aelterman P, Verstraete W. Tubular microbial fuel cell for efficient electricity generation. Environ Sci Technol. 2005;39:8077–82.
3. Bond DR, Holmes DE, Tender LM, Lovley DR. Electrode-reducing microorganisms that harvest energy from marine sediments. Science. 2002; 295(5554):483–5.
4. Liu H, Logan BE. Electricity generation using an air-cathode single chamber microbial fuel cell in the presence and absence of a proton exchange membrane. Environ Sci Technol. 2004;38:4040–6.
5. Lefebvre O, Shen Y, Tan Z, Uzabiaga A, Chang IS, Ng HY. Full-loop operation and cathode acidification of a microbial fuel cell operated on domestic wastewater. Bioresour Technol. 2011;102:5841–8.
6. Lefebvre O, Nguyen TT, Al-Mamun A, Chang IS, Ng HY. T-RFLP reveals high β-Proteobacteria diversity in microbial fuel cells enriched with domestic wastewater. J Appl Microbiol. 2010;109(3):839–50.
7. Logan BE, Hamelers B, Rozendal R, Schröder U, Keller J, Freguia S. Microbialfuel cells: methodology and technology. Environ Sci Technol. 2006;40:5181–92.
8. Lefebvre O, Al-Mamun A, Ooi WK, Tang Z, Chua DH, Ng HY. An insight into cathode options for microbial fuel cells. Water Sci Technol. 2008;57(12): 2031–7.
9. Tran HT, Ryu JH, Jia YH, Oh SJ, Choi JY, Park DH, et al. Continuous bioelectric-ity production and sustainable wastewater treatment in a microbial fuel cell constructed with non-catalyzed granular graphite electrodes and permeable membrane. Water Sci Technol. 2010;61:1819–27.
10. Rabaey K, Lissens G, Siciliano SD, Verstraete W. A microbial fuel cell capable of converting glucose to electricity at high rate and efficiency. Biotechnol Lett. 2003;25(18):531–1535.
11. Rabaey K, Ossieur W, Verhaege M, Verstraete W. Continuous microbial fuel cells convert carbohydrates to electricity. Water Sci Technol. 2005;52(1–2): 515–23.
12. Rhoads A, Beyenal H, Lewandowski Z. Microbial fuel cell using anaerobic respiration as an anodic reaction and biomineralized manganese as a cathodic reactant. Environ Sci Technol. 2005;39:4666–71.
13. Shantaram A, Beyenal H, Raajan R, Veluchamy A, Lewandowski Z. Wireless sensors powered by microbial fuel cells. Environ Sci Technol. 2005;39:5037–42.
14. Lefebvre O, Al-Mamun A, Ng HY. A microbial fuel cell equipped with a biocathode for organic removal and denitrification. Water Sci and Tech. 2008;58(4):881–5.
15. Jang JK, Pham TH, Chang IS, Kang KH, Moon H, Cho KS, et al. Construction andoperation of a novel mediator- and membrane-less microbial fuel cell. Process Biochem. 2004;39:1007–12.
16. Clauwaert P, Aelterman P, Pham TH, Schamphelaire LD, Carballa M, Rabaey K, et al. Minimizing losses in bio-electrochemical systems: the road to applica-tions. Appl Microbiol Biotechnol. 2008;79:901–13.
17. Li Z, Yao L, Kong L. Electricity generation using a baffled microbial fuel cell convenient for stacking. Bioresour Technol. 2008;99:1650–5.
18. Lefebvre O, Ooi WK, Tang Z, Al-Mamun A, Chua D, Ng HY. Optimization of a Pt-free cathode suitable for practical applications of microbial fuel cells:. Bioresour Technol. 2009;1000:4907–10.
19. Virdis B, Rabaey K, Yuan Z, Rozendal RA, Keller J. Electron fluxes in a microbial fuel cell performing carbon and nitrogen removal. Environ Sci Technol. 2009;43:5144–9.
20. B.H. Kim and G.M. Gadd, Bacterial Physiology and Metabolism, Cambridge University Press, 2007, pp. 302–303.
21. Fujita K, Dooley DM. Insights into the mechanism of N_2O reduction by reductively activated N_2O reductase from kinetics and spectroscopic studies of pH effects. Inorg Chem. 2007;46(9):613–5.
22. Schulthess RV, Kuhni M, Gujer W. Release of nitric and nitrous oxides from denitrifying activated sludge. Water Res. 1995;29:215–26.
23. Shiskowski DM, Mavinic DS. The influence of nitrite and pH (nitrous acid) on aerobic-phase, autotrophic N_2O generation in a wastewater treatment bioreactor. J Environ Eng Sci. 2006;5:273–83.
24. Itokawa H, Hanaki K, Matsuo T. Nitrous oxide production in high-loading biological nitrogen removal process under low COD/N ratio condition. Water Res. 2001;35:657–64.
25. Dendooven L, Erson JM. Dynamics of reduction enzymes involved in the denitrification process in pasture soil. Soil Biol Biochem. 1994;26:1501–6.
26. Gorelsky SI, Ghosh S, Solomon E. Mechanism of N_2O reduction by the μ(4)-S tetranuclear Cu-z cluster of nitrous oxide reductase. J Am Chem Soc. 2006; 128:278–90.
27. Anthonisen AC, Loehr RC, Prakasam TBS, Shinath EG. Inhibition of nitrification by ammonia and nitrous acid. J Water Pollut Control Fed. 1976; 48:835–52.
28. Zhou Y, Pijuan M, Yuan ZG. Free nitrous acid inhibition on anoxic phosphorus uptake and denitrification by poly-phosphate accumulating organisms. Biotechnol Bioeng. 2007;98:903–12.
29. Zhou Y, Ganda L, Lim M, Zhiguo Y, Taffan K, Ng WJ. Free nitrous acid (FNA) inhibition on denitrifying poly-phosphate accumulating organisms (DPAOs). Appl Microbiol Biotechnol. 2010;88:359–69.
30. Vadivelu VM, Yuan Z, Fux C, Keller J. The inhibitory effects of free nitrous acid on the energy generation and growth processes of an enriched nitrobacter culture. Environ Sci Technol. 2006;40:4442–8.
31. Logan B, Cheng S, Valerie W, Garett E. Graphite fiber brush anodes for increased power production in air cathode microbial fuel cells:. Environ Sci Technol. 2007;41:3341–6.

32. Cheng S, Logan BE. Ammonia treatment of carbon cloth anodes to enhance power generation of microbial fuel cells:. Electrochemical Communications. 2007;9:492–6.

33. Ghasemi M, Wan Daud WR, Ismail M, Rahimnejad M, Ismail AF, Leong JX, et al. Effect of pre-treatment and biofouling of proton exchange membrane on microbial fuel cell performance. Int J Hydrog Energy. 2013;38(13):5480–4.

34. Clauwaert P, Rabaey K, Aelterman P, De Schamphelaire L, Ham TH, Boeckx P, et al. Biological denitrification in microbial fuel cells. Environ Sci Technol. 2007;41(9):3354–60.

35. Balch WE, Fox GE, Magrum LJ, Woese CR, Wolfe RS. Methanogens: reevaluation of a unique biological group. Microbiol Rev. 1979;43(2):260–96.

36. Lovley DR, Greening RC, Ferry JG. Rapidly Growing Rumen Methanogenic Organism That Synthesizes Coenzyme-M And Has A High-Affinity For Formate. Appl Environ Microbiol. 1984;48(1):81–7.

37. Lovley DR, Phillips EJP. Novel mode of microbial energy metabolism: organic carbon oxidation coupled to dissimilatory reduction of iron or manganese. Appl Environ Microbiol. 1988;54(6):1472–80.

38. Virdis B, Rabaey K, Yuan Z, Keller J. Microbial fuel cells for simultaneous carbon and nitrogen removal. Water Res. 2008;42(12):3013–24.

39. Jiang G, Gutierrez O, Yuan Z. The strong biocidal effect of free nitrous acid on anaerobic sewer biofilms. Water Res. 2011;45:3735–43.

40. Vadivelu VM, Yuan Z, Fux C, Keller J. Stoichiometric and kinetic characterization of nitrobacter in mixed culture by decoupling the growth and energy generation processes: Biotechnol. Bioeng. 2006

41. Zhou Y, Pijuan M, Zeng RJ, Yuan Z. Free nitrous acid inhibition on nitrous oxide reduction by a denitrifying enhanced biological phosphorus removal sludge. Environ Sci Technol. 2008;42(22):8260–5.

23

Assessment of honking impact on traffic noise in urban traffic environment of Nagpur

Ritesh Vijay[1*], Asheesh Sharma[1], Tapan Chakrabarti[2] and Rajesh Gupta[2]

Abstract

Background: In context of increasing traffic noise in urban India, the objective of the research study is to assess noise due to heterogeneous traffic conditions and the impact of honking on it.

Method: Traffic volume, noise levels, honking, road geometry and vehicular speed were measured on national highway, major and minor roads in Nagpur, India.

Results: Initial study showed lack of correlation between traffic volume and equivalent noise due to some factors, later identified as honking, road geometry and vehicular speed. Further, frequency analysis of traffic noise showed that honking contributed an additional 2 to 5 dB (A) noise, which is quite significant. Vehicular speed was also found to increase traffic noise. Statistical method of analysis of variance (ANOVA) confirms that frequent honking ($p < 0.01$) and vehicular speed ($p < 0.05$) have substantial impact on traffic noise apart from traffic volume and type of road.

Conclusions: The study suggests that honking must also be a component in traffic noise assessment and to identify and monitor "No Honking" zones in urban agglomerations.

Keywords: Honking, Traffic noise, Vehicular speed, Traffic volume

Background

Noise pollution, a by-product of urbanization and industrialization, is now recognized as a major problem in urban areas with many adverse health effects [1-4]. The most important factors raising noise pollution in urban areas are vehicular traffic, railway and air traffic [5,6]. Vehicular traffic contributes to about 55% of the total urban noise [7-9]. The need for studies regarding urban noise pollution and its consequences on the environment has motivated various researchers in several counties including India [10-12]. Most cities in India have been facing serious noise pollution problems in the last few decades due to substantial growth in the number of vehicles, expansion of road network, industrialization and urbanization [13-15].

Assessment of traffic noise pollution is not easy and varies with types and physical conditions of vehicles, speed, honking and road geometry [16,17]. Estimation of traffic noise is more difficult in Indian cities considering the heterogeneity in traffic conditions including mixed vehicle types, congestion, road conditions, frequent honking and lack of traffic sense [18-20]. Honking is a common occurrence in India, irrespective of road types and condition, traffic etc. [21]. Driving attitude which includes impatience, over accelerating, sudden braking, abiding traffic rules etc. may also aggravate honking. Kalaiselvi and Ramachandraiah found that horn noise events increase equivalent noise level (L_{eq}) 2 to 13 dB(A) [18,21]. Therefore, there is a need to consider such diverse factors in monitoring and assessment of traffic noise as well as planning of noise abatement measures. The objective of the study is to assess and quantify traffic noise and the impact of honking on it in the urban environment of Nagpur, India. The study will help in defining new 'No Honking' zones in addition to assessing traffic noise and existing horn prohibited areas.

Material and method

The methodology of the present study is elaborated in following sections.

* Correspondence: r_vijay@neeri.res.in
[1]Environmental System Design and Modeling Division, CSIR-NEERI, Nagpur 440020, Maharashtra, India
Full list of author information is available at the end of the article

Study area

Traffic volume, noise levels, spot speed and honking were measured at three sampling locations in the study area during March 2010 – December 2010. The study area lies between 21° 7' 0" to 21° 7' 45" N latitude and 79° 4' 0" to 79° 4' 45" E longitude in Nagpur City, Maharashtra, India (Figure 1). The study area comprises of three main roads namely Wardha road, South-Ambazari road and NEERI road. These are classified as national highway, major and minor roads respectively. Road details including geometry, category, number of traffic lanes and road conditions are considered in the study. The width of national highway, major and minor roads is 21 m, 15 m and 7 m respectively. Road conditions were almost same for all roads with asphalt surface and footpaths on both sides. Road divider separates the flow of mixed traffic at highway (six lanes) and major

road (four lanes) whereas minor road doesn't have any divider.

Data collection

Traffic volume studies were conducted to determine the number, movements, and classification of vehicles at a given location and sampling period. Traffic volume was recorded using video camera and vehicles were counted by viewing recorded footages from cameras on computer system. Vehicles were classified as heavy (truck, bus, bulldozer, trailer, dumper), medium (car, jeep, auto-rickshaw, loading rickshaw) and light (motorcycle, scooter) based on their size and noise emission level. Auto-rickshaw is a three wheeler used as a common means of transportation in India. Noise emitted by traffic vehicles was measured as per standard methods [22,23] using sound level meter [24]. Sound level meter was mounted on a tripod stand

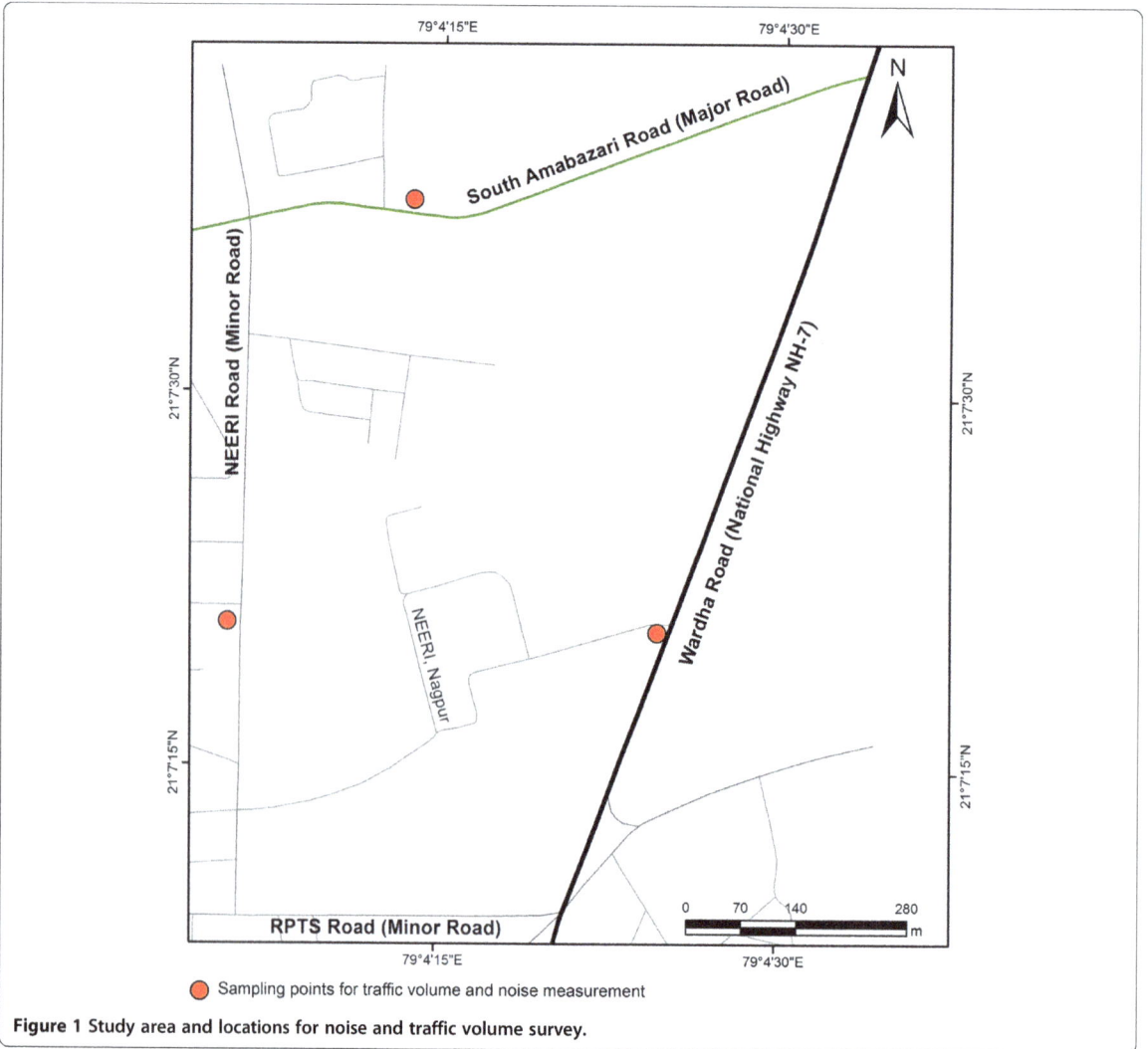

Figure 1 Study area and locations for noise and traffic volume survey.

1.5 m above ground level with slow response mode, frequency weighting "A" and data logging of 1 second time interval. Traffic noise was measured using sound level meter at a distance of 12 m, 10 m and 5 m from the center of national highway, major and minor roads respectively. Similarly, speedometer (Speedet Traffic Radar) was mounted on tripod stand for monitoring speed of

vehicles [25]. Noise emitted from a particular vehicle with corresponding speed was also measured and analyzed for noise-speed response.

Data analysis
An attempt has been made to analyze traffic volume, vehicle speed and honking with their corresponding

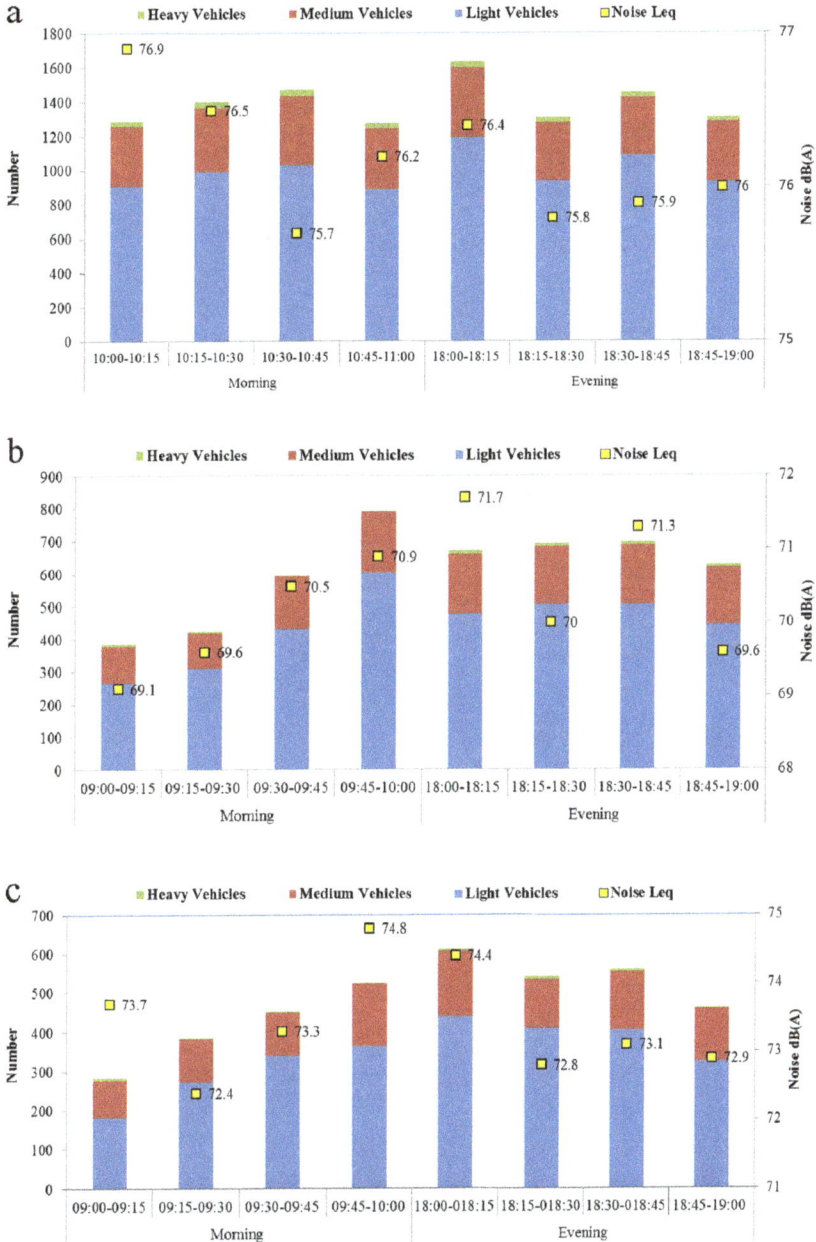

Figure 2 First set of data for traffic and noise during morning and evening peak hours a) National highway b) Major road and c) Minor road.

noise levels. Initially, traffic volume was monitored for 24 hours to identify peak traffic hours in morning and evening. Later, two sets of traffic volume and noise data were monitored during morning and evening peak traffic hours. In the first set of data, traffic and noise levels were measured for 1 hour with 15 minutes time interval while in the second set, honking along with traffic and noise level were measured for 15 minutes with time interval of 1 minute duration. Measured noise data in two sets of readings were analyzed for equivalent (L_{eq}), minimum (L_{min}) and maximum (L_{max}) noise levels. L_{eq} was further analyzed in each time step to assess the impact of honking using frequency component of traffic noise recorded in sound level meter [26]. A statistical analysis was performed to assess the impact of diverse conditions on traffic noise based on the relationship between traffic volume, road geometry and noise data [27]. For this, analysis of variance (ANOVA) and correlation

analysis were carried out to quantify the dependence of traffic volume - equivalent noise, honking - equivalent noise and vehicular speed - corresponding noise level.

Results

Based on the analysis of 24-hour traffic volume, peak traffic flows were observed between 10:00 and 11:00 in case of highway and between 9:00 and 10:00 for major and minor roads in the morning. The number of light, medium and heavy vehicles passing through the highway were 3605, 1427 and 171, respectively during morning peak hour. The observed light, medium, and heavy vehicles on major road were 2338, 612 and 11, respectively while on minor road these values were 1587, 585 and 9, respectively. Similarly, peak traffic flow was observed between 18:00 and 19:00 for all categories of roads in the evening. Number of light, medium and heavy vehicles were 3552, 1663 and 138 at highway, 1861, 754 and

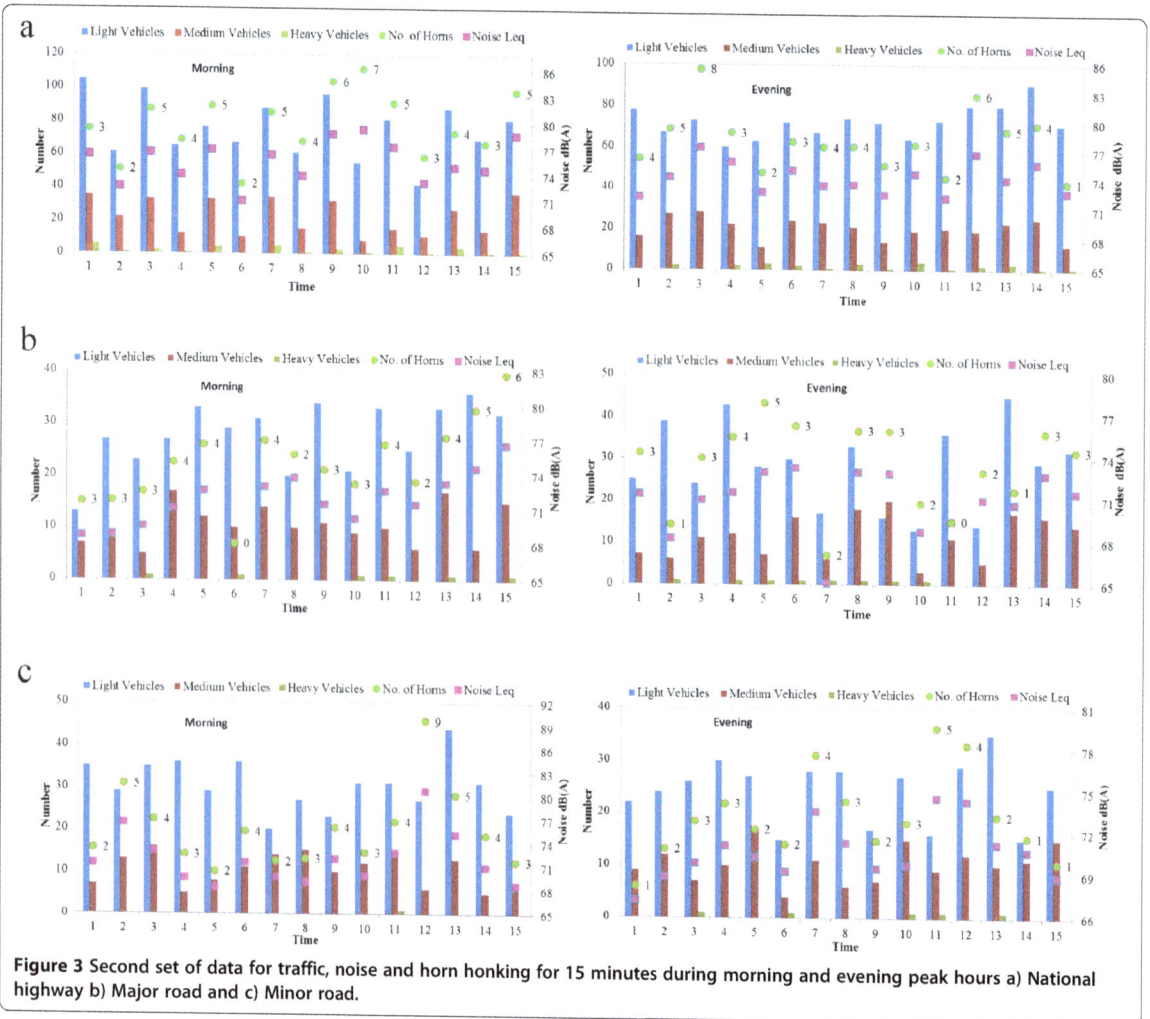

Figure 3 Second set of data for traffic, noise and horn honking for 15 minutes during morning and evening peak hours a) National highway b) Major road and c) Minor road.

27 at major road and 1528, 611 and 8 at minor road, respectively.

To assess the impact of traffic on noise levels, peak hour's traffic and noise levels were measured for 15 minutes interval (Figure 2a, b and c) in first set of data. As per reviewed literature, noise is directly proportional to traffic volume which means that traffic noise increases with increase in traffic volume [28]. However some conflicting results were observed in the present study. For example, at highway, lowest L_{eq} [75.7 dB(A) during 10.30 to 10.45] was noted corresponding to maximum traffic volume and highest L_{eq} [76.9 dB(A) during 10.00 to 10.15] was not corresponding to maximum traffic volume during morning hour (Figure 2a); at major road, highest L_{eq} did not correspond to maximum traffic volume in evening peak hour (Figure 2b) and at minor road, lowest L_{eq} did not correspond to minimum traffic volume in morning and evening (Figure 2c). However, L_{eq} observations conformed to literature findings at highway for highest and lowest L_{eq} during evening peak hour (Figure 2a), at major road for highest L_{eq} during morning and lowest L_{eq} during evening (Figure 2b) and at minor road for highest L_{eq} in morning and evening (Figure 2c). These results show mixed trends between traffic volume and equivalent noise.

As per aforementioned discussion, no statistical relationship could be found between traffic volume and noise level. This suggests that besides traffic volume, other factors are also responsible for contributing noise

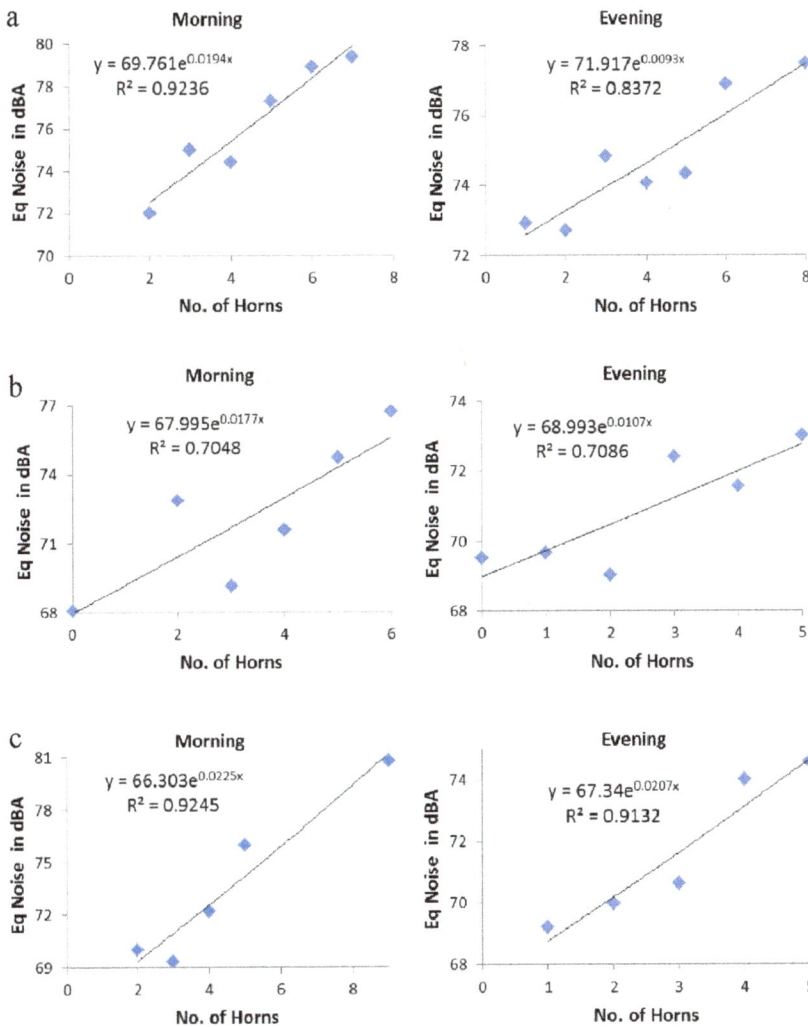

Figure 4 Relationship between horn honking and equivalent traffic noise during morning and evening peak hours a) National highway b) Major road and c) Minor road.

Table 1 Analysis of variance for honking and type of road on traffic noise

Parameters	Degree of freedom	Sum of squares	Mean square	F	p
Honking	1	42.2	42.2	72.8	0.001
Type of road	2	36.8	18.4	31.8	0.001
Interaction	2	0.4	0.2	0.3	0.735
Error	6	3.5	0.6		
Total	11	82.8			

$S = 0.7$; $R^2 = 95.8\%$.

[19]. To identify factors responsible for traffic noise assessment, a second set of data comprising of equivalent noise, traffic volume and honking was collected. These data were collected for 15 minutes duration with one minute time interval in peak traffic hours (Figure 3). Highest L_{eq} [79.4 dB(A)] was observed in 10th minute for least number of vehicles (Figure 3a) during morning at highway. This was due to maximum number of honking recorded. The maximum traffic volume was recorded in 1st minute even though its L_{eq} [76.4 dB(A)] was not the highest. Although traffic volume recorded in 15th minute was lesser than 1st minute, noise level was more due to more number of honking. Further, for same number of horns, noise level in 2nd minute was more than 6th minute due to heavy vehicle. Similar results were observed in the case of 1st and 14th minutes. For same traffic volume in 6th and 8th minutes, L_{eq} was higher in 8th minute due to combined effect of heavy vehicle and honking. Similar scenario of traffic noise was observed in evening peak hour at highway. Highest L_{eq} [77.5 dB(A)] was observed due to eight honking incidents recorded in 3rd minute although traffic volume was not maximum (Figure 3a).

In case of major road, highest L_{eq} [76.7 dB(A)] was observed for 15th minute with most honking while lowest L_{eq} [68.07 dB(A)] was observed in 6th minute with no honking in the morning (Figure 3b). For same number of honks and traffic volume in 4th and 11th minutes, L_{eq} in 11th minute was more due to presence of heavy vehicle. Though L_{eq} levels during 5th and 7th minutes were different, same number of horn incidents and traffic volume was observed. This variation may have been due to vehicle type, its physical condition and speed. Some contrasting results were observed at 5th and 6th minutes

during evening (Figure 3b). For example, highest L_{eq} was observed in 6th minute even though horn incidents were not recorded maximum.

Traffic and noise data on minor road during morning indicate that highest L_{eq} [80.8 dB(A)] was observed in 12th minute with maximum number of horn incidents although traffic volume was not maximum (Figure 3c) while lowest L_{eq} [68.4 dB(A)] was observed in 5th minute with least number of horn incidents. Further, noise level was more in 11th minute as compared to 6th minute with same number of honking and traffic volume due to the presence of heavy vehicle. In evening peak hour, highest L_{eq} [74.6 dB(A)] was observed at 11th minute with maximum number of horns (Figure 3c).

Discussion

Second set of data suggests that honking and heavy vehicles moving on the roads have significant impact on traffic noise as compared to light and medium vehicles. In order to assess the impact of honking on traffic noise, L_{eq} and number of horns were plotted for each category of roads (Figure 4a to c). Average equivalent noise was calculated where equal number of horns was observed for every time step. A strong correlation was observed at highway and minor road while moderate relationship was observed at major road. The correlation coefficients were in the range of 0.84 to 0.97 ($p < 0.05$) suggesting that honking has significant impact on traffic noise, besides traffic volume. Some contradictory results were observed for some time steps where lesser number of horns produced more noise. This requires further analysis.

Further, for quantification of sound level due to honking, Type-I sound level meter was used to

Table 2 Equivalent noise without honking as per statistical and frequency analysis

Type of road	Traffic noise L_{eq} dB(A)		Honking (no)		L_{eq} dB(A) without honking			
					Statistical		Frequency	
	Morn	Even	Morn	Even	Morn	Even	Morn	Even
National Highway	76.6	74.7	63	57	69.8	71.9	72.2	72.1
Major	72.4	71.4	50	37	68.0	69.0	68.1	68.2
Minor	73.6	71.2	57	38	66.3	67.3	69.4	68.9

Morn – morning, even - evening.

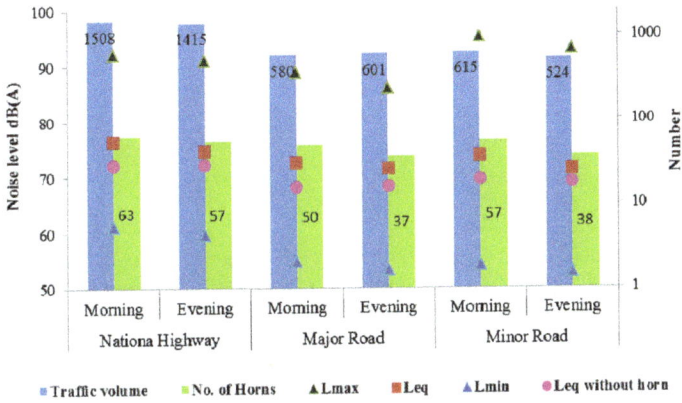

Figure 5 Summarized 15-minutes traffic volume, horn and noise levels.

measure traffic noise under different frequency components distributed in the highest and lowest octaves at 16 Hz and 16 kHz in eleven octaves. Response of honking was observed mostly in the octaves of 500 Hz, 2 kHz and 4 kHz. The logarithmic addition of eleven octaves except these three octaves provides the traffic noise without honking [25]. Honking is responsible for an additional noise of 4 to 5 dB(A) during morning and 2 to 4 dB(A) in evening hour over and above traffic noise for each category of roads which was confirmed using two-way ANOVA as per Table 1. $p < 0.01$ for both independent variables i.e. honking and types of road, indicates that there is a significant impact of these variables on the response of traffic noise level [Confidence Interval (CI) 95%]. A comparison for traffic noise with and without honking was carried out based on the statistical data analysis (Figure 4) and data estimated using frequency analysis as presented in Table 2. Both the analyses confirm the impact of honking on traffic noise.

The results of 15 minutes traffic and noise measurements including traffic volume, number of horns, noise levels L_{min} L_{max} and L_{eq} for with and without honking cases are summarized in Figure 5. The noise values are plotted on primary y-axis and traffic volume with number of horns is plotted on secondary y-axis while timings of measurement are represented on x-axis for each category of roads. Highest L_{eq} [76.3 dB(A)] was observed at highway with maximum traffic volume (1508) and most number of horn incidents (63 nos.). The L_{eq} at minor road is observed more than major road though traffic volume and number of horns are nearly same during peak hours. This variation is mostly due to lesser width of minor road (7 m) as compared to major road (15 m) resulting in reduction of distance between center line of road and position of sound level meter. Moreover, minor road doesn't have divider to control the mixed traffic flow.

A separate study was carried out to estimate the impact of vehicle type and speed on traffic noise level

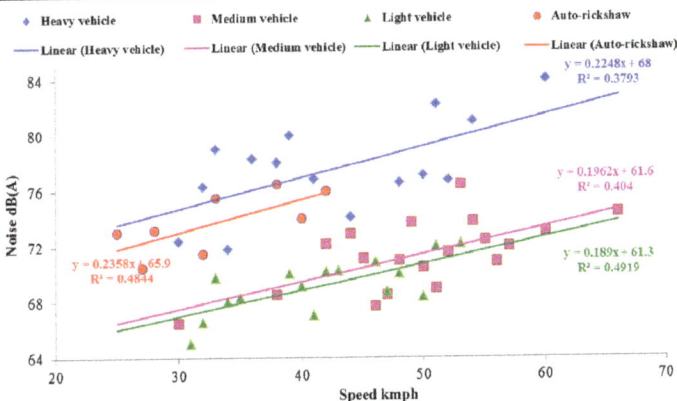

Figure 6 Relationship between vehicular speed and noise level.

Table 3 Analysis of variance for vehicular type and speed on traffic noise

Parameters	Degree of freedom	Sum of squares	Mean square	F	p
Vehicle type	3	156.9	52.3	22.38	0.001
Vehicular speed	3	31.1	10.4	4.44	0.035
Error	9	21.0	2.3		
Total	15	209.1			

$S = 1.5$; $R^2 = 89.94\%$.

(Figure 6). The speed of light, medium and heavy vehicles varied in the range of 40–45 kmph, 50–56 kmph and 30–38 kmph, respectively. For all categories of vehicles, noise level varies linearly with speed. Impact of heavy vehicles and auto-rickshaw on traffic noise is comparatively more than light and medium vehicles. An increase in speed from 35 to 55 kmph, increases the noise level by nearly 4–5 dB(A) except for auto-rickshaw. While in case of auto-rickshaw, increase in speed from 25 to 40 kmph increases noise by nearly 4 dB(A). A statistical analysis using two-way ANOVA was performed to assess the significance of individual vehicle type and speed on traffic noise (Table 3). $p < 0.01$ for vehicular type indicates that there is significant difference in the type of vehicle on the response of traffic noise level (CI - 95%). Similarly, $p < 0.05$ for vehicular speed signifies the impact of variation in speed on traffic noise level (CI 95%). The analysis suggests that type of vehicle (i.e. heavy, medium, light and auto) is more dominant than vehicular speed.

Conclusions

Monitoring and assessment of traffic noise in urban environment is complex due to various influencing factors such as traffic volume, honking, vehicular speed, road geometry etc. Traffic noise was assessed in the urban agglomeration of Nagpur, India considering above factors. Impact of heavy vehicles on traffic noise was more as compared to light and medium vehicles. Honking is a frequent phenomenon in Indian road context therefore it was observed that honking has significant impact on traffic noise besides traffic volume and vehicular speed. Previous studies also confirmed the effect of honking on traffic noise [18,21,26,29,30] and used as one of the input parameter in traffic noise prediction [31,32]. These studies do not provide quantification of honking noise in heterogeneous traffic while present research provides quantification of noise due to honking based on frequency analysis of traffic noise. This was also confirmed by statistical analysis considering traffic noise and honking data. Using this, it was found that honking induced an additional 2 to 5 dB(A) noise over and above traffic noise. Further, increase in vehicular speed from 35 to 55 kmph also increases traffic noise by 4 to 5 dB(A) for all types of vehicles. The present study suggests that

honking must also be a component, apart from monitoring of traffic volume and vehicular speed in traffic noise assessment. Additionally, the study will help in assessing existing horn prohibited areas and defining new 'No Honking zones.

Competing interests

The authors declare that they have no competing interests.

Authors' contributions

All of the authors have the equal contribution. All authors read and approved the final manuscript.

Acknowledgements

Authors are thankful to the Director, CSIR-NEERI for providing encouragement, necessary infrastructural support, and kind permission for publishing the research article. Department of Science and Technology, New Delhi is also acknowledged for providing the financial support to carry out this study. Authors also acknowledge Er. Ankit Gupta for assistance in statistical analysis and Ms. Trupti Mardikar for editing language corrections.

Author details

[1]Environmental System Design and Modeling Division, CSIR-NEERI, Nagpur 440020, Maharashtra, India. [2]Civil Engineering Department, VNIT, Nagpur 440010, Maharashtra, India.

References

1. Zannin PHT, Calixto A, Diniz F, Ferreira JA. A survey of urban noise annoyance in a large Brazillian city: the importance of subjective analysis in conjunction with an objective analysis. Environ Impact Assess. 2003;23:245–55.
2. Lisa Gonise RN, Louis Hagler MD. Noise pollution: a modern plague. South Med J. 2007;100(3):287–94.
3. Nathaniel MM. Noise pollution: the sound behind heart effects. Environ Health Perspect. 2007;115(11):A536–7.
4. Pathak V, Tripathi BD, Mishra VK. Evaluation of traffic noise pollution & attitudes of exposed individuals in working place. Atmos Environ. 2008;42(16):3892–8.
5. Oyedepo OS, Saadu AA. Evaluation and analysis of noise levels in Ilorin metropolis, Nigeria. Environ Monit Assess. 2010;160:563–77.
6. Sazegar Niya A, Bahrayni Tossi M, Moradi H. Noise pollution and traffic noise index at some Mashhad main streets in high traffic hours of summer. Iran J Med Phys. 2005;2(8):21–30.
7. Amrah A, Al-Omari A, Sharabi R. Evaluation of traffic noise pollution in Amman, Jordan. Environ Monit Assess. 2006;120:499–525.
8. Omidvari M, Nouri J. Effects of noise pollution on traffic policemen. Int J Environ Res. 2009;3(4):645–52.
9. Martin MA, Tarrero MA, Gonzalez A, Machimbarrena M. Exposure–effect relationships between road traffic noise annoyance and noise cost valuations in Valladolid, Spain. Appl Acoust. 2006;67(10):945–58.
10. Al-Mutairi NZ, Al-Attar MA, Al-Rukaibi FS. Traffic-generated noise pollution: exposure of road users and population in Metropolitian Kuwait. Environ Monit Assess. 2011;183:65–75.

11. Banerjee D, Chakraborty SK, Bhattacharyya S, Gangopadhyay A. Attitudinal response towards road traffic noise in the industrial town of Asansol, India. Environ Monit Assess. 2009;151(1–4):37–44.

12. Mansouri N, Pourmahabadian M, Ghasenkhani M. Road traffic noise in downtown area of Tehran. Iran J Environ Health Sci Eng. 2006;3(4):267–72.

13. Reddy RB, Ramachandaiah A. Traffic noise in some typical urban lanes. J Acoust Soc India. 1995;23(4):53–7.

14. Vidya Sagar T, Nageswara Rao G. Noise pollution levels in Vishakhapatnam city (India). J Environ Sci Eng. 2006;48(2):139–42.

15. Singh D, Praksh A, Srivastava AK, Kumar K, Jain VK. The effects of meteorological parameters in ambient noise modelling studies in Delhi. Environ Monit Assess. 2013;2013(185):1873–82.

16. Ma G, Tian Y, Ju T, Ren Z. Assessement of traffic noise pollution from 1989 to 2003 in Lanzhou city. Environ Monit Assess. 2006;123:413–30.

17. Sharma A, Vijay R, Sadar KV, Sohony RA, Gupta A. Development of noise simulation model for both stationary and mobile source: A GIS based approach. Environ Model Assess. 2010;15(3):189–97.

18. Kalaiselvi R, Ramachandraiah A. Environmental noise mapping study for heterogeneous traffic conditions. Proceedings of 20th International Congress on Acoustics, ICA 2010 23–27 August 2010, Sydney, Australia, PACS: 43.50.LJ, 43.50.RQ. 1–6.

19. Wani KA, Jaiswal YK. Assessment of noise pollution in Gwalior M.P. India. Adv Biores. 2010;1(1):54–60.

20. Gosh PK. Epidemiological study of the victims of vehicular accidents in Delhi. J Indian Med Assoc. 1992;90(12):309–12.

21. Kalaiselvi R, Ramachandraiah A. A noise mapping study for heterogeneous road traffic conditions considering horn sounds. J Acoust Soc Am. 2011;129:2380.

22. Indian Standard, IS-3098. Noise emitted by moving road vehicles, measurement, Indian Standards Institution, 1980, Manak Bhavan, 9 Bahadur Shah Zafar Marg, New Delhi 110002.

23. Indian Standard, IS-10399. Noise emitted by stationary road vehicles, methods of measurement, Indian Standards Institution, 1982, Manak Bhavan, 9 Bahadur Shah Zafar Marg, New Delhi 110002.

24. SoundPro 2011. Sound Level Meter (Type I and Type II) SoundPro SP-DL2, 3 M make, St. Paul, MN 55144–1000, USA.

25. Speedet. User Hand Book for Hand Held Traffic Radar Speedet. Hyderabad India: Hindustan Aeronautics Limited (HAL); 1990.

26. Vijay R, Kori C, Kumar M, Chakrabarti T, Gupta R. Assessment of traffic noise on highway passing from urban agglomeration. Fluct Noise Lett. 2014;13(4):1450031. 12 pages.

27. Minitab Inc. Meet Minitab 15. State College, PA: Minitab; 2007.

28. Mitchell P. Speed and road traffic noise, the role that lower speeds could play in cutting noise from traffic. Chatham, Kent, UK: A report commissioned by the UK Noise Association; 2009.

29. Bin J, Rui J, Qing-Song W, Mao-bin H. Honk effect in the two lane cellular automaton model for traffic Flow. Physica A. 2005;348:544–52.

30. Tyagi V, Kalyanaraman S, Krishnapuram R. Vehicular traffic density state estimation based on cumulative road acoustic. IEEE Trans Intell Transport Syst. 2012;13(3):1156–66.

31. Sharma A, Vijay R, Bodhe GL, Malik LG. Adoptive neuro-fuzzy inference system for traffic noise prediction. Int J Comput Appl. 2014;98(13):14–9.

32. Vijay R, Sharma A, Kumar M, Shende V, Chakrabarti T, Gupta R. GIS based noise simulation open source software: N-GNOIS. Fluct Noise Lett. 2015;14 (1):1550005. 12 pages.

24

Mathematical modeling in municipal solid waste management: case study of Tehran

Mohsen Akbarpour Shirazi[1], Reza Samieifard[2*], Mohammad Ali Abduli[2] and Babak Omidvar[2]

Abstract

Background: Solid Waste Management (SWM) in metropolises with systematic methods and following environmental issues, is one of the most important subjects in the area of urban management. In this regard, it is regarded as a legal entity so that its activities are not overshadowed by other urban activities. In this paper, a linear mathematical programming model has been designed for integrated SWM. Using Lingo software and required data from Tehran, the proposed model has been applied for Tehran SWM system as a case study.

Results: To determine the optimal status of the available system for Tehran's Solid Waste Management System (SWMS), a novel linear programming model is applied. Tehran has 22 municipal regions with 11 transfer stations and 10 processing units. By running of the model, the transfer stations and processing units are decreased to 10 and 6 units, respectively.

Conclusions: The proposed model is an alternative method for improvement the SWMS by decreasing the transfer stations and processing units.

Keywords: Mathematical modeling, Municipal solid waste management, Tehran, Lingo

Background

Solid Waste Management (SWM) is a set of consistent and systematic regulations related to control generation, storage, collection, transportation, processing and land-filling of wastes according to the best public health principles, economy, preservation of resources, aesthetics, other environmental requirements and what the public attends to [1]. Many countries are facing problems in managing these problems and need comprehensive and practical solutions. Therefore, the optimization of conditions for sustainable approach to SWM is a key factor for managers and planners of government. Currently, the planners and decision-makers in the area of integrated SWM are confronting increased complexity, uncertainty, and multi-objective of this issue [2]. At the beginning, the process of decision-making on SWM was simple. It is because of the decisions were made only through simple comparison of some options out of the available options. However, different combinations of

various components of this system were gradually propounded considering different factors in integrated SWM resulting in complexity of decision-making process.

At this period of time and regarding the complexities in the integrated solid waste management, decision-makers should distinguish between optimal, good, and fortuitous decision-making. In the optimal decision-making, one can solve the optimal problem using the techniques available in other fields. In this solution method, generally some constraints (criteria) are considered, where the function(s) is to be optimized through applying some methods. Good decision-making is done based on experience, trial and error or comparison between different options of the integrated SWM. Although it is possible to choose decisions close to the optimal state using this decision-making method, today these methods are not applicable due to increased number of different combinations in the decision-making process. In the fortuitous decision-making, since decisions are made with no scientific base, so the results are not acceptable [3].

In Iran, ever-increasing rate of population growth and constant development of cities, on the one hand, and

---

* Correspondence: r_samifard@yahoo.com
[2]Department of Environmental Engineering, Graduate Faculty of Environment, University of Tehran, Azin alley, Ghods St. Enghelab Ave., Tehran, Iran
Full list of author information is available at the end of the article

proliferation and development of industrial, commercial, and service activities, on the other hand, have resulted in generation of large amounts of solid waste in cities. In the majority of cases, this has caused numerous problems considering shortage of facilities and budget. One of these problems is environmental pollution. Therefore, today a key factor of environmental pollution is mismanagement of different types of waste. Over 3.5 million tons of solid waste is generated daily around the world, where 80 % is recycled to the consumption cycle in developed countries and the rest is disposed or incinerated in a hygienic way [4].

The Tehran city is located in the north of Iran with a population of 8154051 (from the last census in 2011) [5], which currently generates around 6629 tons of municipal solid waste on a daily bases [6]. In 2004, with the support of the World Bank, extensive studies were initiated on the integrated SWM strategy of Tehran. The reports of these studies were finally delivered to the Organization of Waste Recycling & Composting (OWRC) in 2005 [7, 8]. However, the plan of integrated solid waste management in Tehran was never planned and pursued in the form of an executive program. The aim of presented study is the planning of the current status of SWM in Tehran with a scientific method.

The use of computational systems, due to absorption of a large number of resources and having great impacts on the environment, one can help decision-makers to achieve significant savings in costs and improve recovery of wastes [9]. Yu et al. [10] conducted a study on optimization of the long-term performance of municipal SWM system in the form of a bi-targeted mathematical model. In this paper, a linear and dynamic programming bi-targeted model is proposed for decision-making on supporting long-term operations of municipal SWM system. The proposed mathematical model simultaneously deals with calculation of the economic productivity and environmental pollution from the municipal SWM system within several time periods. Optimal exchange across the entire studied time horizon indicates the accuracy of this model. The proposed model has been calculated and solved by Lingo Software. The proposed model provided an effective solution for the long-term operational planning of the municipal SWMS.

Khaiwal et al. focused on the analysis of municipal SWM system and the methods to minimizing it in Chandigarh in India. This city is situated in the north part of India [11]. The information related to the SWM methods in Chandigarh was investigated for conductance of the mentioned research. The key information was collected from stakeholders through interviews and information of the registry related to transportation and solid waste disposal. This study has emphasized that the solid waste recycling system in this city is poor, where

negligence of this important section has followed negative consequences. Thus, for environmentally-friendly SWM systems, a serious decision-making process and adjustment of operational activities are required. They suggested that the new frameworks should be framed on the properly design integrated solid waste management system with high recovery rates cost-effectiveness, and other environmental impacts.

Sie et al. [12] studied the optimal processing network for municipal SWM in Iskandar, Malaysia. In this paper, a mathematical programming model including integration of four principal consuming technologies has been presented to facilitate the optimal processing of the network. The mentioned model is able to predict the best combination out of the technologies of solid waste disposal, the procedure of solid waste recovery, prediction of product generation, estimation of the capacity of facilities, prediction of greenhouse gases (GHG) emitted from the system and eventually generation of the optimal and cost-effective solution for municipal SWM. The Mixed-Integer Linear Programming (MILP) model presented for profitability has been used as a model studied from the municipal SWM system. Based on results, the best mix of solid waste utilization technologies based on solid waste allocation to value added products was found to be landfill LFG capture (43.19 %), incineration (8.34 %), recycling (48.44 %) and composting (0.03 %).

Hui et al. [13] investigated the heat value of municipal solid waste in China considering the solid waste properties (physiochemical compounds). In this research, after investigation of the physiochemical properties of the municipal solid waste, the statistical indices including the mean, standard deviation, coefficient of changes, and physical analyses were used for determination of the heat value of municipal waste. The results showed that the chemical characteristics should be considered for thermal conversion process of Chinese SWM. Lohri et al. [14] examined the fiscal stability in municipal SWM, costs, and revenues of SWM in Bahir Dar City, Ethiopia. This research conducts a cost-income analysis based on data from July 2009 until June 2011. The analysis indicated that the overall costs in the SWM system in Bahir Dar has increased dramatically within this period due to increased costs related to the transportation of solid waste including the cost of solid waste collection from the households, commercial companies and institutions. The results of this research showed that existence of a structure for accurate analysis of costs and income from the SWM system was paramount to increase productivity and the cost and revenue balance in relation to cost. The obtained results revealed that a strong alliance between the municipality and private enterprise is an appropriate solution for improvement the financial sustainability of a SWM system.

Soltani et al. explored the numerous stakeholders in multi-criteria decision-making in municipal SWM. Municipal SWM is a complex process that includes several environmental, social, and economic criteria [15]. In addition, it also includes decision-making in solid waste management problems such as finding suitable sites for solid waste disposal or strategies commonly requiring various stakeholders such as governments, municipalities, industries, experts, and even the public. Results proved that the Analytical Hierarchy Process is the most common approach in consideration of multiple stakeholders.

Arena & Di Gregorio studies solid waste management planning based on analysis of the flow of the materials. This paper describes the results obtained from a municipal solid waste management planning [16]. This paper has investigated different components of SWM using analysis method of the flow of materials together with the results of the evaluation studies of the integrated solid waste management cycle. They found that the combination of material and substance flow analysis with an environmental assessment method is an alternative tool-box for comparing solid waste management technologies and scenarios.

Several models have been designed for integrated SWM systems. Although these models are a proper means for helping decision-makers and engineers in the integrated SWM planning process, each of them has considered only a certain portion of the SWM sections. In the present study, a mathematic model was used to optimize the current system of SWM in Tehran and identify the number of proper sites for transferring and processing of waste.

Methods

In this research, a novel linear programming model is presented for investigation of the current status of municipal SWM in Tehran and for its optimization. This mathematical model tries to optimize the current system of SWM in Tehran and identify the number of proper sites for transferring and processing of waste. Furthermore, this model determines the extent of recovery and disposal of wastes using each of the recovery, landfilling, and composting methods. Another capability of the presented models is SWM based on the type and properties of the generated solid waste in every municipal region. Based on this ability, given the type of solid waste generated in the region, for each of them the best method of management is selected. The objective function in this model is increasing the profitability of the entire system of SWM.

Tehran municipality has placed presentation of urban services such as management and planning for organizing municipal wastes at the top of its agenda. Relevant officials of the solid waste recovery in 22 regions of

Tehran were approached in order to collect data about the municipal solid waste generation through interviewing, filling out questionnaires, conducting field visits from Aradkooh landfilling and processing complex and collecting information on disposal and destiny of solid wastes. Based on the available data, Tehran's wastes can be categorized into three groups of municipal wastes, organizational wastes together with the wastes generated by towns, hospitals, and animals. The municipality of each region is responsible for collection and transportation of the wastes generated in that region. The major part of the municipal wastes along with a part of the organizational and town wastes, are transferred to the transfer stations available throughout Tehran after collection. The wastes transferred to transfer stations are transported to the Aradkooh landfilling and processing complex, situated 32 km away from the south east of Tehran. The following conceptual representation is used as a base for Tehran municipal SWM modeling (Fig. 1).

Conceptual representation of the model

As can be observed, the regions are connected only to the transfer stations. Under the current state, no processing is carried out at the transfer stations, but it can be provided in the future in these stations, accordingly, this property has been considered in the presented model. At the next stage, after collection of wastes at the transfer stations, the wastes are transported to other units called processing units to be processed. In these units, after processing and separation of wastes into three groups of dry valuable, dry non-valuable and wet wastes, they are transferred to other units. The separated solid waste is guided to one of the recovery, composting, or landfilling stages based on the type of developed optimal productivity. The proposed model has the ability of separating and allocating wastes to the mentioned units, bringing about the maximum profitability. Furthermore, based on the policies of SWM development in Tehran, a number of other facilities including incinerator, biogas, and electricity generation have also been considered in the model. Locating of the transfer stations and processing units is another property of this model. Accordingly, in this paper, a linear mathematical programming model is developed based on the solid waste flow process in Tehran. The list of symbols, parameters and variables concerning the mathematical model has been shown in Appendix I.

Model constraints

In case $Y_j = 1$ therefore the station j will be established as a transfer station and if the $Y_j' = 1$ therefore the station j will be established as the transfer station and processing unit:

$$Y_j + Y_j' \leq 1 \tag{1}$$

Each region connects to one station:

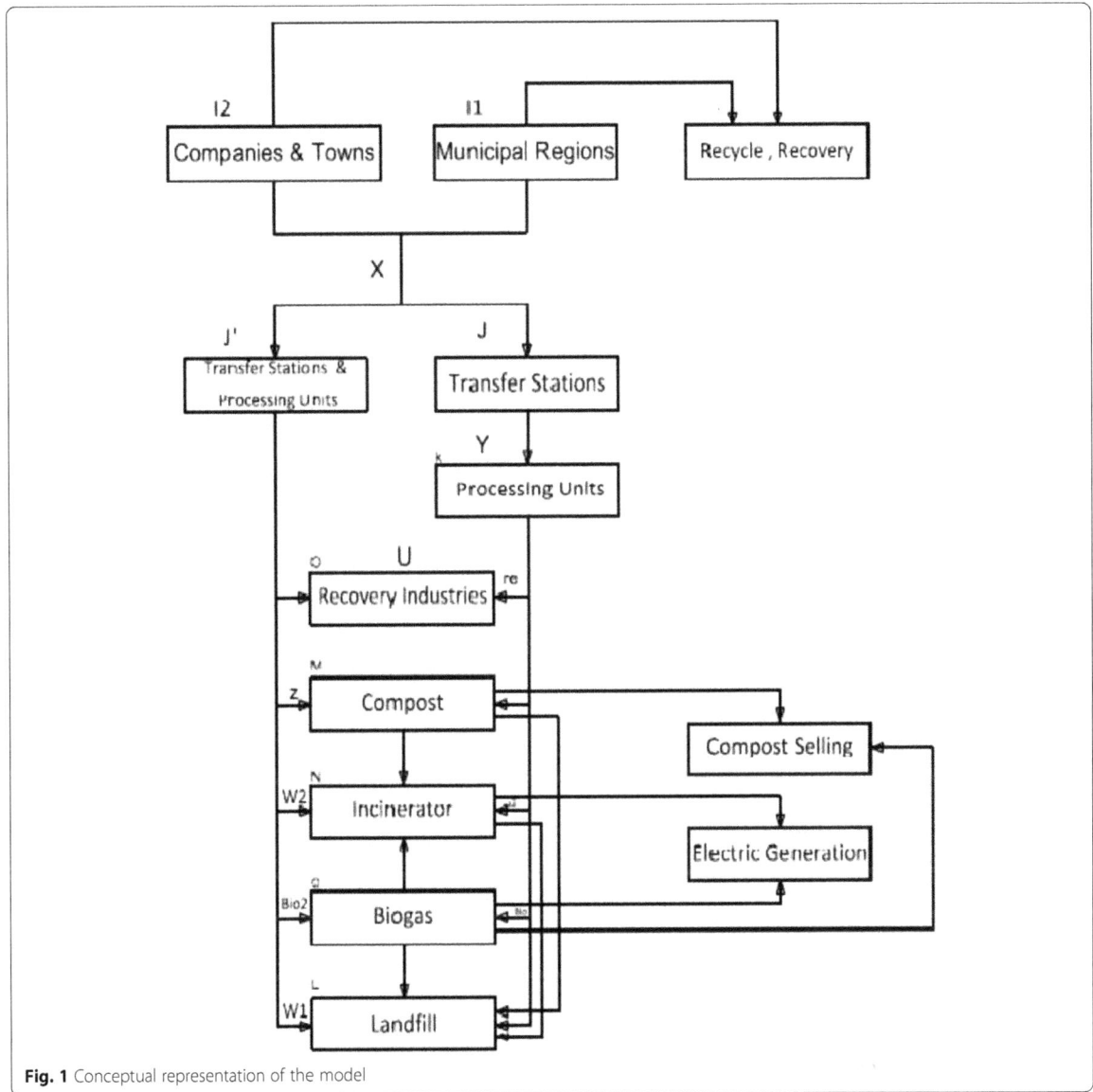

Fig. 1 Conceptual representation of the model

$$\sum_j x_{ij} + \sum_j x'_{ij} = 1 \qquad (2)$$

If $Y_j = 1$, the x_{ij} can get value:

$$x_{ij} \leq Y_j \qquad (3)$$

If $Y'_j = 1$, the x'_{ij} can get value:

$$x'_{ij} \leq Y'_j \qquad (4)$$

The capacity constraint of transfer station j: the amount of unseparated (mixed) solid waste from region i to transfer station j has to less than the capacity of that station:

$$\sum_i D_i(1-\alpha_i)x_{ij} \leq Cap_j^s \quad \forall j = 1, 2, \ldots \qquad (5)$$

The capacity constraint of transfer station and processing unit j: the amount of unseparated (mixed) solid waste from region i to transfer station and processing unit j has to less than the capacity of that station:

$$\sum_i D_i(1-\alpha_i)x'_{ij} \leq Cap_j^{'s} \quad \forall j = 1, 2, \ldots \qquad (6)$$

If the j is transfer station, the transferred solid waste from the regions will be sent to the processing units:

$$D_i(1-\alpha_i)x_{ij} = \sum_K Y_{ijK} \qquad (7)$$

The capacity of the processing units has to more than the incoming solid waste from transfer stations:

$$\sum_i \sum_j Y_{ijK} \leq Cap_K^{pr} LK_k \qquad (8)$$

The allocated solid waste to the transfer station and processing unit j, after transformation to waste P_2, will be sent to one of the compost, recovery, biogas, incinerator or landfill facilities:

$$\sum_{P_1} \sum_i D_i(1-\alpha_i)x'_{ij_1}\delta_{iP_1}\gamma_{P_1P_2} = \sum_M \beta_{P_2}^1 Z_{jMP_2}^2 + \sum_g \beta_{P_2}^5 Bio_{jgP_2}^1$$
$$+ \sum_O \beta_{P_2}^4 rej_{OP_2} + \sum_N \beta_{P_2}^3 W_{3jNP_2}$$
$$+ \sum_L \beta_{P_2}^2 W_{4jLP_2} \qquad (9)$$

In the following, the left side of the equation shows the amount of dry valuable waste, dry non-valuable solid waste and wet solid waste after the processing unit. The flow of the mentioned wastes after the processing units could be the incinerator, landfill, compost, recovery, biogas facilities or combination of these facilities:

$$\sum_{P_1} \sum_j \sum_i Y_{ijK}\delta_{iP_1}\gamma_{P_1P_2} = \sum_N \beta_{P_2}^3 W_{2KNP_2} + \sum_L \beta_{P_2}^2 W_{1KLP_2}$$
$$+ \sum_M \beta_{P_2}^1 Z_{KMP_2} + \sum_O \beta_{P_2}^4 U_{KOP_2}$$
$$+ \sum_g \beta_{P_2}^5 Bio_{KgP_2}^2 \qquad (10)$$

The constraints relating to the capacity of facilities
Capacity of recovery facilities:

$$\sum_j \sum_{P_2} rej_{OP_2} + \sum_K \sum_{P_2} U_{KOP_2} \leq Cap_O^1 \times LO_O \ \forall O \qquad (11)$$

Capacity of compost facilities:

$$\sum_{P_2} \sum_j Z_{jMP_2}^2 + \sum_K \sum_{P_2} Z_{KMP_2} \leq Cap_M^2 \times LM_M \ \forall M \qquad (12)$$

Capacity of landfill facilities:

$$\sum_j \sum_{P_2} W_{4jLP_2} + \sum_N SL_{NL}^2 + \sum_L SL_{ML}^1 + \sum_K \sum_{P_2} W_{1KLP_2}$$
$$+ \sum_g SL_{gL}^3 \leq Cap_L^4 \times LL_L \ \forall L \qquad (13)$$

Capacity of incinerator facilities:

$$\sum_j \sum_{P_2} W_{3jNP_2} + \sum_K \sum_{P_2} W_{2KNP_2} + \sum_M IN_{MN}^1 \qquad (14)$$
$$+ \sum_g IN_{gN}^2 \leq Cap_N^3 \times LN_N \ \forall N$$

Capacity of biogas facilities:

$$\sum_j \sum_{P_2} Bio_{jgP_2}^1 + \sum_{P_2} \sum_K Bio_{KgP_2}^2 \leq Cap_g \times Lg_g \ \forall g \qquad (15)$$

Mass balance constraints
Mass balance relating to the compost facilities:

$$\sum_j \sum_{P_2} Z_{jMP_2}^2 + \sum_K \sum_{P_2} Z_{KMP_2} = FE_M^1 + \sum_L SL_{ML}^1$$
$$+ \sum_N IN_{MN}^1 \qquad (16)$$

Mass balance relating to the biogas facilities:

$$\sum_j \sum_{P_2} Bio_{jgP_2}^1 + \sum_{P_2} \sum_K Bio_{KgP_2}^2 = EL_g^1 + FE_g^2 + \sum_N IN_{gN}^2$$
$$+ \sum_L SL_{gL}^3 \qquad (17)$$

Mass balance relating to the incinerator facilities:

$$\sum_j \sum_{P_2} W_{3jNP_2} + \sum_{P_2} \sum_K W_{2KNP_2} + \sum_g IN_{gN}^2 \qquad (18)$$
$$+ \sum_M IN_{MN}^1 = EL_N^2 + \sum_L SL_{NL}^2$$

Environmental constraints
Environmental constraints relating to the transfer station:

$$FEA_{jpe}\sum_i D_i(1-\alpha_i)\left(x'_{ij} + x_{ij}\right) \leq Vcap_{pe} \qquad (19)$$

Environmental constraints relating to the processing unit(s):

$$FEm_{kpe}\sum_i \sum_j Y_{ijK} \leq Mcap_{pe} \qquad (20)$$

Environmental constraints relating to the compost facilities:

$$FEw_{mpe}\left(\sum_k \sum_{p_2} Z_{kmp_2} + \sum_{p_2} \sum_j Z_{mjp_2}^2\right) \leq Wcap_{pe} \qquad (21)$$

Environmental constraints relating to the landfill facilities:

$$VL\left(\sum_k \sum_{p_2} W_{1klp_2} \times VL_{p2} + \sum_j \sum_{p_2} W_{4jlp_2} \times VL_{p2}\right) \leq Scap \qquad (22)$$

Objective function

$$max \ (benefit-cost) \qquad (23)$$

Table 1 Amount and composition of municipal solid waste generated in regions in 2014

Region no	Waste amount	Used bread	Plastic	PET	Plaster	Talc	Foam	Paper	Cardboard	Ferrous metal	Non-ferrous metal	Textile	Glass	Wood	Rubber	Soil	Special	Putrescible material	Other
1	448	0.04	0.04	0.02	0.02	0.01	0.00	0.08	0.08	0.01	0.01	0.00	0.03	0.00	0.00	0.00	0.01	0.65	0.00
2	516	0.04	0.03	0.01	0.05	0.01	0.00	0.08	0.03	0.02	0.01	0.01	0.03	0.00	0.00	0.00	0.03	0.65	0.00
3	305	0.02	0.03	0.02	0.04	0.00	0.00	0.07	0.04	0.01	0.00	0.01	0.01	0.01	0.00	0.00	0.02	0.69	0.00
4	703	0.05	0.03	0.01	0.04	0.01	0.00	0.04	0.06	0.02	0.00	0.01	0.02	0.01	0.01	0.01	0.02	0.69	0.00
5	534	0.04	0.03	0.01	0.03	0.01	0.00	0.07	0.04	0.01	0.01	0.01	0.02	0.01	0.00	0.00	0.03	0.68	0.00
6	278	0.03	0.03	0.01	0.03	0.00	0.00	0.11	0.05	0.01	0.01	0.01	0.03	0.00	0.00	0.00	0.05	0.62	0.01
7	278	0.05	0.03	0.01	0.03	0.00	0.00	0.06	0.03	0.03	0.01	0.01	0.04	0.00	0.00	0.01	0.04	0.66	0.00
8	273	0.05	0.03	0.02	0.03	0.01	0.00	0.06	0.04	0.02	0.00	0.03	0.02	0.00	0.00	0.00	0.02	0.68	0.00
9	130	0.05	0.04	0.01	0.04	0.01	0.00	0.05	0.04	0.03	0.00	0.02	0.02	0.02	0.00	0.01	0.03	0.62	0.00
10	202	0.04	0.02	0.01	0.02	0.01	0.01	0.04	0.04	0.03	0.01	0.03	0.01	0.03	0.03	0.02	0.06	0.59	0.01
11	247	0.06	0.03	0.02	0.02	0.00	0.01	0.05	0.03	0.04	0.00	0.01	0.02	0.00	0.00	0.00	0.01	0.71	0.00
12	339	0.03	0.04	0.02	0.03	0.00	0.00	0.06	0.05	0.02	0.00	0.02	0.01	0.00	0.00	0.00	0.04	0.68	0.01
13	163	0.05	0.02	0.01	0.04	0.01	0.00	0.05	0.04	0.03	0.01	0.01	0.02	0.00	0.00	0.01	0.04	0.65	0.00
14	321	0.04	0.02	0.01	0.02	0.00	0.00	0.04	0.07	0.02	0.00	0.01	0.01	0.00	0.00	0.00	0.01	0.75	0.00
15	470	0.06	0.02	0.01	0.04	0.01	0.00	0.05	0.04	0.03	0.00	0.02	0.02	0.00	0.00	0.00	0.03	0.68	0.00
16	229	0.06	0.01	0.01	0.04	0.01	0.00	0.03	0.05	0.02	0.00	0.01	0.01	0.00	0.00	0.01	0.04	0.71	0.00
17	183	0.05	0.03	0.01	0.03	0.01	0.00	0.05	0.03	0.02	0.00	0.01	0.01	0.00	0.00	0.00	0.01	0.73	0.00
18	282	0.03	0.03	0.01	0.02	0.01	0.00	0.06	0.01	0.02	0.00	0.01	0.02	0.00	0.01	0.02	0.02	0.73	0.00
19	200	0.05	0.03	0.02	0.04	0.00	0.00	0.04	0.04	0.02	0.00	0.02	0.01	0.00	0.00	0.00	0.04	0.69	0.00
20	309	0.03	0.01	0.01	0.03	0.00	0.00	0.03	0.03	0.02	0.00	0.01	0.02	0.00	0.01	0.01	0.03	0.77	0.00
21	119	0.05	0.03	0.02	0.04	0.01	0.00	0.06	0.04	0.03	0.01	0.01	0.01	0.00	0.01	0.00	0.02	0.67	0.00
22	100	0.04	0.02	0.01	0.03	0.01	0.00	0.05	0.04	0.03	0.00	0.02	0.02	0.00	0.00	0.00	0.01	0.71	0.00

$$\begin{aligned}
\text{cost} = &\sum_i \sum_j D_i(1-\alpha_i)\left(x_{ij} + x'_{ij}\right)C_{ij}^1 + \sum_j \sum_K \sum_i Y_{ijK}C_{jK}^2 + \sum_{P_2}\sum_j \sum_O re_{jOP_2}C_{jO}^3 + \sum_{P_2}\sum_K \sum_O U_{KOP_2}C_{KO}^4 \\
&+ \sum_{P_2}\sum_K \sum_M Z_{KMP_2}C_{KM}^5 + \sum_{P_2}\sum_j \sum_M Z_{jMP_2}^2 C_{jM}^6 + \sum_j \sum_N \sum_{P_2} W_{3jNP_2}C_{jN}^7 + \sum_K \sum_{P_2}\sum_N W_{2KNP_2}C_{KN}^7 \\
&+ \sum_M \sum_N IN_{MN}^1 C_{MN}^8 + \sum_N \sum_g IN_{gN}^2 C_{gN}^9 + \sum_{P_2}\sum_j \sum_g Bio_{jgP_2}^1 C_{gj}^{10} + \sum_{P_2}\sum_K \sum_g Bio_{KgP_2}^2 C_{Kg}^{11} \\
&+ \sum_M \sum_L SL_{ML}^1 C_{ML}^{12} + \sum_N \sum_L SL_{NL}^2 C_{NL}^{13} + \sum_g \sum_L SL_{gL}^3 C_{gL}^{14} + \sum_{P_2}\sum_K \sum_L W_{1KLP_2}C_{KL}^{15} + \sum_{P_2}\sum_j \sum_L W_{4jLP_2}C_{jL}^{16} \\
&+ \sum_o CO_o LO_o + \sum_k CK_k LK_k + \sum_m CM_m LM_m + \sum_l CL_l LL_l + \sum_n CN_n LN_n + \sum_g Cg_g Lg_g + Cj_j^1 Y_j + Cj_j^2 Y'_j \\
&+ \left(\sum_M \sum_L SL_{ML}^1 + \sum_N \sum_L SL_{NL}^2 + \sum_g \sum_L SL_{gL}^3 + \sum_K \sum_L \sum_{P_2} W_{KLP2}^1 + \sum_j \sum_L \sum_{P2} W_{jLP2}^4\right) Cdis
\end{aligned}$$

$$(24)$$

$$\begin{aligned}
Benefit = &\left(\sum_{P_2}\left(\sum_j \sum_O re_{jOP_2} + \sum_K \sum_O U_{KOP_2}\right)B_{P_2}^1\right) \\
&+ \alpha_i D_i B^2 + \sum_g EL_g^1 B_g^3 + \sum_N EL_N^2 B_N^4 \\
&+ \sum_M FE_M^1 B_M^5 + \sum_g FE_g^2 B_g^6
\end{aligned}$$

$$(25)$$

Solving model with Lingo software

The process of solving a math program requires a large number of calculations and is, therefore, best performed by a computer program. Lingo is a mathematical modeling language designed particularly for formulating and solving a wide variety of optimization problems including linear programing. Lingo optimization software uses branch and bound methods to solve problems of this type.

The obtained model is solved by Lingo 13.0 optimization software.

Results and discussion
Case study

The presented model has been applied for a case study related to the SWM system in Tehran. Tehran contains 22 regions in which the amount of daily municipal solid waste generation is 6629 tons. Moreover, the wastes generated in the regions by citizens and organizations have been categorized in 20 different groups. These groups are further converted to three groups of dry valuable (type 1), dry non-valuable (type 2) and wet (type 3) wastes at processing units. Table 1 provides the amount of generated solid waste in the regions as well as its composition in terms of

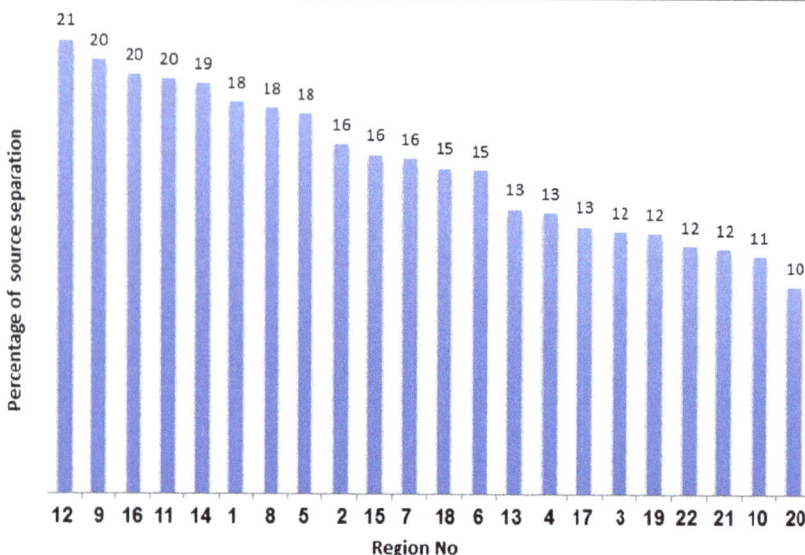

Fig. 2 Status of solid waste separation at source for each region in 2014

Table 2 Specifications of municipal solid waste transfer stations in Tehran in 2014

Transfer station name	Covered region(s)	Nominal capacity (ton/day)	Receiving waste (ton/day)
Darabad	1,3	1299	540–570
Zanjan	2,10	1037	1000
Banihashem	4,8	485	400–450
Hakimiyeh	4,8	485	780–790
Kuhak	5,21,22	2227	750
Beyhaghi	3,6,7	864	750–800
Harandi	11,12	691	600–650
Azadegan	13,14,15	1484	1100
Yaran	9,10,17,18	1484	700
Jehad	16,19	1484	500–530
Shahid Avini	20	323	250–300

percentage in 2014 [6]. The second column of the mentioned table indicates the amount of daily generated solid waste (in terms of ton). The third column itself and onward indicate the ratio of each of the solid waste types out of the total generated solid waste (in terms of percentage).

Amount and composition of municipal solid waste generated in regions in 2014

Furthermore, some of the solid waste generated in the point of origin and before the collection are separated. These wastes are called dry solid waste separated at source. The following figure represents the percentage of solid waste separated at source for every region [6] (Fig. 2).

Status of solid waste separation at source for each region in 2014

There are 11 transfer stations in the SWM system of Tehran. In all of these 11 stations, only collection and transfer of solid waste take place with no processing operation. Table 2 shows the nominal capacity of all of these stations and the current status of the regions allocated to them together with the amount of solid waste sent to them [6].

Specifications of municipal solid waste transfer stations in Tehran in 2014

On the other hand, in Tehran's SWM system, there are 10 solid waste processing units available, all of which are located in the Aradkooh landfilling and processing complex in Tehran. All of the collected wastes are transferred to this complex. The following figure indicates the solid waste processing units' capacity (in terms of ton) per day [6] (Fig. 3).

Name and nominal capacity of solid waste processing units in Aradkooh

In the current situation, there is no possibility for generation of electricity using biogas and incineration facilities, since these facilities are still under construction. Therefore, only generation of compost, recovery of dry materials and landfilling of solid waste are done in the Aradkooh complex. Currently, there are two compost generation facilities with the capacities of 2200 and 1800 tons/day, one center of solid waste recovery with a capacity of 50000 tons/day, and one solid waste landfilling center with a capacity of 60000 tons/day are available.

In order to determine the optimal status of the available system for Tehran's SWM system, after extraction

Fig. 3 Name and nominal capacity of solid waste processing units in Aradkooh

Table 3 Model result relating the regions' allocation to the transfer stations (X_{ij})

Transfer station name	Region(s) to be covered
Darabad	1
Zanjan	2
Banihashem	-
Hakimiyeh	8
Kuhak	5,21,22
Beyhaghi	3,6,7
Harandi	11,12,16
Azadegan	13,14,15
Yaran	9,10,17,18
Jehad	4,19
Shahid Avini	20

of information and run of the model, the following results were obtained. The results indicated a cost amounting to 577179 US $. It means in the best scenario, the existing system incurs a cost of 577179 US $ per day for the municipal SWM. Accordingly, the current facilities and systems are unprofitable.

The results obtained from the optimization of the current situation indicate that all of the 11 available transfer stations are not processing units and they act in the form of transfer stations only. Therefore, no region is directly connected to the processing units and the solid waste of all of the 22 regions of Tehran is first moved to transfer stations, and then goes to the processing units. The following table summarizes the way the 22 regions of Tehran are covered by transfer stations as output of the model. As can be seen, the transfer station 3 (Banihashem) is established neither as a transfer nor a processing center. Thus the transfer station 3 can be closed based on Lingo optimization results (Table 3).

Model result relating the regions' allocation to the transfer stations (X_{ij})

As output of the model, the amount of solid waste allocated from each region to the processing unit is shown in Table 4. The first column in this table represents the name of the available processing units, while the other columns indicate the name of the region allocated to the processing unit and the amount of solid waste sent to them. Based on the obtained results, the processing units of M3, cargo, Plant, and S3 should not be reopened. Moreover, five different regions send their wastes to the processing units of S10 and M1-M2.

Model result concerning regions' solid waste allocation to the processing units

Table 5 indicates that each transfer station is allowed to send its collected solid waste to which processing unit. Some of the transfer stations are allowed to transfer their wastes to multiple processing units such as the transfer station 6 connected to three processing units.

Model result concerning transfer stations' allocation to the processing units

Furthermore, two composts, landfilling, and recovery facilities are also established. The amount of compost generated from the compost facility 1 is 2200 tons and from facility 2 is 1800 tons. As previously stated, the solid waste Type 1 is not allowed to be sent to the compost facility and thus is directly transferred to the recovery facilities, where only solid waste of Type 2 and 3 are sent to the compost facilities. Meanwhile, the processing units of 1 and 2 and 7–10 are also opened. Furthermore, most of the solid waste Type 2 has been transferred to the landfilling facility and landfilled, where solid waste Type 3 has been sent for compost. The whole solid waste of the Type 1 has been

Table 4 Model result concerning regions' solid waste allocation to the processing units

Processing unit	Regions to be covered									
	Region	Waste amount	Region	Waste amount	Region	Waste amount	Region	Waste amount	Region	Waste amount
M1-M2	3	226.3	4	107	12	267.8	17	159.2	18	239.7
S9	2	433.4	4	164.7	13	141.8	14	260	-	-
M3	-	-	-	-	-	-	-	-	-	-
Cargo	-	-	-	-	-	-	-	-	-	-
Plant	-	-	-	-	-	-	-	-	-	-
S3	-	-	-	-	-	-	-	-	-	-
S1-S2	1	367.3	11	158.6	22	88	20	278.1	-	-
S10	6	236.3	7	233.5	8	223.9	9	104	10	179.8
S4-S5	3	42	15	394.8	19	176	21	104.7	-	-
S6-S7	4	339.9	5	437.9	11	39	16	183.2	-	-

Table 5 Model result concerning transfer stations' allocation to the processing units

Processing unit	Allocated transfer stations
M1-M2	6,7,9,10
S9	2,8,10
M3	-
Cargo	-
Plant	-
S3	-
S1-S2	1,5,7,11
S10	4,6,9
S4-S5	6,8,5,10
S6-S7	5,7,10

sent to the recovery facilities and recovered. Table 6 indicates the amount of different types of wastes sent from different processing units to the compost, landfill and recovery facilities.

Model result concerning solid waste flow after the processing units

At the end, the results obtained from the model solution indicate the decreased number of the available transfer stations and processing units. From among the transferring stations, station 3 has the potential to be closed due to proximity to the other transfer stations. Similarly, the processing units of M1-M2, S9, S1-S2, S10, S4-S5 and S6-S7 have the potential to respond to the solid waste sent by the transfer stations, thus there is no need to re-open other processing units. The amount of solid waste sent to the compost facilities of 1 and 2 is equal to their entire capacity necessitating them to operate with their full capacity. Therefore, the combination of optimal transfer and processing units can generate the most acceptable daily solid waste collection and transport cycle. Recent studies indicated that adopting integrating solid waste treatment technologies like composting with effective source separation for organic fraction, solid waste recovery or recycling can help in achieving economic

and environmental benefits [17]. Sharholy et al. reviewed the municipal solid waste management in Indian cities. They found that the management of MSW requires proper infrastructure, maintenance and upgrade for four activities, i.e., solid waste generation, collection, transportation, and disposal [18]. However, the optimization of SWM due to consideration of the interaction of various factors (e.g. economic development, social policy, environmental quality) and parameters (e.g. operation cost, transportation cost, and waste- generation rate), is difficult [19]. Therefore, multi- criteria decision analysis (MCDA) is an alternative method to optimize the MSWM. Xi et al. used constrained mixed-integer linear programming (ICCMILP) method. There are 6 transfer stations with the total capacity of 5500 ton/d, charged with the classification and pretreatment of MSW so as to increase the utilization rate and volume reduction efficiency. It was estimated that through compression and dehydration, most of the leachate would be removed, and the compression ratio would be 2.4–4 times of solid waste compactors.

Conclusion

In this paper, a linear programming model has been presented for optimization of the transferring and processing units of solid waste in Tehran. Tehran has 22 municipal regions in which 20 types of solid waste are generated. Similarly, there are 11 transfer stations and 10 processing units in Tehran. After running of the model, the results have indicated that the optimal situation for Tehran in order to maximize the profitability was presence of 10 transfer station and 6 processing units. Furthermore, the amount of solid waste sent to the compost facilities of 1 and 2 is equal to their entire capacity necessitating them to operate with their full capacity. Future improvement can be focused on two aspects. First, determination of physic- chemical composition of solid waste in Tehran to predict the proximate and ultimate analysis and heating value from physical composition. Second, the optimization of incineration, recycling and composting after completing facilities in Tehran.

Table 6 Model result concerning solid waste flow after the processing units

Facility name	Waste type	M1-M2	S9	S1-S2	S10	S4-S5	S6-S7
Compost No.1	Dry non-valuable (Type 2)	-	-	-	-	-	187.3
	Wet (Type 3)	702	682	628.2	-	-	-
Compost No.2	Type 2	-	-	-	-	-	-
	Type 3	-	-	-	621.8	488.3	689.9
Landfill	Type 2	33.4	28.6	15.2	42.6	23.8	31.6
	Type 3	-	-	-	-	-	-
Recovery	Dry valuable (Type 1)	264.7	289.5	248.6	313	205.5	278.5

Appendix 1 List of symbols and variables

Indices

i: region's solid waste (I1) and companies and industrial parks (I2)

j: urban service station (transfer station)

K: processing unit

L: Landfill unit

M: compost

N: incinerator

O: Recovery industries

P_1: Solid waste type

P_2: dry valuable waste, dry non- valuable waste, wet waste

g: biogas

pe: different types of environmental pollutions

Parameters and Variables

Cap_j^s: urban service station capacity (transfer station)

$Cap_j^{'s}$: urban service station capacity, in case the station is the transfer station and processing unit

α_i: percentage of the solid waste of the region i recovered directly (source separation)

D_i: the amount of solid waste generated in the region i

δ_{iP_1}: the amount (percentage) of solid waste P_1 generated in the region i

Cap_K^{pr}: the capacity of the processing unit K

$\gamma_{P_1P_2}$: the amount (percentage) of solid waste P_1 transformed into solid waste P_2 $\sum_{P_2}\gamma_{P_1P_2}=1$

$\beta_{P_2}^1$: if the solid waste P_2 sent to the compost the value will be 1, otherwise 0

$\beta_{P_2}^2$: if the solid waste P_2 sent to the landfill the value will be 1, otherwise 0

$\beta_{P_2}^3$: if the solid waste P_2 sent to the incinerator the value will be 1, otherwise 0

$\beta_{P_2}^4$: if the solid waste P_2 sent to the recovery the value will be 1, otherwise 0

$\beta_{P_2}^5$: if the solid waste P_2 sent to the biogas the value will be 1, otherwise 0

Cap_g: the capacity of the biogas g

Cap_O^1: the capacity of the recovery O

Cap_M^2: the capacity of the compost M

Cap_N^3: the capacity of the incinerator N

Cap_L^4: the capacity of the landfill L

FEA_{jpe}: the amount of pe generated in transfer station j

$Vcap_{pe}$: the maximum allowed pe in transfer station j

$Mcap_{pe}$: the maximum allowed pe in processing unit

$Wcap_{pe}$: the maximum allowed pe in compost unit

Scap: the maximum allowed Leachate in landfill unit

FEm_{kpe}: the amount of pe generated in processing unit K

FEw_{mpe}: the amount of pe generated in compost unit M

VL_{p2}: the amount of generated Leachate as per landfilled solid waste P_2

x_{ij}: if the solid waste of region i allocated to transfer station j the value is 1, otherwise 0

x'_{ij}: if the solid waste of region i allocated to transfer and processing station j the value is 1, otherwise 0

Y_{ijK}: the amount of the demand of the solid waste of region i accumulated in the transfer station j and sent to the processing unit K

W_{1KLP_2}: the amount of solid waste P_2 sent from processing unit K to landfill L

W_{2KNP_2}: the amount of solid waste P_2 sent from processing unit K to incinerator N

W_{3jNP_2}: the amount of solid waste P_2 sent from transfer and processing station j to incinerator N

W_{4jLP_2}: the amount of solid waste P_2 sent from transfer and processing station j to landfill L

Z_{KMP_2}: the amount of solid waste P_2 sent from processing unit K to compost M

U_{KOP_2}: the amount of solid waste P_2 sent from processing unit K to recovery O

$Z_{jMP_2}^2$: the amount of solid waste P_2 sent from transfer station j to compost M

$Bio_{jgP_2}^1$: the amount of solid waste P_2 sent from transfer station j to biogas g

re_{jOP_2}: the amount of solid waste P_2 sent from transfer station j to recovery O

φ_i: fraction of generated solid waste in region i to be separated in source

$Bio_{KgP_2}^2$: the amount of solid waste P_2 sent from processing unit K to biogas g

FE_M^1: the amount of the generated compost in compost M

FE_g^2: the amount of the generated compost in biogas g

EL_g^1: the amount of the generated electricity in biogas g

EL_N^2: the amount of the generated electricity in incinerator N

SL_{ML}^1: the amount of the solid waste (compost) sent from compost M to landfill L

SL_{NL}^2: the amount of the solid waste (ash) sent from incinerator N to landfill L

SL_{gL}^3: the amount of the solid waste sent from biogas g to landfill L

IN_{MN}^1: the amount of the solid waste (compost) sent from compost M to incinerator N

IN_{gN}^2: the amount of the solid waste sent from biogas g to incinerator N

Lg_g: 1 = if the biogas g selected, otherwise = 0

LK_k: 1 = if the processing K selected, otherwise = 0

LN_N: 1 = if the incinerator N selected, otherwise = 0

LL_L: 1 = if the landfill L selected, otherwise = 0

LM_M: 1 = if the compost M selected, otherwise = 0

LO_O: 1 = if the recovery O selected, otherwise = 0

Y_j: in case j is transfer station

Y'_j: in case j is transfer station and processing unit

C^1_{ij}: transfer cost from region i to transfer station j

C^2_{jK}: transfer cost from transfer station j to processing unit K

C^3_{jO}: transfer cost from transfer station j to recovery industry O

C^4_{KO}: transfer cost from processing unit K to recovery industry O

C^5_{KM}: transfer cost from processing unit K to compost M

C^6_{Mj}: transfer cost from transfer station j to compost M

C^7_{jN}: transfer cost from transfer station j to incinerator N

C^8_{MN}: transfer cost from compost M to incinerator N

C^9_{gN}: transfer cost from biogas g to incinerator N

C^{10}_{gj}: transfer cost from transfer station j to biogas g

C^{11}_{Kg}: transfer cost from processing unit K to biogas g

C^{12}_{ML}: transfer cost from compost M to landfill L

C^{13}_{NL}: transfer cost from incinerator N to landfill L

C^{14}_{gL}: transfer cost from biogas g to landfill L

C^{15}_{KL}: transfer cost from processing unit K to landfill L

C^{16}_{jL}: transfer cost from transfer station j to landfill L

$Cdis$: landfill cost per unit waste

CO_o: establishment cost of recovery O

CK_k: establishment cost of processing unit K

CM_m: establishment cost of compost M

CL_l: establishment cost of landfill L

CN_n: establishment cost of incinerator N

Cg_g: establishment cost of biogas g

$Cj^1_{jY_j}$: establishment cost of transfer station j

$Cj^2_{jY'_j}$: establishment cost of transfer station and processing unit j

$B^1_{P_2}$: the benefit from the recovery of one unit of solid waste type P_2

B^2: the benefit from the separated solid waste at source

B^3_g: the benefit from the sell of one unit of generated electricity in biogas g

B^4_N: the benefit from the sell of one unit of generated electricity in incinerator N

B^5_M: the benefit from the sell of one unit of compost from compost M

B^6_g: the benefit from the sell of one unit of compost from biogas g

Competing interests
The authors declare that they have no competing interests.

Authors' contribution
Authors participated in this research including design, data analysis and manuscript preparation. All authors read and approved the final manuscript.

Acknowledgements
The authors are most grateful to the relevant officials of Tehran municipality, Iran, for their collaboration in this research.

Author details
[1]Department of Industrial Engineering and Management Systems, Amirkabir University of Technology, Tehran, Iran. [2]Department of Environmental Engineering, Graduate Faculty of Environment, University of Tehran, Azin alley, Ghods St. Enghelab Ave., Tehran, Iran.

References
1. Powell JC. The evaluation of waste management options. Waste Manag Res. 1996;14(6):515–26.
2. Kang HY, Schoenung JM. Electronic waste recycling: a review of U.S. infrastructure and technology options, Resources. Conserv Recycl. 2005;45(4):368–400.
3. Sufian MA, Bala BK. Modeling of urban solid waste management system: the case of Dhaka city. Waste Manag. 2007;27(7):858–68.
4. Abduli MA, Naghib A, Yonesi M, Akbari A. Life cycle assessment (LCA) of solid waste management strategies in Tehran: landfill and composting plus landfill. Environ Monit Assess. 2010;178:487–98.
5. Hesamizadeh K, Sharafi H, Keyvani H, Alavian SM, Najafi-Tireh Shabankareh A, Sharifi Olyaie R, Keshvari M. Hepatitis A virus and hepatitis E virus seroprevalence among blood donors in Tehran, Iran. Hepat Mon. 2016;16(1):32215–9.
6. Abduli M, Akbarpour Shirazi M, Omidvar B, Samieifard R. A survey of municipal solid waste generation in 22 regions of Tehran with solid waste reduction approach. TB. 2015;14(2):23–33.
7. OWRC. Statistical report on 2002–2005. Organization for Waste Recycling and Composting (OWRC). Iran: Tehran Municipality; 2005.
8. OWRC. Statistics report on 2005. Organization for waste recycling and composting. Iran: Tehran Municipality; 2006.
9. Ghiani G, Laganà D, Manni E, Musmanno R, Vigo D. Operations research in solid waste management: a survey of strategic and tactical issues. Comput Oper Res. 2014;44:22–32.
10. Yu H, Solvang WD, Li S. Optimization of long-term performance of municipal solid waste management system: a bi-objective mathematical model. Int J Energy Environ. 2015;6(2):153–64.
11. Ravindra K, Kaur K, Mor S. System analysis of municipal solid waste management in Chandigarh and minimization practices for cleaner emissions. J Clean Prod. 2015;89:251–6.
12. Tan ST, Lee CT, Hashim H, Ho WS, Lim JS. Optimal process network for municipal solid waste management in Iskandar Malaysia. J Clean Prod. 2014;71:48–58.
13. Zhou H, Meng AH, Long YQ, Li QH, Zhang YG. An overview of characteristics of municipal solid waste fuel in China: physical, chemical composition and heating value. Renew Sustain Energy Rev. 2014;36:107–22.
14. Lohri CR, Camenzind EJ, Zurbrügg C. Financial sustainability in municipal solid waste management – costs and revenues in Bahir Dar, Ethiopia. Waste Manag. 2014;34(2):542–52.
15. Soltani A, Hewage K, Reza B, Sadiq R. Multiple stakeholders in multi-criteria decision-making in the context of municipal solid waste management: a review. Waste Manag. 2015;35:318–28.
16. Arena U, Di Gregorio F. A waste management planning based on substance flow analysis. Res Conserv Recycl. 2014;85:54–66.
17. Herva M, Neto B, Roc E. Environmental assessment of the integrated municipal solid waste management system in Porto (Portugal). J Clean Prod. 2014;70:183–93.
18. Sharholy M, Ahmad K, Gauhar M, Trivedi RC. Municipal solid waste management in Indian cities – a review. Waste Manag. 2014;34:542–52.
19. Xi BD, Su J, Huang GH, Qin XS, Jiang YH, Huo SL, Ji DF, Yao B. An integrated optimization approach and multi-criteria decision analysis for supporting the waste-management system of the City of Beijing, China. Eng Appl Artif Intel. 2010;23:620–31.

Permissions

All chapters in this book were first published in JEHSE, by BioMed Central; hereby published with permission under the Creative Commons Attribution License or equivalent. Every chapter published in this book has been scrutinized by our experts. Their significance has been extensively debated. The topics covered herein carry significant findings which will fuel the growth of the discipline. They may even be implemented as practical applications or may be referred to as a beginning point for another development.

The contributors of this book come from diverse backgrounds, making this book a truly international effort. This book will bring forth new frontiers with its revolutionizing research information and detailed analysis of the nascent developments around the world.

We would like to thank all the contributing authors for lending their expertise to make the book truly unique. They have played a crucial role in the development of this book. Without their invaluable contributions this book wouldn't have been possible. They have made vital efforts to compile up to date information on the varied aspects of this subject to make this book a valuable addition to the collection of many professionals and students.

This book was conceptualized with the vision of imparting up-to-date information and advanced data in this field. To ensure the same, a matchless editorial board was set up. Every individual on the board went through rigorous rounds of assessment to prove their worth. After which they invested a large part of their time researching and compiling the most relevant data for our readers.

The editorial board has been involved in producing this book since its inception. They have spent rigorous hours researching and exploring the diverse topics which have resulted in the successful publishing of this book. They have passed on their knowledge of decades through this book. To expedite this challenging task, the publisher supported the team at every step. A small team of assistant editors was also appointed to further simplify the editing procedure and attain best results for the readers.

Apart from the editorial board, the designing team has also invested a significant amount of their time in understanding the subject and creating the most relevant covers. They scrutinized every image to scout for the most suitable representation of the subject and create an appropriate cover for the book.

The publishing team has been an ardent support to the editorial, designing and production team. Their endless efforts to recruit the best for this project, has resulted in the accomplishment of this book. They are a veteran in the field of academics and their pool of knowledge is as vast as their experience in printing. Their expertise and guidance has proved useful at every step. Their uncompromising quality standards have made this book an exceptional effort. Their encouragement from time to time has been an inspiration for everyone.

The publisher and the editorial board hope that this book will prove to be a valuable piece of knowledge for researchers, students, practitioners and scholars across the globe.

List of Contributors

Edris Bazrafshan, Leili Mohammadi and Alireza Ansari-Moghaddam
Health Promotion Research Center, Zahedan University of Medical Sciences, Zahedan, Iran

Amir Hossein Mahvi
Department of Environmental Health Engineering, School of Public Health, Tehran University of Medical Sciences, Tehran, Iran
Center for Solid Waste Research, Institute for Environmental Research, Tehran University of Medical Sciences, Tehran, Iran
National Institute of Health Research, Tehran University of Medical Sciences, Tehran, Iran

Jalil Jaafari, Alireza Mesdaghinia and Ramin Nabizadeh
School of Public Health, Tehran University of Medical Sciences, Tehran, Iran

Mohammad Hoseini
Department of Environmental Health Engineering, School of Public Health, Shiraz University of Medical Sciences, Shiraz, Iran

Hossein kamani
Health Promotion Research Center, Zahedan University of Medical Sciences, Zahedan, Iran

Amir Hossein Mahvi
School of Public Health, Tehran University of Medical Sciences, Tehran, Iran
Center for Solid Waste Research, Institute for Environmental Research, Tehran University of Medical Sciences, Tehran, Iran
National Institute of Health Research, Tehran University of Medical Sciences, Tehran, Iran

Hamidreza Heidari and Farideh Golbabaei
Department Occupational Health, School of Public Health, Tehran University of Medical Sciences, Tehran, Iran

Aliakbar Shamsipour
Department Physical Geography, School of Geography, University of Tehran, Tehran, Iran

Abbas Rahimi Forushani
Department Epidemiology and Biostatistics, School of Public Health, Tehran University of Medical Sciences, Tehran, Iran

Abbasali Gaeini
Department Sport Physiology, School of Physical Education and Sport Science, University of Tehran, Tehran, Iran

Ahmad Moarefian
Department of Chemical Engineering, College of Science, Shahrood Branch, Islamic Azad University, Shahrood, Iran

Hossein Alizadeh Golestani
Department of Chemical Engineering, Quchan Branch, Islamic Azad University, Quchan, Iran

Hooman Bahmanpour
Department of Environment, College of Science, Shahrood Branch, Islamic Azad University, Shahrood, Iran

Mohammad Delnavaz
Civil and Environmental Engineering Faculty, Tarbiat Modares University, Tehran, Iran
Civil Engineering Department, Faculty of Engineering, Kharazmi University, Tehran, Iran

Bita Ayati and Hossein Ganjidoust
Civil and Environmental Engineering Faculty, Tarbiat Modares University, Tehran, Iran

Sohrab Sanjabi
Material Engineering Department, Nano Materials Division, Tarbiat Modares University, Tehran, Iran

Aliakbar Hedayati and Elaheh Hassan Nataj Niazie
Department of Fisheries, Faculty of Fisheries and Environment, Gorgan University of Agricultural Sciences and Natural Resources, Gorgan, Iran

Tahereh Sadeghi Yekta, Mohammad Khazaei and Ahmad Reza Yari
Research Center for Environmental Pollutants, Qom University of Medical Sciences, Qom, Iran

Ramin Nabizadeh
Department of Environmental Health Engineering, School of Public Health, Tehran University of Medical Sciences, Poursina St, Keshavarz Blvd, PO BOX: 6446-14155, Tehran, Iran

Amir Hossein Mahvi
Department of Environmental Health Engineering, School of Public Health, Tehran University of Medical Sciences, Poursina St, Keshavarz Blvd, PO BOX: 6446-14155, Tehran, Iran
Center for Solid Waste Research, Institute for Environmental Research, Tehran University of Medical Sciences, Tehran, Iran

Simin Nasseri
Department of Environmental Health Engineering, School of Public Health, Tehran University of Medical Sciences, Poursina St, Keshavarz Blvd, PO BOX: 6446-14155, Tehran, Iran
Center for Water Quality Research, Institute for Environmental Research, Tehran University of Medical Sciences, Tehran, Iran

Rishi Pal Chauhan and Amit Kumar
Department of Physics, National Institute of Technology, Kurkshetra 136119, India

Seyed Vali Hosseini
Department of Fisheries, College of Agriculture & Natural Resources, University of Tehran, Karaj, Iran

Soheil Sobhanardakani
Department of the Environment, College of Basic Sciences, Hamedan Branch, Islamic Azad University, Hamedan, Iran

Hamed Kolangi Miandare
Department of Fisheries, Gorgan University of Agricultural Sciences and Natural Resources, Gorgan, Iran

Mohammad Harsij
Department of Natural Resources, Gonbad Kavous University, Gonbad Kavous, Iran

Joe Mac Regenstein
Department of Food Science, Cornell University, Ithaca, New York, USA

Lakmal Jayarathna
Material Technology Section, Industrial Technology Institute, No 363, Bauddhaloka Mawatha, Colombo 07, Sri Lanka
Chemical and Environmental System Modeling group, Institute of Fundamental Studies, Hanthana Road, Kandy, Sri Lanka

Athula Bandara
Department of Chemistry, University of Peradeniya, Peradeniya, Sri Lanka

W.J. Ng
Nanyang Environment and Water Research Institute, Singapore, Singapore

Rohan Weerasooriya
Department of Soil Science, University of Peradeniya, Peradeniya, Sri Lanka

Nandhini Palanisamy, Jayaprakash Ramya, Srilakshman Kumar and NS Vasanthi
Department of Biotechnology, Bannari Amman Institute of Technology, Sathyamangalam, Tamil Nadu, India

Preethy Chandran
CeNTAB, School of Chemical and Biotechnology, SASTRA University, Thanjavur 613401, Tamil Nadu, India

Sudheer Khan
Department of Biotechnology, Bannari Amman Institute of Technology, Sathyamangalam, Tamil Nadu, India
CeNTAB, School of Chemical and Biotechnology, SASTRA University, Thanjavur 613401, Tamil Nadu, India

Fathollah Aalipour
Food and Drug Administration, Sharekord University of Medical Sciences, Shahkord, Iran

Maryam Mirlohi, Mohammad Jalali and Leila Azadbakht
Food Security Research Center, School of Nutrition and Food Sciences, Isfahan University of Medical Sciences, Hezargrib Street, Isfahan, Iran

Hossein Hazrati, Jalal Shayegan and Seyed Mojtaba Seyedi
Department of Chemical and Petroleum Engineering, Sharif University of Technology, Tehran, Iran

Masuma Moghaddam Arjmand and Abbas Rezaee
Department of Environmental Health Engineering, Faculty of Medical Sciences, Tarbiat Modares University, Tehran, Iran

Simin Nasseri
Department of Environmental Health Engineering, School of Public Health, and Center for Water Quality Research, Institute for Environmental Research, Tehran University of Medical Sciences, Tehran, Iran

Said Eshraghi
Department of Pathobiology, School of Public Health, Tehran University of Medical Sciences, Tehran, Iran

Gholam Hossein Safari and Mahmood Alimohammadi
Department of Environmental Health Engineering, School of Public Health, Tehran University of Medical Sciences, Tehran, Iran

Simin Nasseri
Department of Environmental Health Engineering, School of Public Health, Tehran University of Medical Sciences, Tehran, Iran Center for Water Quality Research, Institute for Environmental Research, Tehran University of Medical Sciences, Tehran, Iran

Amir Hossein Mahvi and Kamyar Yaghmaeian
Department of Environmental Health Engineering, School of Public Health, Tehran University of Medical Sciences, Tehran, Iran

Center for Solid Waste Research, Institute for Environmental Research, Tehran University of Medical Sciences, Tehran, Iran

Ramin Nabizadeh
Department of Environmental Health Engineering, School of Public Health, Tehran University of Medical Sciences, Tehran, Iran Center for Air Pollution Research, Institute for Environmental Research, Tehran University of Medical Sciences, Tehran, Iran

Ali D Boloorani
Department of Remote Sensing & GIS & Geoinformatics Research Institute (GRI), University of Tehran, Tehran, Iran

Seyed O Nabavi
Geoinformatics Research Institute (GRI), University of Tehran and Department of Geography and Regional Research, University of Vienna, Vienna, Austria

Hosain A Bahrami
Department of Soil Science, Tarbiat Modares University, Tehran, Iran

Fardin Mirzapour
Faculty of Electrical Engineering, Sadra Institute of Higher Education, Isfahan, Iran

Musa Kavosi, Esmail Abasi and Rasoul Azizi
Geoinformatics Research Institute (GRI), University of Tehran, Tehran, Iran

Vali Alipour
Department of Environmental Health Engineering, School of Public Health, Tehran University of Medical Sciences, Tehran, Iran

Simin Nasseri, Ramin Nabizadeh Nodehi and Amir Hossein Mahvi
Department of Environmental Health Engineering, School of Public Health, Tehran University of Medical Sciences, Tehran, Iran Center for Water Quality Research (CWQR), Institute for Environmental Research (IER), Tehran University of Medical Sciences, Tehran, Iran

Alimorad Rashidi
Nanotechnology Research Center, Research Institute of Petroleum Industry (RIPI), Tehran, Iran

Ovidiu Murărescu, George Murătoreanu and Mădălina Frînculeasa
Department of Geography, Valahia University of Târgoviște, Târgoviște, Romania

Mahboobeh Ghoochani, Masoud Yunesian and Shahrokh Nazmara
Department of Environmental Health Engineering, School of Public Health, Tehran University of Medical Sciences, Tehran, Iran

Sakine Shekoohiyan
Department of Environmental Health Engineering, School of Public Health, Hormozgan University of Medical Sciences, Bandar- Abbas, Iran

Amir Hossein Mahvi
Department of Environmental Health Engineering, School of Public Health, Tehran University of Medical Sciences, Tehran, Iran
Center for Solid Waste Research, Institute for Environmental Research, Tehran University of Medical Sciences, Tehran, Iran
National Institute of Health Research, Tehran University of Medical Sciences, Tehran, Iran

Mehran Mohammadian Fazli, Negin Soleimani and Mohammadreza Mehrasbi
Department of Environmental Health Engineering, Zanjan Universiry of Medical Sciences, Zanjan, Iran

Sima Darabian and Jamshid Mohammadi
Medical Entomology and Mycology Department, School of Medicine, Zanjan Universiry of Medical Sciences, Zanjan, Iran

Ali Ramazani
Biotechnology Departments, School of Pharmacy, Zanjan University of Medical Sciences, Zanjan, Iran

Hamid Forootanfar
Department of Pharmaceutical Biotechnology, Faculty of Pharmacy, Kerman University of Medical Sciences, Kerman, Iran

Shahla Rezaei, Hamed Tahmasbi, Mehdi Mogharabi and Mohammad Ali Faramarzi
Department of Pharmaceutical Biotechnology, Faculty of Pharmacy and Biotechnology Research Center, Tehran University of Medical Sciences, P.O. Box 14155-6451, Tehran 1417614411, Iran

Hamed Zeinvand-Lorestani
Department of Pharmacology and Toxicology, Faculty of Pharmacy, Tehran University of Medical Sciences, P.O. Box 14155-6451, Tehran 1417614411, Iran

Alieh Ameri
Department of Medicinal Chemistry, Faculty of Pharmacy, Kerman University of Medical Sciences, Kerman, Iran

Abdullah Al-Mamun
Department of Civil & Architectural engineering, Sultan Qaboos University, Al-Khodh, P.C. 123, P.O. Box 33, Muscat, Sultanate of Oman

Mahad Said Baawain
Centre for Environmental Studies and Research, Sultan Qaboos University, Al-khodh, P.C. 123, Muscat P.O. Box 17, Muscat, Sultanate of Oman

Ritesh Vijay and Asheesh Sharma
Environmental System Design and Modeling Division, CSIR-NEERI, Nagpur 440020, Maharashtra, India

Tapan Chakrabarti and Rajesh Gupta
Civil Engineering Department, VNIT, Nagpur 440010, Maharashtra, India

Mohsen Akbarpour Shirazi
Department of Industrial Engineering and Management Systems, Amirkabir University of Technology, Tehran, Iran

Reza Samieifard, Mohammad Ali Abduli and Babak Omidvar
Department of Environmental Engineering, Graduate Faculty of Environment, University of Tehran, Azin alley, Ghods St. Enghelab Ave., Tehran, Iran

Index

www.ingramcontent.com/pod-product-compliance
Lightning Source LLC
Chambersburg PA
CBHW061946190326
41458CB00009B/2803